The Chemistry, Biology and Medical Applications of Hyaluronan and its Derivatives

Related titles published by Portland Press:

Keratan Sulphate: Chemistry, Biology, Chemical Pathology
edited by H. Greiling and J.E. Scott
1989 ISBN 0 90449 825 5

Dermatan Sulphate Proteoglycans: Chemistry, Biology,
Chemical Pathology
edited by J.E. Scott
1993 ISBN 1 85578 051 8

Interstitium, Connective Tissue and Lymphatics
edited by R.K. Reed, N.G. McHale, J.L. Bert, C.P. Winlove and
G.A. Laine
1995 ISBN 1 85578 073 9

Connective Tissue Biology: Integration and Reductionism
edited by R.K. Reed and K. Rubin
1998 ISBN 1 85578 118 2

WENNER-GREN INTERNATIONAL SERIES

VOLUME 72

The Chemistry, Biology and Medical Applications of Hyaluronan and its Derivatives

Edited by

T.C. Laurent

PORTLAND PRESS
London and Miami

Published by Portland Press Ltd, 59 Portland Place, London W1N 3AJ, U.K.
Tel: (+44) 171 580 5530; e-mail: edit@portlandpress.co.uk

In North America orders should be sent to Ashgate Publishing Co., Old Post Road,
Brookfield, VT 05036-9704, U.S.A.

ISBN 1 85578 119 0

British Library Cataloguing in Publication Data
A catalogue record for this book is available from the British Library

Typeset by Portland Press Ltd
Printed in Great Britain by Cambridge University Press, Cambridge, UK

Front cover shows: an atomic force microscope view of a hyaluronan
molecule in a sample of hyaluronan (weight-average molecular weight 2×10^6)
dissolved in a magnesium chloride solution and placed on a mica surface. The
dimension of the molecular domain is approximately 500nm. Courtesy of M.
Cowman and E.A. Balazs.

Contents

Medical applications of hyaluronan and derivatives of hyaluronan

Endogenous hyaluronan as a clinical marker

Mark E. Adams
Arthritis Society Chair in Rheumatic Diseases/Rheumatology, Department of Medicine, and McCaig Centre for Joint Injury and Arthritis Research, The University of Calgary, Canada

Endre A. Balazs
Matrix Biology Institute, 65 Railroad Avenue, Ridgefield, NJ 07657, U.S.A.

Philip A. Band
Biomatrix, Inc., 65 Railroad Avenue, Ridgefield, NJ 07657, U.S.A.

Carlos Belmonte
Instituto de Neurociencias, Universidad Miguel Hernandez, San Juan de Alicante, 03550 San Juan de Alicante, Spain

Sven Björnsson
Department of Clinical Chemistry, Lund University, Sweden

Anna M. Blom
Department of Medical and Physiological Chemistry, University of Uppsala, Biomedical Center, Box 575, S-751 23 Uppsala, Sweden

Hege Bothner Wik
Pharmacia & Upjohn, S-751 82 Uppsala, Sweden

T.J. Brown
Department of Biochemistry and Molecular Biology, Monash University, Clayton, Victoria 3168, Australia

Anthony Calabro
Department of Biomedical Engineering, Wb3 Research Institute, Cleveland Clinic Foundation, Cleveland, OH 44195, U.S.A.

Mary K. Cowman
Department of Chemical Engineering, Chemistry, and Materials Science, Polytechnic University, Six Metrotech Center, Brooklyn, NY 11201, U.S.A.

Anthony J. Day
Department of Biochemistry, University of Oxford, South Parks Road, Oxford, OX1 3QU, U.K.

Janet L. Denlinger
Biomatrix, Inc., 65 Railroad Aveune, Ridgefield, NJ 07657, U.S.A.

Anna Engström-Laurent
Department of Internal Medicine, University Hospital of Umeå, S-901 85 Umeå, Sweden

J.R.E. Fraser
Department of Biochemistry and Molecular Biology, Monash University, Clayton, Victoria 3168, Australia

Erik Fries
Department of Medical and Physiological Chemistry, University of Uppsala, Biomedical Center, Box 575, S-751 23 Uppsala, Sweden

Csaba Fulop
Department of Biomedical Engineering, Wb3 Research Institute, Cleveland Clinic Foundation, Cleveland, OH 44195, U.S.A.

Nicola Goodstone
Department of Biomedical Engineering, Wb3 Research Institute, Cleveland Clinic Foundation, Cleveland, OH 44195, U.S.A.

Stefan Gustafson
Department of Medical and Physiological Chemistry, University of Uppsala, Box 575, S-751 23 Uppsala, Sweden

R. Harrison
Cardiovascular Research, The Hospital for Sick Children, 555 University Avenue, Toronto, Ontario M5G 1X8, Canada

Vincent C. Hascall
Department of Biomedical Engineering, Wb3 Research Institute, Cleveland Clinic Foundation, Cleveland, OH 44195, U.S.A.

Dick Heinegård
Department of Cell and Molecular Biology, Lund University, PO Box 94, S-221 00 Lund, Sweden

Paraskevi Heldin
Department of Medical and Physiological Chemistry, Biomedical Center, Uppsala University, Box 575, S-751 23 Uppsala, Sweden

Daniel M. Hittner
Department of Chemical Engineering, Chemistry, and Materials Science, Polytechnic University, Six Metrotech Center, Brooklyn, NY 11201, U.S.A.

Michael Hogg
Department of Biomedical Engineering, Wb3 Research Institute, Cleveland Clinic Foundation, Cleveland, OH 44195, U.S.A.

Maureen R. Horton
Department of Medicine, Division of Pulmonary and Critical Care Medicine, Johns Hopkins University School of Medicine, 858 Ross Research Building, 720 Rutland Avenue, Baltimore, MD 21205, U.S.A.

G. Hou
Cardiovascular Research, The Hospital for Sick Children, 555 University Avenue, Toronto, Ontario M5G 1X8, Canada

Jong Soo Kim
Department of Chemical Engineering, Chemistry, and Materials Science, Polytechnic University, Six Metrotech Center, Brooklyn, NY 11201, U.S.A.

Cheryl B. Knudson
Departments of Biochemistry and Pathology, Rush Medical College, Rush-Presbyterian-St. Luke's Medical Center, 1653 West Congress Parkway, Chicago, IL 60612-3864, U.S.A.

Warren Knudson
Departments of Biochemistry and Pathology, Rush Medical College, Rush-Presbyterian-St. Luke's Medical Center, 1653 West Congress Parkway, Chicago, IL 60612-3864, U.S.A.

Nancy E. Larsen
Biomatrix, Inc., 65 Railroad Avenue, Ridgefield, NJ 07657, U.S.A.

Claude Laurent
Department of Oto-Rhino-Laryngology, University Hospital, University of Umeå, S-901 85 Umeå, Sweden

Torvard C. Laurent
Department of Medical and Physiological Chemistry, University of Uppsala, Box 575, S-751 23 Uppsala, Sweden

Ulla B.G. Laurent
Department of Medical and Physiological Chemistry, University of Uppsala, Box 575, S-751 23 Uppsala, Sweden

Jayne Lesley
Cancer Biology Laboratory, The Salk Institute, P.O. Box 85800, San Diego, CA 92186-5800, U.S.A.

Min Li
Department of Chemical Engineering, Chemistry, and Materials Science, Polytechnic University, Six Metrotech Center, Brooklyn, NY 11201, U.S.A.

Z. Lin
Cardiovascular Research, The Hospital for Sick Children, 555 University Avenue, Toronto, Ontario M5G 1X8, Canada

Jianshe Liu
Department of Chemical Engineering, Chemistry, and Materials Science, Polytechnic University, Six Metrotech Center, Brooklyn, NY 11201, U.S.A.

Donald MacCallum
Department of Anatomy, University of Michigan, Ann Arbor, MI 48109, U.S.A.

Dale M. Marecak
Department of Medicinal Chemistry, 308 Skaggs Hall, College of Pharmacy, The University of Utah, Salt Lake City, UT 84112, U.S.A., and Department of Chemistry, University at Stony Brook, Stony Brook, NY 11794-3400, U.S.A.

James F. Marecek
Department of Chemistry, University at Stony Brook, Stony Brook, NY 11794-3400, U.S.A.

Charlotte M. McKee
Department of Medicine, Division of Pulmonary and Critical Care Medicine, Johns Hopkins University School of Medicine, 858 Ross Research Building, 720 Rutland Avenue, Baltimore, MD 21205, U.S.A.

Matthias Mörgelin
Department of Cell and Molecular Biology, Lund University, PO Box 94, S-221 00 Lund, Sweden

Paul W. Noble
Department of Medicine, Division of Pulmonary and Critical Care Medicine, Johns Hopkins University School of Medicine, 858 Ross Research Building, 720 Rutland Avenue, Baltimore, MD 21205, U.S.A.

Ashfaq A. Parkar
Department of Biochemistry, University of Oxford, South Parks Road, Oxford, OX1 3QU, U.K.

Bernard Pessac
CNRS UPR 9035, Développement & Immunité du Système Nerveux Central, 15 rue de l'Ecole de Médecine, 75270 Paris cedex 06, France

Glyn O. Phillips
Research Transfer Ltd., 2 Plymouth Drive, Radyr, Cardiff CF4 8BL, U.K.

Miguel A. Pozo
Instituto de Neurociencias, Universidad Miguel Hernandez, San Juan de Alicante,
03550 San Juan de Alicante, Spain

Glenn D. Prestwich
Department of Medicinal Chemistry, 308 Skaggs Hall, College of Pharmacy,
The University of Utah, Salt Lake City, UT 84112–5820, U.S.A.

Antonia Salustri
Department of Public Health and Cell Biology, University of Rome 'Tor Vergata', 00173
Rome, Italy

J.E. Scott
Chemical Morphology, Chemistry Building, Manchester University, Oxford Road,
Manchester M13 9PL, U.K.

D.M. Shaw
Department of Immunology, Faculty of Medicine, University of Liverpool, Liverpool
L69 3BX, U.K.

Yngve Sommarin
Department of Cell and Molecular Biology, Lund University, PO Box 94, S-221 00
Lund, Sweden

Markku Tammi
Department of Anatomy, University of Kuopio, 70211 Kuopio, Finland

Raija Tammi
Department of Anatomy, University of Kuopio, 70211 Kuopio, Finland

Bryan P. Toole
Department of Anatomy and Cellular Biology, Tufts University School of Medicine,
136 Harrison Avenue, Boston, MA 02111, U.S.A.

E. A. Turley
Cardiovascular Research, The Hospital for Sick Children, 555 University Avenue,
Toronto, Ontario M5G 1X8, Canada

Koen P. Vercruysse
Department of Medicinal Chemistry, 308 Skaggs Hall, College of Pharmacy,
The University of Utah, Salt Lake City, UT 84112, U.S.A., and Department of
Chemistry, University at Stony Brook, Stony Brook, NY 11794-3400, U.S.A.

Charles Weiss
Department of Orthopaedics and Rehabilitation, Mt. Sinai Medical Center, Miami
Beach, FL 33140, U.S.A.

D.C. West
Department of Immunology, Faculty of Medicine, University of Liverpool, Liverpool
L69 3BX, U.K.

Ove Wik
Pharmacia & Upjohn, S-751 82 Uppsala, Sweden

Michael R. Ziebell
Department of Medicinal Chemistry, 308 Skaggs Hall, College of Pharmacy,
The University of Utah, Salt Lake City, UT 84112, U.S.A., and Department of
Physiology and Biophysics, University at Stony Brook, Stony Brook, NY 11794-3400,
U.S.A.

ADH	Adipic dihydrazide
AE	Adverse events
AF	Aggregation factors
AFM	Atomic force microscopy
BEHAB	Brain-enriched hyaluronan-binding protein
bFGF	Basic fibroblast growth factor
BK	Bradykinin
CAM	chorioallantoic membrane
COC	Cumulus cell–oocyte complex
CP MAS	Cross-polarization magic-angle spinning
crg-2	Cytokine-responsive gene 2
CS	Chondroitin sulphate
CSF	Cerebrospinal fluid
DMSO	Dimethyl sulphoxide
DSC	Differential scanning calorimetry
ECM	Extracellular matrix
EDCI	Ethyl(N,N-dimethylaminopropyl)carbodiimide
EGF	Epidermal growth factor
EHL	Extensor hallucis longus
ESR	Electron spin resonance; erythrocyte sedimentation rate
FAK	Focal adhesion kinase
FSH	Follicle-stimulating hormone
GAG	Glycosaminoglycan
GAPDH	Glyceraldehyde 3-phosphate dehydrogenase
GH	Growth hormone
GHAP	Glial hyaluronate binding protein
GI	Gastrointestinal
HA	Hyaluronan; hyaluronic acid
HAas	Hyaluronidase
HABD	Hyaluronan binding domain
HABP	Hyaluronan binding protein
HABR	Hyaluronan binding region
HAS	Hyaluronan synthase
5-HT	5-Hydroxytryptamine
HYA	Hyaluronan
IαI	Inter-α-inhibitor
ICAM-1	Intercellular adhesion molecule-1
IFN	Interferon
IGF-1	Insulin-like growth factor-1
IL-1	Interleukin-1
iNOS	Inducible nitric oxide synthase
LALLS	Low-angle laser-light scattering
LCV	Linear crystallization velocity
LEC	Liver endothelial cells
LP	Link protein
LPS	Lipopolysaccharide
MAN	Medial articular nerve

MIP	Macrophage inflammatory protein
MOT	Mouse ovarian ascites tumour
N-CAM	Nerve cell adhesion molecule
NHM	Normal human mesothelial
NIF-NaHA	Non-inflammatory fraction of sodium hyaluronate
NMR	Nuclear magnetic resonance
NOE	Nuclear Overhauser effect
NSAID	Non-steroidal anti-inflammatory drugs
OA	Osteoarthritis
ODFR	Oxygen-derived free radical
PαI	Pre-α-inhibitor
PDGF	Platelet-derived growth factor
PDHA	Partially depolymerized hyaluronan
PG	Prostaglandin/prostacyclin
PKA	Protein kinase A
PKC	Protein kinase C
PMC	Pericellular molecular cage
PMN	Polymorphonuclear leucocytes
pp125[FAK]	Focal adhesion protein tyrosine kinase
RA	Rheumatoid arthritis
RANTES	Regulated on activation, normal T-cell expressed and secreted
RGD	Arginine-glycine-aspartic acid
RHAMM	Receptor for hyaluronan-mediated motility
SHAP	Serum-derived hyaluronan-associated protein
TGF-β	Transforming growth factor-β
TNF	Tumour necrosis factor
TNM	Tetranitromethane
TSG-6	Tumour necrosis factor-stimulated gene-6
VEGF	Vascular endothelial growth factor

Introduction

T.C. Laurent

Department of Medical and Physiological Chemistry, University of Uppsala, Box 575, S-751 23 Uppsala, Sweden

The Wenner-Gren Foundations, located at Wenner-Gren Center in Stockholm, support scientific exchange worldwide. As part of their activity a number of international scientific symposia are held every year at the Wenner-Gren Center and the proceedings are published in the Wenner-Gren International Series.

Hyaluronan (hyaluronic acid) is a polysaccharide found in all vertebrate organs and fluids that have been analysed to date, with the highest concentrations in the extracellular matrix of soft connective tissues. Karl Meyer gave the name hyaluronic acid to a major component of the vitreus of the eye after he discovered that the polysaccharide, known until then as hyalomucin or mucoitin-sulphate, did not have any sulphate groups [1]. He also correctly identified that mucoitin-sulphate, previously described in the umbilical cord, synovial fluid and other tissues, is hyaluronic acid. The term hyaluronan is an adjustment of the name derived from the accepted rules of polysaccharide nomenclature [2]. Interest in hyaluronan has increased dramatically since about 1980, when major clinical applications in ophthalmology and in the treatment of joint disease, coupled to industrial production of the polymer, were introduced. Furthermore, it has been realized that hyaluronan plays an important cell biological role through its interactions with cells.

As an expression of the unique and important role played by hyaluronan in the family of connective tissue polysaccharides, three international symposia have been organized in which hyaluronan has been the sole topic. The first was held in St. Tropez in 1985 and the second in London in 1988. The proceedings of the latter symposium have been published [3]. The third symposium, published in this volume, was organized by the Wenner-Gren Foundations in Stockholm and was held in September 1996. Between the two last symposia there have been major breakthroughs on the structure of hyaluronan; on its biosynthesis, including the discovery of the hyaluronan synthase; on the molecular interaction between hyaluronan and proteins, including the determination of the three-dimensional structure of a hyaluronan-binding protein; on the biological effects of cell–hyaluronan interactions; and on the different biological activities of high molecular weight hyaluronan and hyaluronan fragments. These aspects are covered by different contributors. However, a strong feature of the symposium was the emphasis placed on the production of new derivatives of hyaluronan and on the various medical applications of hyaluronan and these derivatives.

I would like to thank all the contributors to the symposium for valuable contributions and for co-operation in finishing this publication. I would also like to thank Ms Sarah Harrison and Portland Press for kind collaboration.

Dr Endre A. Balazs introduced hyaluronan as a treatment of clinical disorders more than 25 years ago and he has been the leading authority on this matter ever since. A year before the 1996 symposium he reached the age of 75. The symposium was held in his honour and it gives me great pleasure to dedicate this volume to my teacher, Bandi.

References

1.   Meyer, K. and Palmer, J.W. (1934) J. Biol. Chem. 107, 629–634
2.   Balazs, E.A., Laurent, T.C. and Jeanloz, R.W. (1986) Biochem. J. **235**, 903
3.   Evered, D. and Whelan, J., eds. (1989) The Biology of Hyaluronan, Ciba Foundation Symposium No. 143, Wiley, Chichester

Endre A. Balazs – a pioneer in hyaluronan research

T.C. Laurent

Department of Medical and Physiological Chemistry, University of Uppsala, Box 575, S-751 23 Uppsala, Sweden

Endre Balazs (pictured above) is known among all his numerous friends and colleagues as Bandi. He became my scientific mentor in September 1949. We have thus known each other for almost half a century and I have never stopped feeling a great admiration for him.

Bandi was born in Budapest in 1920. He received his medical education and scientific training in Hungary during the Second World War and his first scientific publication came in 1939. In 1943 he wrote a paper on the formation of synovial fluid that became the introduction to his continuing interest in connective tissue and hyaluronan. Bandi attended his first international congress in Stockholm in 1947. He lectured on the effect of the extracellular matrix on growth of fibroblasts, and came in contact with Professor Hjalmar Holmgren from the Department of Experimental Histology, Karolinska Institute, Sweden, who was interested in connective tissue polysaccharides. Bandi decided to move to Sweden and did so the same year, before the communists took over in Hungary. Bandi stayed in Sweden for three years and established many contacts, which would turn out to be important both for his own work and for Swedish medical science. At the end of 1950, when the Korean war threatened world peace, he emigrated to Boston, U.S.A.

When I met Bandi I was 18 years old and in my second year of medical school. I had been offered an unpaid junior instructorship in histology and Bandi immediately engaged me. All available time between the medical courses was spent on hyaluronan research and when I look back on my life, this first year that I spent with Bandi was decisive to my future career. I had met a warm, charismatic and hard-working person full of original ideas and enterprise, who could arouse enthusiasm in his collaborators. During this year I became a co-author of three papers dealing with the polyelectrolyte viscosity of hyaluronan; the degradation of hyaluronan by irradiation (free radicals); and the biological activity of sulphated hyaluronan.

In Boston, Bandi was employed to develop an eye research laboratory, the Retina Foundation, affiliated with the Massachusetts Eye and Ear Infirmary, the eye and ear department of Harvard Medical School. The enterprise started in an old apartment house in Chambers Street, to be demolished ten years later, and ended up in a new modern laboratory building with the name Boston Biomedical Research Institute at Staniford Street. In these laboratories Bandi assembled scientists from all over the world, but perhaps with a dominance of Swedes and Hungarians during a period of 25 years. I worked together with my wife Ulla for two periods (1953–54 and 1959–61) at the Retina Foundation. Bandi concentrated his efforts on the chemistry of the vitreus because it represented the simplest intercellular matrix and because of its possible role in the aetiology of retinal detachment. Owing to his work this is probably the most well-characterized of all connective tissues. He collaborated with many experts and he was never afraid of employing new techniques, from animal experimentation — he even rebuilt a farm for experiments — to advanced physical methods; the laboratory was equipped with all possible instruments, including an analytical ultracentrifuge, Tiselius electrophoresis apparatus, electron microscope and one of the first NMR machines for biological work.

From the beginning Bandi searched for practical applications of his research. I believe that in Budapest during the war he had a patent for making an egg white substitute out of synovial fluid. In his work on the vitreus he searched for a natural replacement for injured tissue and he was already experimenting in the late 1950s with implants of collagen gels imbibed with hyaluronan. He realized that hyaluronan could have great potential in implants and he developed techniques to purify the polysaccharide to such an extent that it would not give any tissue reactions due to impurities. In a seminal paper at a conference in Turku in 1972 he summarized his work on the effect of concentrated high-molecular weight hyaluronan on lymphomyeloid cells and fibroblasts and established the foundations for using hyaluronan solutions in medical treatments.

Rydell, Butler and Balazs published a paper in 1970 in which they showed that racehorses with traumatic joints rapidly recovered after treatment with intra-articular injections of concentrated hyaluronan solutions. In 1972 Bandi convinced the Pharmacia Pharmaceutical Company in Uppsala to start production of hyaluronan according to his methods for the treatment of arthritis in humans and horses and for replacing vitreus in retinal detachment.

Some years later hyaluronan solutions were used by David Miller and Robert Stegmann, on the advice of Bandi, to protect the structures in the anterior chamber of the eye during artificial lens implantations. This became a great success and around 1980 a solution of 1% hyaluronan prepared from rooster combs (Healon®) to be used for eye surgery became one of Pharmacia's largest products.

Bandi became professor of Experimental Ophthalmology at Columbia University in 1975 and worked in New York until 1985. During this time he concentrated on the ophthalmological use of hyaluronan. Since then he has devoted his time to his company, Biomatrix Inc in New Jersey, where he is developing new products based on hyaluronan, especially ones for joint treatment. Several chapters in this book will describe new practical applications. Having pioneered the use of hyaluronan for medical use he is now competing with biotechnology companies all over the world who have realized the importance of this polysaccharide.

Bandi is not only a great scientist and entrepreneur, he is also a superb administrator, which is seen not only in his building of laboratories and biotechnology companies but also in starting scientific organizations, arranging international symposia, editing books and publishing journals. We have all benefitted from his proficiency in this respect.

It was difficult to convince Bandi to accept a symposium in his honour and it needed some persuasion. I am glad that I succeeded. I must also admit that he has had a great influence on the programme of the conference, which helped in making it a very successful meeting. All who were present paid their homage to a great man and a great friend.

Chemical morphology of hyaluronan

J.E. Scott

Chemical Morphology, Chemistry Building, Manchester University, Oxford Road, Manchester M13 9PL, U.K.

Introduction

I chose the title of this paper while recalling years of enormously enjoyable discussions with Endre Balazs. Although I believe I have the first Chair of Chemical Morphology, he was a chemical morphologist a decade before I completed my Ph.D. thesis. Our discussions were passionate, prolonged and noisy to an extent that light sleepers rose from their beds to enquire whether police action was necessary. No one had hitherto stimulated me to that extent, and my intuitive feeling, that the only line worth taking was that which led to a bridge between structure and function, was greatly strengthened. From the comparative isolation I had known until then Balazs emerged as a kind of role model, respected, admired and challengeable. Of all the workers I have known in this field, Endre Balazs and Karl Meyer *consistently* surprised and interested me.

When I first met him, in 1958, he was much involved with vitreous [1], which he saw as a model connective tissue. A few years later, working with Robin Stockwell, I interpreted our results on cartilage in terms of a matrix of collagen fibrils inflated by 'mucopolysaccharides', which took the compressive stresses that cartilage is designed to withstand. I was struck by the formal analogy between Balazs' scheme for vitreous and our mechanism for cartilage elasticity. On further development and analysis, a modified and generalized version of Balazs' diagram became the flag under which I marched for well over a decade [2]. Even though it was superseded as further ultrastructural data became available, it remains a valuable introduction to fundamental concepts of extracellular matrix (ECM).

The 'mucopolysaccharide' of human vitreous is hyaluronan (HA), on the distribution, turnover and, later, clinical use of which Balazs has done much very valuable pioneering work. The change from the nomenclature of 'mucopolysaccharide' to 'glycosaminoglycan' (GAG) was proposed by Balazs and Jeanloz, and it is still the most recent (now 30 years old!), generally accepted attempt to bring order into the chaotic terminology of this field.

As 'hyaluronic acid', HA was characterized in the vitreous by Karl Meyer, who went on to identify and elucidate the structures of the different chrondroitin, dermatan (dermochondan) and keratan sulphates. All were polymers of repeating disaccharide units connected via alternating 1→3 and β1→4 glycosidic bonds. HA produced solutions of much higher viscosity than the others and this intrigued biophysicists and physiologists. Balazs understood that the physical properties of HA were fundamental to the behaviour and structure of vitreous. The groups of Balazs, Ogston and Laurent (and others) mounted extensive programmes to investigate the subtleties of hyaluronan and its interactions with other biocolloids.

The conclusions of Laurent [3], that hyaluronan behaves as a random coil with considerable intrinsic stiffness, were based on bulk solution properties extrapolated back to infinite dilution, as the physical and theoretical models required. At higher concentrations entanglement was postulated to occur. Ogston [4], using similar techniques, considered a 'mutable cage' model, perhaps with evanescent aggregation of HA molecules. Early electron microscopy by Jensen and Carlsen [5] and Rowen and co-workers [6] showed long, thin anastomosing threads of approximately molecular dimensions.

HA, a non-interacting molecular stuffing?

I was intrigued by the problems of defining randomness. It seemed likely that interactions with other biopolymers would be less avid and less specific in the absence of order at some level, because of the probable loss of entropy in moving from a disordered state to a structured complex. The alternative scenario was that HA had developed during evolution to interact with nothing, and this was not unthinkable. As the perfect non-interacting molecular stuffing HA might have a role, e.g. in rooster comb, Wharton's jelly or the tumescent sex skin of female baboons. Possibly even in the vitreous.

Then I obtained evidence that HA was not random, in the sense that it had a secondary structure. Periodate was found to probe the reactivity of the glycol group in the uronate of the repeating disaccharide unit of a series of monomers and polymers. In HA and the chondroitins the glycol environment was quite different from that in other GAGs and model monomers, so that the rate of periodate oxidation was much reduced.

A secondary structure involving intra-residue H-bonds was proposed [7] that was later found to be compatible with X-ray fibre data, when two more H-bonds were added to those already suggested [8]. Intra-residue bonds would be expected to stiffen the HA chain, in accord with the results of Laurent [3], which suggested a statistical unit in the Kuhn random walk treatment of about 30 disaccharides, implying great stiffness.

The presence of all the proposed H-bonds in dimethyl sulphoxide (DMSO) solutions of HA was proven by NMR in a series of papers (see, for example, Scott and co-workers [9]) with Frank Heatley of the Chemistry Department, Manchester University. It appears from this structure (Figure 1a) that almost every polar group (OH, acetamido and -COO^-) is hooked up in intramolecular interactions. Potential interactions with other species are therefore weakened, which is desirable in something that fills space in tissues. It will then not sequester passing molecules, thus avoiding clogging up communication channels [10]. HA did indeed seem to be an entropy-rich, perfect, non-interactive stuffing.

Figure 1

(a)

(b)

Secondary structure of HA

(a) Tetrasaccharide fragment from HA showing four hydrogen bonds (single dashed lines) plus a water bridge between the acetamido NH and the -COO⁻ groups. The pairs of transglycosidic protons (●) are subject to strong NOEs (nuclear Overhauser effects) in water, which establishes the presence of a predominantly twofold helical configuration [13], as shown. (b) Pentasaccharide constructed from Courtauld space-filling atoms, showing hydrogen bonds (arrowed) and the extensive hydrophobic patch (CHs with + signs) stretching across about three sugar residues in the twofold helix as shown in (a). This

In the interim it was found that spin couplings and NH chemical shifts of the 2CH–NH groups in water were incompatible with the all-*trans* arrangement seen in DMSO, but that there was a water bridge instead of a direct NH→COO⁻ H-bond in aqueous solutions, thus bringing the observations into line [11]. Although the effect of a water environment was to weaken the H-bonds [12] by competition, NOEs (nuclear Overhauser effects) showed that the twofold helix was the preferred structure in water as it was in DMSO, not only for HA but also for chondroitins and keratans [13]. In fact, water plays a part in stabilizing the twofold helix, as shown by computer simulation [14].

Hydrophobic bonding

On subsequent examination [11] of the HA secondary structure, which had been published nine years previously, it became clear that a most important aspect had been overlooked. There was a large hydrophobic patch of 8CHs, equivalent to octane, stretching across three sugar units (Figure 1b) that had not previously been noticed [15]. This suggested that completely different, *non-polar*, interactions were likely, and the observations that had begun to accumulate on the biological multipotency of HA, including interactions at high dilution, began to make sense.

Several examples of hydrophobic interactions of biological significance have recently been discovered and interpreted on the basis of this structure. Ghosh and co-workers [16], using NMR, showed that phospholipids, including platelet-activating factor, interacted with HA. They suggested that the latter and similar lipids could be bound to HA in synovial fluid, thus limiting their pharmacological activity. At a higher level, HA was found to increase the fluidity of red

Figure 2

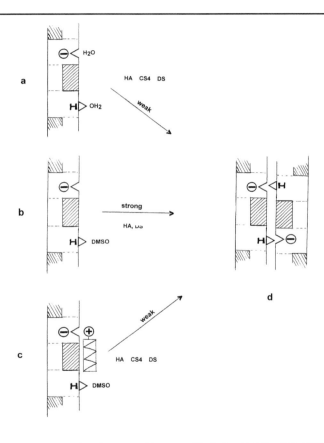

Scheme illustrating the occurrence of hydrophobic and H-bonds in complexes between HA (and also dermatan sulphate, DS) molecules

(a)–(c) Unaggregated forms of GAGs (e.g. HA) in water (a) or DMSO [(b) and (c)]. The complexed forms are shown in (d) in which the hydrophobic (hatched) patches come together, reinforced by H-bonds. In an aqueous environment [form (a)], hydrogen donating and accepting groups are blocked by H$_2$O. In dry DMSO only H-donor groups are blocked by the solvent so that H-bonding and hydrophobic bonding are possible with a maximal increase in viscosity. Decyl or other long aliphatic chains block the hydrophobic patches in form (c), thus reducing interaction avidity and lowering viscosity. See also [13], the text and Table 1. CS4, chrondroitin sulphate 4.

	1 H$_2$O Na$^+$ (η)	2 DMSO Na$^+$, Li$^+$ (η)	3 DMSO C$_{10}$H$_{21}$N$^+$(CH$_3$)$_3$ (η)	4 Computer simulation H$_2$O	5 Electron microscopy H$_2$O	**Table 1**
Hyaluronan	++	++++	++	+++	++++	
Dermatan sulphate	+	++++	+	?	++	
CS-6 sulphate	±	±	ND	++	++	
CS-4 sulphate	±	±	±	−	−	
Keratan sulphate	+	±	±	+	?	

Evidence for aggregation of various salts of anionic GAGs in DMSO or H$_2$O solution from viscosity (η), computer simulation or electron microscopy

The + signs are a qualitative estimate of the strength of the observed interaction. An increase from column 1 to 2 probably indicates increased intermolecular H-bonding, a decrease from column 2 to 3 indicates decreased inter-GAG hydrophobic bonding (see text). The +s in column 1 for dermatan and keratan sulphates are based on light scattering and gel chromatographic data, respectively (Scott [18]). ?, Result unclear; ND, not determined; −, no aggregation. See also Scott and co-workers [13].

blood cell membranes [17]. Thus, HA is a new kind of amphiphile, being simultaneously highly polar while possessing marked abilities to interact in specific ways with non-polar structures.

The interesting question was, could it interact with itself? Several lines of evidence suggested very strongly that it could. Computer simulations using molecular dynamics in 'water' (i.e. with a large number of H$_2$O molecules as the 'environment') implied that aggregation would be favoured [14]. More directly, electron micrographs of rotary-shadowed preparations contained ordered, aligned aggregates of long, thin filaments that could ramify and rejoin a meshwork that resembled an irregular honeycomb [14]. The combination of molecular dynamics and electron microscopy permitted an overall view [18]. Aggregation was driven by hydrophobic and hydrogen bonding (Figure 2), but opposed by electrostatic repulsion. This is analogous to formation of the DNA double helix.

Direct demonstration of this mechanism was not envisaged until I recently recognized that an experimental system had been in the literature for some years that demonstrates the importance of the two interactions, hydrophobic and hydrogen bonds [13]. When the Na$^+$ salts of some anionic ECM GAGs are dissolved in dry DMSO they produce solutions of dramatically high viscosity (almost jellies). This indicates that interactions between these GAGs are very much stronger in dry DMSO than they are in water. These are the GAGs that computer simulation and electron microscopy showed or suggested can aggregate in water. The solution in dry DMSO can then be manipulated in ways that are not open to aqueous solutions.

Starting from the very viscous solutions in dry DMSO: (1) the addition of a small quantity of water (<10% v/v) is accompanied by a sharp drop in viscosity. The simplest explanation is that proton-accepting groups on the GAGs are solvated by the added water, as already demonstrated by NMR [11], thus weakening interactions with proton donors on other GAG molecules (Figure 2).

Figure 3

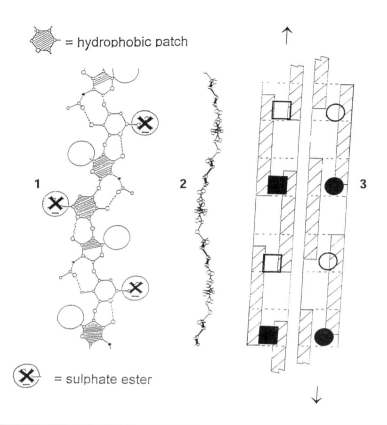

= hydrophobic patch

= sulphate ester

Secondary and tertiary structures of HA

(1) and (2) show the geometry of the ECM GAGs in plan and side view, respectively. They have a strong preference to form twofold helices [13], which have a gentle waveform in the plane of the illustration (1) and also at right angles to this aspect, as shown in (2). When two such molecules are aligned antiparallel as in (3) the two waves complement each other and the hydrophobic patches (hatched areas) come together as shown, (3). Arrows indicate the polarity. That in (3) is a side view of the duplex, with transverse dotted lines delimiting a sugar unit. Simultaneously, the -COO⁻ (□) and acetamido NH (○) groups are able to H-bond to each other. The filled symbols are on the same 'edge' of the tapelike molecules, opposite to that of the open symbols [19].

(2) If the Na^+ or Li^+ cations are replaced by decyltrimethyl ammonium, the GAG solutions are not viscous. The simplest explanation is that the decyl chains align with the hydrophobic patches, competing with the hydrophobic patches on other GAG molecules and thus weakening lyophobic bonding between GAGs.

The presence of bulky aliphatic chains between GAG molecules might sterically hinder H-bonding, thus further weakening interactions between GAGs.

(3) The phenomena are stereospecific, since the marked difference between chrondroitin sulphate 4 and dermatan sulphate would not be expected if only bonding between randomly placed sites was responsible for the interactions.

(4) Both H-bonds and lyophobic bonding are necessary to give stable interactions at room temperature.

Putting together the above facts, and taking into account that a good fit between two GAG molecules can be achieved only if they are oriented antiparallel to each other, the structure shown in Figure 3 is proposed (see also Scott [19]). In this context, Atkins and co-workers [20] suggested an acetamido–carboxylate bond between HA molecules in fibres, based on X-ray diffraction studies.

Meshworks

Important corollaries emerged. Low-molecular-mass HA at low concentrations formed fragmented networks, in contrast to high-molecular-mass HA, where the meshwork was infinite at very low concentrations [14]. It is relevant that Rees and co-workers [21] found that an admixture of low-molecular-mass HA dramatically altered the rheological properties of high-molecular-mass HA, which they tentatively suggested was an aggregation-breaking effect. Their result is difficult to explain in terms of entanglement in the high-molecular-mass HA solution. Molecular-weight-dependent meshwork-forming behaviour explained how low-molecular-mass HA was less able to protect cells against fluxes of free radicals generated by the enzyme glucose oxidase coupled to a Fenton-type system than high-molecular-mass HA at the same concentration ([22], Figure 4). The

(a)   **Figure 4**

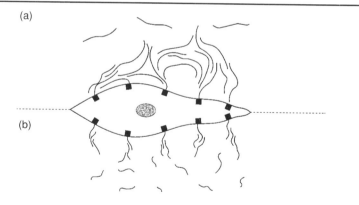

(b)

Cell-anchored HA meshworks that hinder the approach to the cell membrane of particles etc. that may cause damage

In the upper half (a) high-molecular-mass meshworks contain overlapping and reinforcing HA molecules, effectively forming a palisade through which penetration by large particles is prevented. In the lower half (b) low-molecular-mass HA molecules are not long enough to overlap and form coherent lamellae as in (a), leaving channels through which large particles can approach and perhaps damage the cell. A practical demonstration of this possibility used an enzyme (glucose oxidase) as a large particle that generated OH· free radicals in a Fenton-type system. High-molecular-mass HA was much more effective in preventing free-radical damage to the cells than low-molecular-mass HA [22].

overlapping lamellar structure proposed in this picture was speculative at the time, but such structures do occur, as shown in Figure 5 (see below).

Apart from structural implications in the tissues, these results have pharmacological significance. Chang and co-workers [23] discovered that HA solutions that had been heated and then rapidly cooled had a strong inhibitory effect on complement-activated lysis of erythrocytes, which was lacking if the heated solution was cooled slowly. They interpreted this as being due to a reversible denaturation or annealing of a structure present and stable at physiological temperatures. They had no basis on which to formulate this structure in detail. This raises interesting questions about which part of the HA molecule is hidden in the stable native form, and then uncovered by heating and dissociation. The authors suggested that this phenomenon might also extend to other activities. HA might be a reservoir of physiological activity that is released under conditions in which aggregates dissociate or break down.

HA as a tissue organizer

HA is present in large amounts in young and developing ECMs [24,25]. As such it is possible that it plays a part in organizing the tissue, as do other GAGs in the ECM [26]. The simplest cell cultures that produce HA are streptococci, in that no

Figure 5

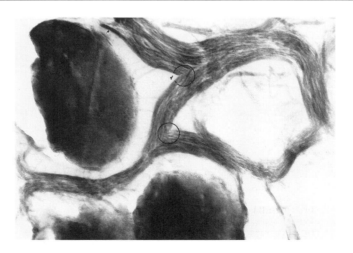

Cupromeronic-Blue-stained (e.g. [26]) streptococci 60 min after beginning to produce HA in culture [27], ×55000

The dark cells are surrounded by a lamellar structure of HA, which is organized with parallel stained elements of HA, about 5 nm in width. Two lamellae anastomose in the (circled) structures. These lamellar structures (somewhat like the conjectural structures in Figure 4) overlap several cells, possibly acting to orient the streptococci in line ahead formation, thus organizing the overall pattern of streptococci in culture. (Scott and Prehm, unpublished work).

further developmental stages of a more complicated kind are produced, in contrast to, for example, mammalian fibroblasts. No clear reasons why streptococci should produce HA have been adduced. We thought that it would be interesting to see if there was any ordered structure or clear production pathway from streptococci, using Cupromeronic Blue for electron microscopic demonstration of HA.

Streptococci were grown and deposited on cellulose acetate membranes after thorough washing and enzymic cleaning. At zero time they had no obvious Cupromeronic-Blue-stained material attached to their exteriors. After 60 min they were accompanied by an ordered sheath of stained HA that was arranged in clear patterns, mainly connecting cells together. There was a lamellar structure similar to that shown in Figure 4, which has potential for amalgamating with and continuing along the direction in which the strands of the lamellae were oriented.

At higher magnification the ordered directed structure with anastomoses is clearly visible (Figure 5). It seems possible the HA is helping to keep the cells together in a linear fashion, which is typical of the way streptococci grow and replicate.

It is tempting to see this as the simplest case of tissue organization, dependent on the specific properties of ECM molecules, to produce an ECM organization in which the cells are oriented and guided. The relevance of this finding to tissue development is an interesting question.

References

1. Balazs, E.A. (1981) Biomed. Foundations Ophthalmol. **1**, 1–16
2. Scott, J.E. (1972) Philos. Trans. R. Soc. London Ser. B **271**, 235–242
3. Laurent, T.C. (1957) Arkiv Kemi **11**, 487–496
4. Preston, B., Davies, M. and Ogston, A.G. (1965) Biochem. J. **96**, 449–471
5. Jensen, C.E. and Carlsen, F. (1954) Acta Chem. Scand. **8**, 1357–1360
6. Rowen, J.W., Brunish, R. and Bishop, F.W. (1956) Biochim. Biophys. Acta **19**, 480–489
7. Scott, J.E. and Tigwell, M.J. (1978) Biochem. J. **173**, 103–114
8. Atkins, E.D.T., Meader, D. and Scott, J.E. (1980) Int. J. Biol. Macromol. **2**, 318–319
9. Scott, J.E., Heatley, F. and Hull, W.E. (1984) Biochem. J. **220**, 197–205
10. Scott, J.E. (1980) Conn. Tissue (Japan) **11**, 111–120
11. Heatley, F. and Scott, J.E. (1988) Biochem. J. **254**, 489–494
12. Sicinska, W., Adams, B. and Lerner, L. (1993) Carbohydr. Res. **242**, 29–51
13. Scott, J.E., Heatley, F. and Wood, B. (1995) Biochemistry **34**, 15467–15474
14. Scott, J.E., Cummings, C., Brass, A. and Chen, Y. (1991) Biochem. J. **274**, 699–705.
15. Scott, J.E. (1989) in The Biology of Hyaluronan (Evered, D. and Whelan, J., eds.), Ciba Foundation Symp. No. 143, pp. 6–20, Wiley, Chichester
16. Ghosh, P. Hutadilok, N. and Lentini, A. (1994) Heterocycles **38**, 1757–1774
17. Clarke, D.N. and Sirs, J.A. (1988) Int. J. Microcirculation Clin. Exp. (Special issue, August), S153
18. Scott, J.E. (1992) FASEB J. **6**, 2639–2645
19. Scott, J.E. (1994) Biochem. J. **298**, 221–222
20. Atkins, E.D.T., Phelps, C.F. and Sheehan, J.K. (1972) Biochem. J. **128**, 1255–1263
21. Welsh, E.J., Rees, D.A., Morris, E.R. and Madden, J.K. (1980) J. Mol. Biol. **138**, 375–382
22. Presti, D. and Scott, J.E. (1994) Cell Biochem. Function **12**, 281–288
23. Chang, N-S., Boackle, R.J. and Armand, G. (1985) Mol. Immunol. **22**, 391–397
24. Toole, B.P., Jackson, G. and Gross, J. (1972) Proc. Natl. Acad. Sci. U.S.A. **69**, 1384–1386
25. Laurent, T.C. and Fraser, J.R. Hyaluronan (1992) FASEB J. **6**, 2397–2404
26. Scott, J.E. (1992) Eye **6**, 553–555
27. Mausolf, A., Jungmann J., Robenek, H. and Prehm, P. (1990) Biochem. J. **276**, 191–196

4

Hyaluronan interactions: self, water, ions

Mary K. Cowman*, Jianshe Liu, Min Li, Daniel M. Hittner and Jong Soo Kim
Department of Chemical Engineering, Chemistry, and Materials Science, Polytechnic University, Six Metrotech Center, Brooklyn, New York, 11201, U.S.A.

Abstract

Hyaluronan (HA) may be visualized by atomic force microscopy (AFM) as either isolated chains or a network-like matrix of chains. The network structures illustrate the tendency of HA to self-associate on a mica surface where the local concentration is high. Using a combination of light scattering and polyacrylamide-gel electrophoresis, weak intermolecular association of HA segments can also be demonstrated in aqueous solutions containing NaCl but not KCl. The HA networks, formed by interpenetration and by self-association, have significant effects on the nucleation and growth of ice crystals. These data show the importance of understanding the effect of HA on water structure and dynamics in relation to the formation of networks of interacting chains.

Results and discussion

It has been well established that hyaluronan (HA) can self-associate. Both intra- and intermolecular associations can exist under physiological conditions, creating a network-like matrix of HA chains. We have been interested in the structure, stabilization and consequences of such self-associations.

Electron microscopic investigations of HA have shown isolated HA chains, as well as networks of HA chains in which inter-chain association is apparent [1,2]. Recently, we have begun to examine HA by tapping-mode AFM. One advantage of AFM is that the resulting image is three-dimensional, allowing access to height information. Sample preparation is also simple. In tapping-mode AFM, the HA sample may be applied to a mica surface as a solution, rinsed and dried under nitrogen and then imaged in air through the remaining water layer, which is a few nanometres thick. When HA is applied to mica as a very dilute (1 μg/ml) aqueous solution containing 10 mM $MgCl_2$ (to minimize electrostatic repulsion between the HA chains and the mica surface), isolated chains may be observed (Figure 1). The contour length of the rooster comb HA chain shown is approximately 4 μm. Since the projected length of an HA disaccharide is generally 0.8–1.0 nm in favoured conformations, the contour length probably corresponds to a chain containing 4000–5000 disaccharides, or a molecular weight of NaHA of $1.6–2.0 \times 10^6$. This is in good agreement with the expected average molecular

Figure 1

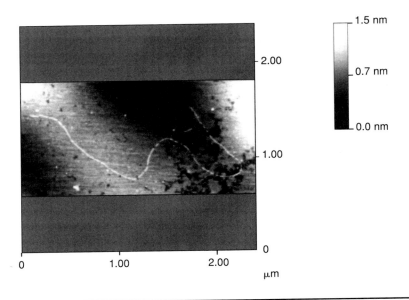

Tapping-mode atomic force microscope image of rooster comb HA, applied to mica as a 1 μg/ml solution in 10 mM MgCl$_2$

The sample was rinsed with H$_2$O, dried under a gentle stream of N$_2$ and imaged in air at ambient temperature and humidity. Height range was 1.5 nm.

weight of HA in this sample. The height of the HA chain was measured as 0.5–0.6 nm, which is also in good agreement with the diameter of a HA chain. If the same HA is applied to mica as a 100-fold higher concentration solution in physiological phosphate-buffered saline, the AFM image no longer shows isolated HA chains. Rather, a network of chains is seen (Figure 2). Some chains appear to be crossing, with an apparent build-up of material at the intersections. Other chains appear to be diverging from double-stranded segments. It is important to note that a solution of HA at the stated concentration and molecular weight is well below the coil overlap concentration. HA chains in such a solution can be subjected to chromatographic or electrophoretic separation, or analysed by physicochemical methods, usually without serious complication due to intermolecular self-association. Thus the network seen by AFM is largely a consequence of the collapse of chains on to the mica surface. It does, however, illustrate the tendency of HA chains to form associations as the local concentration increases.

What stabilizes self-association of HA? For high-molecular-weight HA, it has been reported that self-association is present in aqueous solution containing NaCl, but not KCl [3]. This report has been disputed by other researchers [4]. Because studies on high-molecular-weight HA can be complicated by competition between intra- and inter-molecular associations, we have addressed this question using short HA segments, generated by enzymic digestion of the polymer. In previous studies [5], performed in 0.15 M NaCl, we found that HA segments containing approximately 35 or more disaccharides can exhibit intramolecular association (chain folding or coiling). Shorter HA segments,

Figure 2

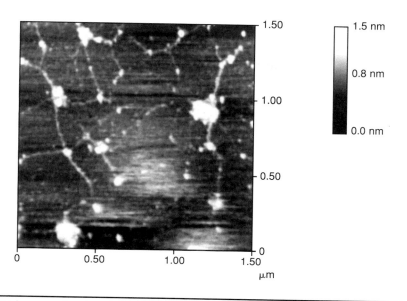

Tapping-mode atomic force microscope image of rooster comb HA, applied to mica as a 100 μg/ml solution in phosphate-buffered saline
Other conditions were as given in Figure 1.

approximately 7–35 disaccharides in length, can exhibit only intermolecular association. A comparison between the effects of NaCl and KCl in promoting self-association of HA segments is shown in Figure 3. Low-angle laser light scattering analysis was performed on HA segments whose molecular weights were independently determined by polyacrylamide-gel electrophoresis (PAGE). When studied in 0.15 M KCl, HA segments averaging either 27 or 40 disaccharides in length give good linear extrapolations to yield molecular weights slightly below those determined by PAGE (and adjusted for the disaccharide residue weight of K-HA). We could find no strong evidence for intermolecular self-association. In 0.15 M NaCl, the larger HA segments also gave a good linear extrapolation to yield a molecular weight slightly below that determined by PAGE for Na-HA. The shorter HA segments in NaCl solution are less well-fitted to a linear extrapolation, which, if done, would yield a molecular weight higher than that found by PAGE (13 200 versus 10 800). If a concentration-dependent monomer–dimer equilibrium is used as a model to fit the data (with the monomer molecular weight set to that found by PAGE), we can fit the observed concentration dependence using an association constant of 2.4×10^3 M^{-1}. This is approximately one-quarter the association constant found by Fransson and co-workers [6] for dermatan sulphate in 0.5 M sodium acetate. Thus we see a relatively weak association of short HA segments in NaCl, but not KCl. Longer HA segments in NaCl can satisfy the self-association tendency by intramolecular interaction. Future models for HA self-association should take into account the ion specificity. It is possible that counter-ion bridges are involved in inter-chain

Figure 3

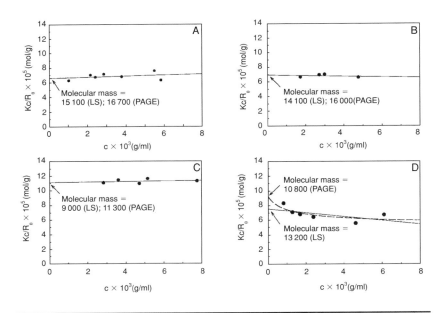

Low-angle laser light scattering analysis of HA segments in 0.15 M NaCl or 0.15 M KCl

The solid lines represent linear least-squares fits to the data. The dashed line is the best fit to a monomer–dimer equilibrium. (A) HA averaging 40 disaccharides in length, in 0.15 M KCl; (B) HA averaging 40 disaccharides in length, in 0.15 M NaCl; (C) HA averaging 27 disaccharides in length, in 0.15 M KCl; (D) HA averaging 27 disaccharides in length, in 0.15 M NaCl. Kc, optical constant of 1.71×10^{-7} mol · cm^2 · g^{-2}; c, HA concentration; R_θ, excess Rayleigh factor.

associations. Alternatively, differences in water structure in the two salt solutions may result in different degrees of stabilization for associated regions.

While water structure may influence HA self-association, the converse is certainly true; HA self-association influences water structure and dynamics. One method we have used to examine water in HA solutions is to measure the linear crystallization velocity (LCV) of an ice front moving through a narrow tube containing pure solvent or HA solution. In this experiment, a U-shaped tube containing the liquid sample is placed in a cooling bath whose temperature is controlled to within 0.1 °C. The bath is then cooled to a chosen temperature below the freezing point. This is the undercooling temperature, and it affects the rate of ice growth. Ice formation is nucleated by adding a small frozen piece of the same solution to one end of the tube. Growth of the ice proceeds across the tube, and the rate at which the ice front moves is measured. For pure water or saline solution, the LCV increases as the degree of undercooling increases. The effect of HA at a concentration of 1 mg/ml on the LCV in 0.15 M NaCl solution is shown in Figure 4. Several different HA samples, ranging in molecular weight from 1.5×10^4 to 6×10^6, were tested. In all cases, the LCV was lower in HA-containing solutions than for solvent alone. Similar observations have been made

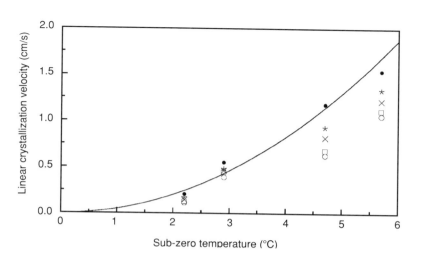

Figure 4

Linear crystallization velocity of ice in 0.15 M NaCl solutions containing HA at 1 mg/ml

Type of HA and its molecular weight: (●) no HA; (*) HA segment, 2.3 $\times 10^4$; (X) bovine vitreus HA, 2.5 $\times 10^5$; (□) bacterial HA, 2 $\times 10^6$; (○) rooster comb HA, 6 $\times 10^6$.

for other polysaccharides such as alginate, carrageenan, guar gum, locust bean gum and tragacanth gum [7]. However, the magnitude of the HA effect, even at the smallest molecular weight studied, is almost the same as the reported effect for a tenfold higher concentration of the other polysaccharides. We could not test such high concentrations of HA because ice propagation was so slow that spontaneous nucleation within the solution caused the solution to freeze before significant movement of the ice front from the added nucleus occurred. Antifreeze glycoprotein also has a large effect on the LCV of water, of a similar magnitude to that seen for HA [8]. This effect is believed to depend on a specific interaction between the glycoprotein and the prism faces of the growing ice crystal, which then is retarded in its growth. While such an interaction is also possible for HA, it should not be strongly dependent on the HA's molecular weight. In Figure 4, it can be seen that high-molecular-weight HA has a much greater effect than low-molecular-weight HA. We have observed that the reduction in LCV relative to solvent alone is almost linearly dependent on the logarithm of the HA molecular weight. It is also nearly linearly dependent on the logarithm of the HA concentration, or the product of the HA concentration and the intrinsic viscosity, known as the coil overlap parameter (J. Liu and M.K. Cowman, unpublished work). Thus the effect of HA on the growth rate of ice crystals may primarily reflect the influence of the HA network in reducing the rate of diffusion of water molecules to the surface of the growing ice crystal.

HA can also affect the nucleation of ice crystals. Nucleation requires the formation of ice clusters of sufficient size for subsequent propagation to be favourable. We have investigated the effect of HA on ice nucleation by using differential scanning calorimetry. In a scan performed at a cooling rate of 3 °C/min, water alone shows the onset of freezing not at the equilibrium freezing

Figure 5

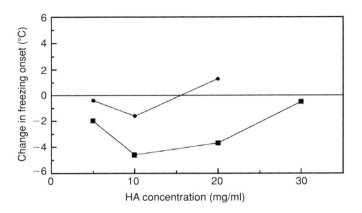

Effect of bacterial HA (molecular weight 2×10^6) on the temperature for the onset of freezing in aqueous solution

Freezing onset was measured relative to pure solvent, by differential scanning calorimetry, cooling at 3 °C/min. (▪)HA in H_2O; (●) HA in 0.15 M NaCl.

point, but at an undercooled temperature of approximately -9 °C. The presence of HA in the solution can further affect the temperature at which the onset of freezing occurs (Figure 5). At a concentration of 5–10 mg/ml, HA inhibits nucleation, which is manifested as a reduction in the temperature for freezing onset in the dynamic scan. Such an effect has also been observed for some cross-linked polymer gels, and found to be dependent on the average pore size in the polymer gel [9]. As the pore size is reduced, the formation of critical size nuclei becomes less likely, and the onset temperature for freezing is reduced. For an HA network, the pore size should decrease as HA concentration increases. Surprisingly, above 1% concentration, HA solutions show a reduced inhibition of nucleation. The change in behaviour is more marked in saline solution than in water. Since self-association of HA is favoured at higher HA concentrations, and in NaCl-containing solution, it appears that self-associated HA is less inhibitory of ice nucleus formation, and may under some circumstances even aid nucleation. It is possible that this effect is due to the formation of associated groups of HA chains, which may act as specific aids to ice nucleation by pre-ordering water molecules. In this respect, associated HA chains may act in a manner similar to bacterial membrane ice nucleation proteins [10].

The effects exerted by HA on ice nucleation and growth indicate a significant influence on water structure and dynamics. We have recently found that HA reduces the enthalpy change associated with the melting of water in 1% HA solutions by approximately 4% (J. Liu and M.K. Cowman, unpublished work). This is much greater than the effect expected on the basis of the small amount of water tightly bound by HA. Such an observation indicates that the structure of water in the vicinity of an HA chain differs from that of bulk water, so that the ice → water transition reflects a different final state than that of bulk water. Theoretical analysis of water organization surrounding simple monosaccharides suggests that this is a phenomenon of general importance for carbohydrates [11].

Experimental methods

Rooster comb HA samples (Pharmacia) were the kind gifts of Drs E.A. Balazs and K. Janné. HA samples derived from human umbilical cord, bovine vitreus and *Streptococcus zooepidemicus* were obtained from Sigma. Molecular weights were determined by agarose-gel electrophoresis [12]. HA segments were isolated as described previously [5], from bovine testicular hyaluronidase digests of human umbilical cord HA, except that fractionation was by size exclusion chromatography on a TSK G4000SW column in 0.15 M NaCl. Molecular weights were determined by PAGE [13]. Atomic force microscopy was performed on a Nanoscope IIIa Multimode scanning probe microscope, equipped with a type EV scanner. Images were obtained in tapping mode, using etched silicon cantilever probes of 125 μm length, at a drive frequency of approx. 300 kHz, and a scan rate of 1 Hz. Low-angle laser light scattering was performed as described previously [5]. Measurements of the linear crystallization velocity of ice were performed by the method of Budiaman and Fennema [7]. Differential scanning calorimetry was performed on a TA Instruments Model 2920, using 10 mg samples in hermetically sealed coated aluminium sample pans, cooled at a rate of 3 °C/min.

This manuscript is dedicated to Dr. Endre A. Balazs, in appreciation of his continuing support, encouragement and interest. From the introduction to this symposium lecture by M.K.C.: "It has been my great pleasure to have worked with Bandi since 1979, when I came to his laboratory at Columbia as a postdoctoral fellow. I had no way of knowing then that such a long-term and enjoyable collaboration would follow. One of my first strong memories of Bandi is the expression on his face when he demonstrated the phenomenon of the viscoelastic putty formation by a 1% hyaluronan solution. This dramatic transition of a viscous solution to a cohesive mass that can be pulled out of a beaker and played with endlessly is by all means a good show. But for Bandi it was more than that. He exhibited a childlike fascination with the mysteries of hyaluronan, the type of wonderment that leads a great scientist to continue working on a problem for the sheer joy of discovery. Over the years we have pondered together many questions about hyaluronan's structure and behaviour. Some of the mysteries have been solved, but many remain. Recently, Bandi acquired an atomic force microscope. The look on his face when he discusses the potential for this method to answer some of the long-standing questions about hyaluronan seems the same today as that I saw 17 years ago. Bandi, it has been a most happy voyage of discovery to work with you, and my co-workers and I dedicate to you these first images recorded for hyaluronan with your latest toy, the atomic force microscope." We thank Dr. Norman Peterson for assistance in data fitting analysis. The thermal analysis work was performed at the Edith Turi Thermal Analysis Laboratory of the Herman F. Mark Polymer Research Institute of Polytechnic University. Part of this work was supported by NIH grant EY 04804.

References

1. Scott, J.E., Cummings, C., Greiling, H., Stuhlsatz, H.W., Gregory, J.D. and Damle, S. (1990) Int. J. Biol. Macromol. **12**, 180–184
2. Scott, J.E., Cummings, C., Brass, A. and Chen, Y. (1991) Biochem. J. **274**, 699–705
3. Sheehan, J.K., Arundel, C. and Phelps, C.F. (1983) Int. J. Biol. Macromol. **5**, 222–228
4. Månsson, P., Jacobsson, Ö. and Granath, K.A. (1985) Int. J. Biol. Macromol. **7**, 30–32
5. Turner, R.E., Lin, P. and Cowman, M.K. (1988) Arch. Biochem. Biophys. **265**, 484–495
6. Fransson, L.A., Nieduszynski, I.A., Phelps, C.F. and Sheehan, J.K. (1979) Biochim. Biophys. Acta **586**, 179–188
7. Budiaman, E.R. and Fennema, O. (1987) J. Dairy Sci. **70**, 534–546
8. Kerr, W.L., Osuga, D.T. and Feeney, R.E. (1987) J. Cryst. Growth **85**, 449–452
9. Arndt, K.F. and Zander, P. (1990) Colloid Polymer Sci. **268**, 806–813
10. Hew, C.L. and Yang, D.S.C. (1992) Eur. J. Biochem. **203**, 33–42
11. Liu, Q. and Brady, J.W. (1996) J. Am. Chem. Soc. **118**, 12276–12286
12. Lee, H.G. and Cowman, M.K. (1994) Anal. Biochem. **219**, 278–287
13. Min, H. and Cowman, M.K. (1986) Anal. Biochem. **155**, 275–285

5

Rheology of hyaluronan

Hege Bothner Wik* and Ove Wik
Pharmacia & Upjohn, S-751 82 Uppsala, Sweden

Introduction

Rheology is the study of the deformation and flow of matter. The rheological properties of hyaluronan have always been of interest, though mostly implicitly. The synovial fluid is viscous, lubricating and forms long threads though it contains only 0.3% hyaluronan. The viscosity of hyaluronan solutions drops dramatically upon degradation, and this fact is used for the determination of the activity of hyaluronidases. The notation 'mucopolysaccharides', previously used for hyaluronan and related polysaccharides, indicates a mucin-like, viscous behaviour.

In this paper we will describe some basic rheological properties of hyaluronan solutions [1].

Hyaluronan size and conformation

It is well-known that hyaluronan is a linear polysaccharide with a relative molecular mass (M_r) in the order of millions. In tissues, the average M_r is about 10^6, while those in pharmaceutical preparations are in the $0.5–5 \times 10^6$ range.

The very large size of hyaluronan is seldom acknowledged as it is hard to grasp the molecular dimensions on a nanometre scale. It may be easier to perceive the dimensions of a hyaluronan molecule by adopting a larger scale, using a hyaluronan molecule with a M_r of 4×10^6 as an example. This molecule contains 10 000 repeating disaccharide units. If the repeating disaccharide unit was 10 cm long, then this hyaluronan molecule would be 10 cm thick and have a length of 1000 m. It may be even easier to perceive these dimensions on a timescale: if the repeating disaccharide unit was 1 s long, the molecule would be 1 s thick and 3 h long. If the repeating disaccharide unit was 1 min, then the molecule would be 1 week long.

The linear hyaluronan molecule adopts in solution a very expanded random coil conformation. This fact was pointed out by Laurent [2]: hyaluronan adopts "a random coil configuration with a certain degree of stiffness". Though John Scott has pointed out that inter- and intramolecular interactions may give a more complicated conformation, the simplified random coil approach is sufficient when discussing and explaining the basic rheology of hyaluronan solutions.

The size of a random coil formed by a hyaluronan molecule with a M_r of 4×10^6 is so large that already at concentrations below 1 mg/ml, i.e. at less than 0.1%, the random coils fill up the whole solution (Figure 1). Therefore, at the

---

* To whom correspondence should be addressed.

Figure 1

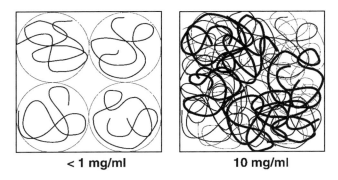

< 1 mg/ml   10 mg/ml

Schematic description of solutions containing hyaluronan random coils at concentrations where the coils fill up the whole solution, and at higher concentration where the coils form a flexible molecular network

concentrations of hyaluronan found in, for example, synovial fluid (3 mg/ml) and in pharmaceutical preparations (typically approx. 10 mg/ml) the hyaluronan random coils form a continuous flexible molecular network in solution, as outlined in Figure 1.

Zero shear viscosity

Viscosity is a measure of the resistance to flow. It is evident that the presence of extended hyaluronan random coils as well as a hyaluronan molecular network will have a profound effect on the flow properties (shear viscosity) of hyaluronan solutions. The viscosity will increase with an increasing degree of entanglement, which in turn depends both on concentration and molecular size.

The intrinsic viscosity is frequently used as a parameter for molecular size. (Please note that the 'intrinsic viscosity' is not a viscosity parameter, but simply is a measure of the volume a polymer occupies in solution in units of volume per weight.) The intrinsic viscosity of hyaluronan is strongly dependent on the dimensions of the analytical equipment used. This becomes especially problematic for hyaluronan samples with a $M_r > 1 \times 10^6$. Therefore we prefer to describe the molecular size by the mass average relative molecular mass, a parameter that currently can be determined by applying absolute techniques.

The bulk viscosity of hyaluronan solutions is, in rheological terms, determined as the zero shear viscosity at steady state. That is, the experiment is performed at such a low flow rate (rheological notation: shear rate) that the conformation of the molecules equals those at rest, i.e. the hyaluronan random coils maintain their spherical shape. In Figure 2 the zero shear viscosity is plotted as a function of the product of the concentration (in mg/ml) and the mass average relative molecular mass. The x-axis is therefore related to the degree of entanglement or coil overlap. Both scales are logarithmic, and the y-axis covers seven decades.

Figure 2

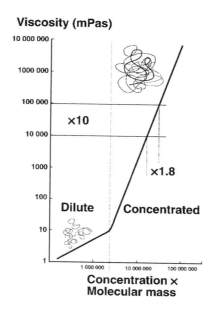

Viscosity (mPas)

×10

×1.8

Dilute   Concentrated

**Concentration ×
Molecular mass**

**Bulk viscosity of hyaluronan solutions determined as the zero shear
viscosity as a function of the product of the concentration and the mass
average relative molecular mass plotted on log–log scale**

At low concentrations or M_r (dilute region) there is no molecular
entanglement, and the hyaluronan molecules behave as separate entities. In this
region there is a slow increase in viscosity with increasing concentration or M_r
from 1 to 10 mPa s. For comparison, the viscosity of water is 1 mPa s.

Entanglement sets in when the product of concentration and M_r gives a
viscosity of about 10 mPa s. In presence of entanglement (concentrated region)
there is a sharp increase in viscosity with concentration or M_r. In this region the
slope in this log–log graph is 3.8, and therefore a less than two-fold increase in
concentration or M_r results in a 10-fold increase in viscosity.

It is often anticipated that there is little difference between hyaluronan
molecules with $M_r > 10^6$. However, the four-fold difference in molecular size
between, for example, hyaluronan of $M_r = 10^6$ and that of $M_r = 4 \times 10^6$ results in an
approximately 100-fold difference in viscosity in the concentrated region. A 1%
solution of hyaluronan with $M_r = 10^6$ will have a viscosity of 3 000 mPa s, while a
solution of hyaluronan with $M_r = 4 \times 10^6$ at the same concentration will have a
viscosity of 400 000 mPa s.

From these data it is also evident that the viscosity of hyaluronan
solutions is very sensitive to degradation of hyaluronan. A hyaluronan molecule
with $M_r = 4 \times 10^6$ contains 20 000 glycosidic bonds. Breakage of this molecule in
half owing to breakage of just one of 20 000 bonds will give $M_r = 2 \times 10^6$, and will
result in a more than 10-fold drop in viscosity in the concentrated region.

Figure 3

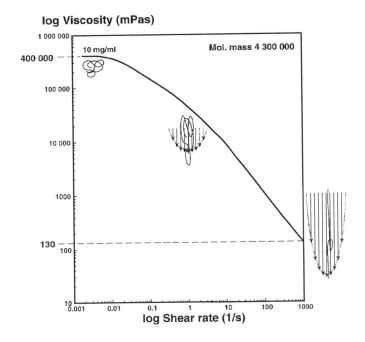

The viscosity of a hyaluronan solution (10 mg/ml, mass average relative molecular mass = 4.3 × 10⁶) as a function of shear rate on log–log scale

The decrease in viscosity with increasing shear rate is attributed to the elongation of random coils in the streamlines of flow.

However, such a drop in bulk viscosity can easily be compensated for by increasing the concentration two-fold.

Shear thinning

When subjected to increasing flow rate (rheological notation: shear rate), the hyaluronan random coils will deform and elongate in the streamlines of flow, as shown in Figure 3. As a result of this elongation the molecules will span fewer streamlines and therefore the deformed coils will affect the flow to a smaller extent. That is, the viscosity of hyaluronan solutions will decrease with increasing shear rate. The viscosity of a 1% hyaluronan ($M_r = 4 \times 10^6$) solution drops from a zero shear viscosity of 400000 to 130 mPa s at a shear rate of 1000 1/s. For an increase in the shear rate of six decades the viscosity drops more than three decades, as shown in Figure 3 where the shear viscosity is plotted as a function of shear rate on a log–log scale.

This shear thinning (pseudoplastic) behaviour is shown in Figure 4 for hyaluronan solutions at constant concentration and in Figure 5 for hyaluronan solutions at constant M_r.

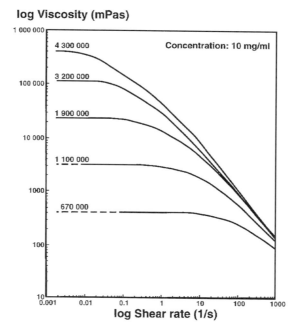

log Viscosity (mPas)

Figure 4

Concentration: 10 mg/ml

log Shear rate (1/s)

Shear dependence of the viscosity of hyaluronan solutions at constant concentration (10 mg/ml) and varying mass average relative molecular mass

At high shear rates the molecules are almost completely elongated, and the viscosity will essentially be a function of the friction between molecular rods. Therefore, at high shear rates, the viscosity will be independent of M_r (Figure 4), but dependent on the spacing between molecules, i.e. concentration. When changing the concentration at constant M_r both the zero shear and the high shear viscosities decrease with decreasing concentration (Figure 5).

This very large shear thinning (pseudoplastic) behaviour is of practical importance since pharmaceutical preparations of hyaluronan usually are delivered in syringes, and the solutions must pass through thin cannulas for proper application of the product. The shear rate in the cannulas is about 1000 1/s or above. Thanks to the shear thinning property of hyaluronan solutions, only moderate force needs to be applied to the syringe piston to expel the hyaluronan solutions through thin cannulas.

Viscoelasticity

When considering the entangled hyaluronan molecular network (Figure 1) we find an appearance that strongly resembles that of a gel. In a gel there are permanent cross-links between molecules, resulting in typical elastic behaviour. The hyaluronan molecular network will exhibit properties combining the elastic

Figure 5

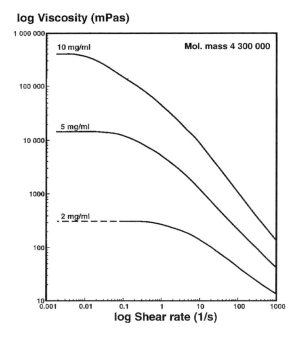

log Viscosity (mPas)

Mol. mass 4 300 000

10 mg/ml

5 mg/ml

2 mg/ml

log Shear rate (1/s)

Shear dependence of the viscosity of hyaluronan solutions at constant molecular mass (4.3 × 10⁶) and varying concentration

behaviour of a gel and the viscous behaviour of a solution. The hyaluronan solutions will be viscoelastic or, as E.A. Balazs has pointed out, more properly, elastoviscous.

The viscoelastic behaviour becomes evident when a hyaluronan solution is subjected to varying speed of impact, or subjected to an oscillating movement. When subjected to slow impact or slow oscillations, the hyaluronan molecules will have time to disentangle and the response is predominantly viscous. The gel-like, elastic character becomes evident upon fast impact or at fast oscillations when there is insufficient time for disentanglement. In these circumstances, the entangled molecular network behaves like a gel, showing an elastic response.

When studying the viscoelastic response of hyaluronan solutions various rheological parameters such as dynamic viscosity, phase angle, and elastic, viscous and complex moduli are obtained as a function of oscillating frequency. In a recent book on basic rheology, Barnes and co-workers [3] pointed out that "by common consent, rheology is a difficult subject". This statement is especially applicable to viscoelastic results. Therefore, E.A. Balazs introduced, in the mid-1980s, an alternative way to present viscoelastic data in a fashion suitable for the non-specialist. The rather complex viscoelastic parameters can be recalculated to percentage elasticity and percentage viscosity as a function of oscillating movement of the solution. This approach has been adopted in Figure 6, where the viscoelastic properties of 1% hyaluronan solutions are presented. The hyaluronan ($M_r = 4.3 \times 10^6$) solution changes from close to viscous behaviour at low

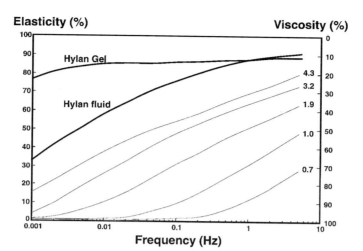

Figure 6

Plotting viscoelastic data according to a proposal by E.A. Balazs

The viscoelastic response is calculated to give percentage elasticity and percentage viscosity. The x-axis is oscillating movement (frequency) in log-scale. Thin lines denote the viscoelastic response of hyaluronan solutions at 10 mg/ml and with different mass average relative molecular mass. The figures denote M_r in millions. The viscoelastic response for modified hyaluronan produced by E.A. Balazs is also included: lightly cross-linked hyaluronan (Hylan fluid) and permanently cross-linked hyaluronan (Hylan gel beads).

frequencies to an almost elastic response at high frequencies. With decreasing molecular size there is a sharp drop in elastic response.

For comparison, the viscoelastic response of E.A. Balazs' modified hyaluronan products Hylan fluid and Hylan gel are also shown in Figure 6. Hylan fluid consists of lightly cross-linked hyaluronan molecules, resulting in an even greater elastic response, and the Hylan gel is a true gel, consisting of permanently cross-linked hyaluronan molecules. This is shown by the fact that the response is constant and almost completely elastic at all frequencies.

Historical review

When summing up previous work on the rheology of hyaluronan solutions in a limited amount of space, we find that Endre A. Balazs' contribution in this field is outstanding.

In the 1950s Balazs studied the shear rate dependence of the viscosity of hyaluronan solutions and synovial fluid [4]. In the 1960s he described the viscoelastic properties of both hyaluronan solutions and synovial fluid [5–7]. Not only did Balazs study the basic properties, but he also put these properties into biological perspective and described and explained the properties of the synovial fluid in health and disease [8]. One can rightfully claim that while Newton and Hooke, 300 years ago, introduced the concepts of viscosity and elasticity, respectively, Balazs was one of the pioneers in the field of synovial fluid and hyaluronan viscoelasticity only 30 years ago.

After the introduction of the non-inflammatory hyaluronan solutions as a surgical aid in ophthalmic surgery, Balazs explained the relationship between rheological properties and surgical performance [9–12]. It is therefore not surprising that the different polymer solutions used in this field commonly are denoted viscoelastics.

Balazs also introduced the concept of 'viscosurgery' [9–12] to describe the use of hyaluronan as well as modified hyaluronan (hylan) solutions and gels in surgery: "Viscosurgery is the use of solutions, putties and gels with viscous, elastic and pseudoplastic properties before, during or after surgery." Thanks to the introduction of this concept, a current topic of the International Organization for Standardization is a standard for 'Ophthalmic Viscosurgical Devices for use in the Anterior Segment'.

This is but a very brief summary of Balazs' outstanding contribution to the field of the rheology of hyaluronan. Both of us, all hyaluronan scientists, as well as ophthalmic surgeons, are indebted to Balazs for his pioneering work on hyaluronan in general and on hyaluronan rheology in particular. Put in perspective, our own work is but a reproduction of Endre A. Balazs' original studies, using modern instruments and extending the range of M_r and concentration.

References

1. Bothner Wik, H. (1991) PhD Thesis, Faculty of Pharmacy, University of Uppsala
2. Laurent, T.C. (1957) Physico-chemical studies on hyaluronic acid. MD Thesis, Karolinska Institute, Stockholm
3. Barnes, H.A., Hutton, J.F. and Walters, K. (1989) An Introduction to Rheology, Elsevier, Amsterdam
4. Balazs, E.A. and Sundblad, L. (1959) Acta Soc. Med. Upsaliensis. **64**, 137–146
5. Balazs, E.A. (1968) Univ. Mich. Med. Center J., Dec. 1968, Special Arthritis Issue, 255–259
6. Gibbs, D.A., Merrill, E.W., Smith, K.A. and Balazs, E.A. (1968) Biopolymers **6**, 777–791
7. Balazs, E.A. and Gibbs, D.A. (1970) in Chemistry and Molecular Biology of the Intercellular Matrix (Balazs, E.A., ed.), vol. 3, pp. 1241–1253, Academic Press, New York
8. Balazs, E.A. (1974) in Disorders of the Knee (Helfet, J., ed.), pp. 63–75, J.P. Lippincott, Philadelphia
9. Balazs, E.A. (1983) in Healon (sodium hyaluronate). A guide to its use in ophthalmic surgery (Miller, D. and Stegmann, R., eds.), pp. 5–28, Wiley, New York
10. Balazs, E.A. (1985) Transplantation Today **2**, 62–64
11. Balazs, E.A. (1986) in Ophthalmic Viscosurgery (Eisner, G., ed.), pp. 3–19, Medicöpea, Montreal
12. Balazs, E.A. (1986) in Treatment of Anterior Segment Ocular Trauma (Miller, D. and Stegmann, R., ed.), pp. 121–128, Medicöpea, Montreal

6

Hyaluronan derivatives: chemistry and clinical applications

Philip A. Band
Biomatrix, Inc., 65 Railroad Avenue, Ridgefield, NJ 07657, U.S.A.

Introduction

The utilization of hyaluronan in medicine has created considerable interest in methods to derivatize hyaluronan and thereby expand its therapeutic applications [1,2]. A great deal of diverse chemistry has been applied, and many new hyaluronan-based materials have been described in the literature. This report will systematically catalogue the chemical strategies employed to modify hyaluronan and the types of derivatives now available. After reviewing the wide range of hyaluronan derivatives created in the laboratory, attention will be focused on those derivatives that are currently used in clinical medicine.

My own personal interest in derivatizing hyaluronan is a direct result of my work with Endre A. Balazs. In 1971, Professor Balazs coined the term *matrix engineering* to describe the use of natural and modified biological matrices derived from hyaluronan to control, direct or augment tissue regenerative processes [3]. He was also the first to recognize the need for water-insoluble, cross-linked derivatives of hyaluronan to be used in medical therapeutics. Biomaterials derived from matrix polymers are intrinsically suited for the control of tissue regeneration because matrix polymers typically play a critical role in all such phenomena. The idea is simple and elegant. By chemically modifying hyaluronan, the physical properties required for new medical applications could be attained without losing hyaluronan's natural biocompatibility. This concept of matrix engineering still serves as a paradigm for the current medical uses of hyaluronan, and provides direction to efforts aimed at derivatizing hyaluronan to make it suitable for new indications.

Why derivatize hyaluronan?

Hyaluronan offers many unique advantages as a starting point for medical products. First and foremost is the ubiquitous distribution of hyaluronan in nature. It is found in virtually every species in the animal kingdom as well as in the capsule of certain microorganisms, and in every tissue in the human body [4]. Moreover, the hyaluronan repeating disaccharide is identical in all species and all tissues and is therefore never itself recognized as immunologically foreign. Thus to prepare a medically usable preparation of hyaluronan the only absolute requirement is to remove any inflammatory fractions. Though this is still a technically difficult process, particularly while maintaining the viscosity and elasticity of the native hyaluronan molecule, methods and specifications have been developed that allow the routine production of highly purified and elastoviscous hyaluronan, generically termed NIF-NaHA, as an acronym for non-inflammatory sodium hyaluronate [5].

The second attribute of hyaluronan that is advantageous in medicine is its unique physicochemical properties. The combination of hyaluronan's polyanionic character, its linear unbranched structure and its high molecular mass make solutions of hyaluronan extremely elastoviscous, yet pseudoplastic enough to be extruded through narrow gauge needles. All of the current medical uses of hyaluronan are based on placing purified hyaluronan matrices directly at sites where either a mechanically functional physical barrier is required (viscosurgery, viscoprotection, viscoseparation) or a mechanically dysfunctional tissue can be supplemented (viscosupplementation, viscoaugmentation) [2].

The third important factor underlying the medical utility of hyaluronan derives from the magnitude and pathways available for systemic hyaluronan metabolism. Large quantities of hyaluronan are turned over on a daily basis, of the order of grams per day in a healthy human adult [4]. This means that endogenous pathways are readily available to metabolize the hyaluronan exogenously applied during medical procedures. Since the capacity for hyaluronan turnover is so high, the small quantities applied therapeutically do not add significantly to the systemic metabolic burden at a given tissue location.

Several goals have driven the extensive recent efforts to derivatize hyaluronan. The primary goal is most often to improve rheological properties or to create new physical forms, such as solids or hydrogels with decreased water solubility. Water solubility controls residence time because the major pathway for hyaluronan elimination is via the flow of aqueous tissue fluid into lymphatics [4]. Experiments using unmodified NIF-NaHA in the form of dried films or concentrated elastoviscous solutions to prevent post-surgical adhesion in animal models demonstrated the need for the development of water-insoluble or slowly dispersable hyaluronan derivatives for tissue separation (see Chapter 27 in this volume). Unmodified hyaluronan is eliminated too rapidly to maintain a long-lasting physical barrier between opposing tissues. By modifying hyaluronan's water solubility via covalent derivatization, its solidity and elimination rate can be controlled to create materials more suitable for adhesion control. The goal is to customize hyaluronan for specific medical needs while maintaining its natural biocompatibility.

Hyaluronan also clearly exerts important biological influences on cells *in vitro* [6,7a,7b]. Though many of these influences are likewise related to the physicochemical properties of hyaluronan, there are at least several cell surface proteins with a specific binding site for hyaluronan that can mediate biological responses *in vitro* [8]. Though therapeutic products based on these observations have yet to emerge, the opportunity exists to exert a therapeutic influence by modifying specific hyaluronan–cell interactions with either hyaluronan itself, fragments of hyaluronan or modified hyaluronan derivatives. Drugs can be covalently attached to hyaluronan or filled in the molecular network of cross-linked hyaluronan derivatives [9,10]. Hyaluronan derivatives themselves can also be created to have pharmacological activities targeting receptors or enzymes. Though these types of hyaluronan products are inherently more difficult to develop than the current generation of products based on hyaluronan's mechanical properties, a far-reaching range of possibilities exists to stimulate ongoing research.

Hyaluronan derivatization: methods and strategies

Attempts to derivatize hyaluronan extend back almost 50 years. During the 1940s Hadidian and Pirie synthesized nitrated and acetylated derivatives of hyaluronan that had anti-hyaluronidase activity [11]. Balazs, Högberg and Laurent synthesized sulphated derivatives of hyaluronan, which they demonstrated to have anti-enzyme activities and to inhibit the proliferation of chicken heart fibroblasts *in vitro* [12]. Jeanloz synthesized a large number of hyaluronan esters in the course of his structural determinations [13]. Laurent, Hellsing and Gelotte used a bis-epoxide reagent to cross-link hyaluronan into a water-insoluble hydrogel similar to the cross-linked dextrans used for gel-permeation chromatography [14]. Carbodiimide-mediated coupling was used by Danishefsky and Siskovic to attach amino acids to hyaluronan [15], and Tengblad coupled hyaluronan to cyanogen-bromide-activated Sepharose® for affinity chromatography and analytical determinations [16]. Though these diverse reactions illustrated the potential of hyaluronan for chemical derivatization, none of the materials synthesized were used in clinical medicine.

The modern era of derivatizing hyaluronan to improve its clinical utility was pioneered during the early 1980s by Endre A. Balazs and his co-workers. The impetus for this development came from the body of clinical, veterinary and laboratory experience that rapidly accumulated during the 1970s after NIF-NaHA became available [3]. Though NIF-NaHA preparations were extremely effective for ophthalmic viscosurgery and for the treatment of equine arthritis, it became clear that indications such as viscosupplementation, adhesion control and tissue augmentation required hyaluronan preparations with a longer residence time or greater solidity and elasticity. Because hyaluronan itself is so water soluble, it was clear that this could only be accomplished by cross-linking or some other type of chemical derivatization. Though several ionic and non-covalent complexes have been described as hyaluronan derivatives, only covalent derivatives will be considered below.

Four groups on the hyaluronan molecule are available for chemical modification: the carboxylate, the acetamido, the reducing end-group and the hydroxy groups. The glycosidic bond is also available in that it is readily hydrolysed to create shorter chains or oligosaccharides. End-group modification also offers some opportunity for derivatization, particularly in short oligosaccharides. Though hyaluronan derivatives via the end-group [17] and acetamido group [18] have been synthesized, the bulk of all efforts to date directed at synthesizing medically useful hyaluronan derivatives have used the hydroxy and/or the carboxylate group as the point of chemical attack.

Two strategies have been employed to chemically modify the physical properties of hyaluronan: cross-linking and attachment of pendant groups. Cross-linking modifies physical properties by increasing molecular size or by forming a continuously cross-linked network of hyaluronan chains that swell in water but no longer dissolve. Attaching pendant groups to hyaluronan modifies its physical properties by adding hydrophobic side chains, reducing polyanionic character or modifying chain aggregation. A significant amount of clinical progress has been made with both classes of hyaluronan derivatives.

Hyaluronan with attached pendant groups

Various pendant groups have been attached to hyaluronan, using a variety of chemical methods. Activated alkyl groups have been esterified to the carboxylate group of hyaluronan [19]. Carbodiimide reagents have also been used to modify the carboxylate group, both on their own [20] and in coupling reactions [21].

The physical properties, biological behaviour and potential medical applications of these hyaluronan derivatives have been extensively explored in laboratory studies. The attachment of a hydrophobic side chain to the carboxylate group of hyaluronan decreases water solubility and therefore increases tissue residence time. The more hydrophobic the group attached, and the more completely the carboxylate group is esterified, the more pronounced these effects are [22]. These desirable changes in physical properties must be balanced against any undesirable changes in biocompatibility, and indeed some of these hyaluronan esters have shown increased activation of complement and tissue reactions that differ from unmodified hyaluronan. The same trends in physical property modification have been observed for hyaluronan derivatives synthesized via carbodiimide activation of the carboxyl group [23].

In addition to attaching pendant groups that modify the physical properties of hyaluronan, several laboratories have synthesized hyaluronan–drug adducts for controlled delivery applications [10,11] and hyaluronan–protein adducts as biomaterials and cell substrates [24,25].

The only hyaluronan with an attached pendant group currently employed in clinical medicine is a composite material used for post-surgical adhesion control. The product is composed of hyaluronan and carboxymethyl cellulose derivatized with carbodiimide (Seprafilm®, Genzyme Inc., Cambridge, MA, U.S.A.). It is supplied as a dry membrane that swells into a hydrated gel when placed as a physical barrier between tissue surfaces. The carbodiimide derivatization is believed to decrease water solubility so that the barrier can remain in place for several days. Little has been published on its metabolism and systemic elimination, but it is reported to remain at the site of implantation for approximately 1 week, and to be eliminated from the body within 1 month. Clinical studies of this derivatized hyaluronan–cellulose composite have found it to reduce post-surgical adhesions in gynaecological and abdominal surgery [26] and this product is now available for clinical practice in Europe and the United States.

Cross-linked hyaluronan

Several different methods have been used to cross-link hyaluronan. By controlling the extent of cross-linking, the type of covalent bond and the hyaluronan functional group involved, it is possible to create a wide range of physically diverse materials. Cross-linking can be so extensive that every hyaluronan molecule is covalently attached within a continuous polymer network and the individual molecules are no longer soluble; or cross-linking can be limited so that only several hyaluronan molecules are covalently bonded and, therefore, the cross-linked hyaluronan remains water-soluble but has an increased average molecular size and, consequently, an increased elastoviscosity.

Most of the reported methodology to cross-link hyaluronan employs polyfunctional reagents such as bis-epoxides [27,28], bis-carbodiimides [29], dihydrazides [30] and divinyl sulphone [31]; the first three reagents attack the carboxylate group preferentially. Divinyl sulphone preferentially attacks the hydroxy group [31], and monofunctional aldehyde reagents have been reported to cross-link hyaluronan via a protein bridge between hydroxy groups [32]. In addition, other polysaccharides and proteins can be cross-linked to hyaluronan to create co-polymers [31].

Generally the reactions reported with these reagents have been used to synthesize continuously cross-linked hyaluronan networks. Reaction conditions have been described using bis-epoxides [33] and monofunctional aldehyde reagents [32] to synthesize soluble cross-linked hyaluronan of increased average relative molecular mass (M_r).

The only cross-linked hyaluronan derivatives currently used in clinical medicine are produced from a class of derivatives generically referred to as hylan polymers [2,31,32,34]. The hylan family of hyaluronan derivatives illustrates how a variety of chemical strategies can be used to customize hyaluronan into physically different materials that are suitable for specific medical indications without losing hyaluronan's unique biocompatibility and rheology. Synvisc® (hylan G-F 20) is a highly elastoviscous synovial fluid supplement used for the treatment of osteoarthritis in a procedure referred to as viscosupplementation. Hylaform® (hylan B) is a solid hydrated gel implanted into soft tissues (viscoaugmentation). Hylashield® (hylan A) is a dilute elastoviscous fluid applied to the surface of the eye for comfort and protection (viscoprotection). (Synvisc®, Hylaform® and Hylashield® are registered trademarks of Biomatrix, Inc., Ridgefield NJ, U.S.A.)

Hylan is a generic term used to refer to a class of hyaluronan derivatives with characteristic features. All hylan polymers are cross-linked via the hydroxy groups of the hyaluronan chain, leaving the carboxylate and the acetamido groups unreacted. The retention of the carboxylate group is especially important because the polyanionic character of hyaluronan is critical to its physicochemical and biological properties. Hylan polymers are available in physical forms ranging from highly elastoviscous solutions to viscoelastic gels and solids (membrane, tubes). This allows the physical properties and residence time of hyaluronan to be controlled without an unacceptable loss in biocompatibility.

Figure 1

Solution of 1% hylan A in the form of a viscoelastic putty

The solution has sufficient coherence and elasticity to provide physical support, yet it still meets the physicochemical definition of a solution because it can be diluted, it will flow and will adopt the shape of its container.

The hylan polymers currently used in medicine are designated hylan A and hylan B. Hylan A is synthesized *in situ* by treating hyaluronan-rich tissue sources with aldehydes prior to extraction [32]. The aldehyde activates the hydroxy groups of the hyaluronan, forming a protein-mediated cross-link and creating soluble hyaluronan polymers in the subsequent extracts that have average

Figure 2

Solid hydrated gel containing 0.5% hylan B at equilibrium hydration

Note that this material is not an aqueous solution; it is a solid that retains a shape and sharp edge, and will not dissolve in water.

M_rs of up to 24×10^6. Hylan B is synthesized by reacting hyaluronan or hylan A with divinyl sulphone under mild alkaline conditions [31]. With this reagent sulphonyl bis-ethyl cross-links are formed, producing an infinite hyaluronan network that is no longer water soluble. The reaction conditions can be varied to create materials that range from soft deformable gels to solid membranes and tubes, with prolonged or permanent residence times. Examples of hylan A solutions and hylan B solids are illustrated in Figures 1 and 2, respectively.

The rheological properties and physical form of hylan A, hylan B and hyaluronan are compared in Table 1. Though hylan A remains water soluble, like hyaluronan, both its viscosity and elasticity are enhanced by an order of magnitude at an equivalent concentration. Hylan B is no longer in the physical form of a solution, but it is hydrated to a viscoelastic gel or gel slurry at an equilibrium polymer concentration of 0.3% or greater. Hylan B is a highly elastic polymer at a wide range of frequencies (0.001 to 10 Hz). This elasticity is 10-fold higher than that of a hylan A of the same concentration. This table illustrates how two different cross-linking reactions change the physical form and rheological properties of hyaluronan.

	Physical form	Low shear viscosity in Pa s (0.001 s^{-1})	Modulus of elasticity (G') in Pa (0.01 Hz)
1% hyaluronan*	aqueous solution†	242	4
1% hylan A‡	aqueous solution†	2190	22
0.5% hylan B§	hydrated† gel slurry	NA	69

*NIF-NaHA (Healon®, Pharmacia, Upjohn), average Mr is 3×10^6. †The aqueous media consisted of phosphate-buffered physiological saline solution. ‡Biomatrix, Inc.; average Mr is 6×10^6. §Biomatrix, Inc.; infinite molecular network, water insoluble.

Table

Comparison of the rheological properties and physical form of hyaluronan and hylan

The effect of cross-linking on residence time is illustrated in Table 2, using hylan B to demonstrate how dramatically water solubility affects residence time. The three tissues compared – the knee, skin and vitreus – differ with respect to their rates of endogenous hyaluronan turnover, presumably because they are subjected to different mechanical forces. Thus the rate of turnover is much lower

	Comparative half-life	
	1% hyaluronan	0.5% hylan B
Knee	12 h*	9 days†
Skin	24 h‡	>9 months§
Vitreus	7 days¶	>10 years**

*Rabbit knee [35]. †Rabbit knee [38]. ‡Rabbit skin [36]. §Guinea pig skin [39]. ¶Owl monkey vitreus [37]. **Owl monkey vitreus [40].

Table 2

Effect of cross-linking on residence time in tissue; comparison of hylan B and hyaluronan

in a mechanically protected tissue like the vitreus than in musculoskeletal tissues like the knee, which are constantly subjected to mechanically induced fluid flow. In all cases the residence time of hylan B is significantly longer than that of hyaluronan, though it is still proportional to the mechanical forces on the tissue. In a mechanically protected compartment like the vitreus, hylan B can remain resident for years because no mechanical forces are present to degrade the hylan B polymer.

Functional improvements due to cross-linking are illustrated in Table 3. Both hylan A and hylan B interact more strongly with water than hyaluronan, as shown by differential scanning calorimetry [41]. In addition, hylan A is 3-fold more stable than hyaluronan to free-radical degradation [42].

The biocompatibility of hylan polymers has been extensively evaluated and found to be virtually identical to that of hyaluronan [43,44]. This illustrates an important paradigm. Chemical modification of hyaluronan should always be limited to the extent necessary to achieve functionally important characteristics. Hylan polymers illustrate how small levels of chemical derivatization can modify the functional characteristics of hyaluronan in medically useful ways.

Table 3		Wc* (g of water per g of polymer at saturation)	G value†
	Hyaluronan	3.0	6
	Hylan A	10.0	2
	Hylan B	15.8	Not determined

*[41]. †Number of chain breaks per unit radiation input.

Comparative functional characteristics of hyaluronan and hylan [42]

Conclusion

The extensive efforts directed at creating hyaluronan derivatives useful in medicine attests to both the current and future importance of hyaluronan as a starting point for therapeutic products. The main lesson of the work to date is that less is more. The least modification necessary to fill a specific need is always the place to start. In expanding the medical uses of hyaluronan beyond the capabilities of the unmodified molecule, chemical derivatization should be performed as gently and as minimally as possible. Modification of the native molecule should be limited to the minimum change required to bring about the customization required for a particular indication.

The work reviewed in this report demonstrates many important structure–function relationships within a wide range of hyaluronan derivatives. As the utility of such hyaluronan derivatives become established in medicine, many new indications will emerge in what will be a fertile field of research for many years to come.

The hylan polymers described in this paper were developed in the laboratories of Biomatrix under the leadership of Endre A. Balazs with the participation of Ed

Leshchiner, Adele Leshchiner, Nancy Larsen, Phil Band, Janet Denlinger, Betty Morales and Arnold Goldman.

In 1995 Ed Leshchiner passed away prematurely. On behalf of Dr. Balazs and the other members of the Biomatrix development team, I would like to acknowledge the seminal contributions of Ed Leshchiner to the field of hyaluronan derivatization. His brilliance had an impact on his co-workers and on the field to which he contributed so richly.

References

1. Balazs, E.A. and Denlinger, J.L. (1989) in The Biology of Hyaluronan (Evered, D. and Whelan, J., eds.), Ciba Foundation Symp. No. 143, pp. 265–280, Wiley, New York
2. Balazs, E.A., Leshchiner, E., Larsen, N.E. and Band, P. (1995) in Handbook of Biomaterials and Applications (Wise, D.L., ed.), pp. 2719–2741, Marcel Dekker, New York
3. Balazs, E.A. (1971) in Hyaluronic Acid and Matrix Implantation, p. 131, Biotrics Inc., Arlington, MA
4. Laurent, T.C. and Fraser, J.R.E. (1992) FASEB J. **6**, 2397–2404
5. Balazs, E.A. (1979) Ultrapure Hyaluronic Acid and the Use Thereof. U.S. Pat. No. 4 141 973.
6. Balazs, E.A. and Darzynkiewicz, Z. (1973) In Biology of Fibroblast (Kulonen, E. and Pikkarainen, J., eds.), pp. 237–252, Academic Press, London
7. (a) Forrester, J.V. and Balazs, E.A. (1980) Immunology **40**, 435–446; (b)Toole, B.P. (1991) in Cell Biology of Extracellular Matrix (Hay, E.D., ed.), pp. 305–341, Plenum Press, New York
8. Knudson, C.B. and Knudson, W. (1993) FASEB J. **7**, 1233–1241
9. Balazs, E.A., Leshchiner, A., and Larsen, N.E. (1990) Drug Delivery Systems Based on Hyaluronans. Derivatives Thereof and Their Salts and Methods of Producing Same. U.S. Pat. No. 5 128 326
10. Pouyani, T. and Prestwich, G.D. (1994) Bioconjug. Chem. **5**, 339–347
11. Hadidian, Z. and Pirie, N.W. (1947) Biochem. J. **42**, 466
12. Balazs, E.A., Högberg, B. and Laurent, T.C. (1951) Acta Physiol. Scand. **23**, 168–178
13. Jeanloz, R.W. (1952) J. Biol. Chem. **197**, 141–150
14. Laurent, T.C., Hellsing, K. and Gelotte, B. (1964) Acta Chem. Scand. **18**, 274–275
15. Danishefsky, I. and Siskovic, E. (1971) Carbohydr. Res. **16**, 199–201
16. Tengblad, A. (1979) Biochim. Biophys. Acta **578**, 281–289
17. Raja, R.H., LeBoeuf, R.D., Stone, G.W. and Weigel, P.H. (1984) Anal. Biochem. **139**, 168–177
18. Dahl, L.B., Laurent, T.C. and Smedsrød, B. (1988) Anal. Biochem. **175**, 397–407
19. della Valle, F. and Romeo, A. (1986) Esters of Hyaluronic Acid. U.S. Pat. No. 4 851 521
20. Hamilton, R., Fox, E.M., Acharya, R.A. and Walts, E.A. (1987) Water Insoluble Derivatives of Hyaluronic Acid. U.S. Pat. No. 4 937 270
21. Kuo, J-W., Swann, D.A. and Prestwich, G.D. (1991) Bioconjug. Chem. **2**, 232–241
22. Cortivo, E., Brun, P., Rastrelli, A. and Abatangelo, G. (1991) Biomaterials **12**, 727–730
23. Kuo, J-W., Swann, D.A. and Prestwich, G.D. (1992) Water-Insoluble Derivatives of Hyaluronic Acid and Their Methods of Preparation and Use. U.S. Pat. No. 5356883
24. Glass, J.R., Dickerson, K.T., Stecker, K. and Polarek, J.W. (1996) Biomaterials **17**, 1101–1108
25. Rhee, W.M. and Berg, R.A. (1993) Glycosaminoglycan-Synthetic Polymer Conjugates. U.S. Pat. No. #5 510 418
26. Becker, J.M., Dayton, M.T., Fazio, V.W., Beck, D.E., Stryker, S.J., Wexner, S.D., Wolff, B.G., Rowers, P.L., Smith, L.E., Moore, M. and Beart, R.W. (1995) American College of Surgeons 81st Annual Meeting, New Orleans
27. Mälson, T. (1987) Material of Polysaccharides Containing Carboxyl Groups, and a Process for Producing Such Polysaccharides. U.S. Pat. No. 4 963 666
28. della Valle, F. and Romeo, A. (1987) Cross-linked Esters of Hyaluronic Acid. U.S. Pat. No. 4 957 744
29. Kuo, J-W., Swann, D.A. and Prestwich, G.D. (1994) Water-insoluble Derivatives of Hyaluronic Acid and Their Methods of Preparation and Use. U.S. Pat. No. 5 502 081
30. Pouyani, T., Harbisoa, G.S. and Prestwich, G.D. (1994) J. Am. Chem. Soc. **116**, 7515–7522
31. Balazs, E.A. and Leshchiner, A. (1984) Cross-Linked Gels of Hyaluronic Acid and Products. U.S. Pat. No. 4 582 865
32. Balazs, E.A., Leshchiner, E., Leshchiner, A. and Band, P. (1985) Chemically Modified Hyaluronan Acid and Preparation and Method of Recovery Thereof From Animal Tissues. U.S. Pat. No. 4 713 448

33. Sakurai, K., Ueno, Y. and Okayama, T. (1985) Crosslinked Hyaluronic Acid and Its Use. U.S. Pat. No. 4716224

34. Balazs, E.A. and Leshchiner, E.A. (1989) in Cellulosics Utilization: Research and Rewards in Cellulosics (Inagaki, H. and Phillips, G.O., eds.), pp. 233–241, Elsevier Applied Sciences, New York

35. Brown, T.J., Laurent, U.B.G. and Fraser, J.R.E. (1991) Exp. Physiol. **76**, 125–134

36. Laurent, U.G.B., Dahl, L.B. and Reed, R.K. (1991) Exp. Physiol. **76**, 695–703

37. Denlinger, J.L., El-Mofty, Aly A.M. and Balazs, E.A. (1980) Exp. Eye Res. **31**, 101–117

38. Band, P., Goldman, A., Barbone, A., Reiner, K. and Balazs, E.A. (1995) Materials Research Society, Spring Meeting, April 17-21, 1995, San Francisco, CA, 433 (abstract)

39. Larsen, N.E., Leshchiner, E., Pollack, C.T., Balazs, E.A. and Piacquadio, D. (1995) In Polymers in Medicine and Pharmacy: Proceedings of the Materials Research Society Spring Meeting, 17–21 April 1995, San Francisco, CA (Mikos, A.G., Leong, K.W., Radomsky, M.L., Tamada, J.A. and Yaszemski, M.J., eds.), pp. 193–197, Mater. Res. Soc., Pittsburgh, PA

40. Balazs, E.A., Vadasz, A. and Goldman, A.I. (1990) Proceedings of the IX Int. Congress Eye Research, 29 July to 4 August 1990, Helsinki, Finland (Uusitalo, H., Palkama, A. and Mahlberg, K., eds.), vol VI University Press, 165 (abstract)

41. Takigami, S., Takigami, M. and Phillips, G.O. (1995) Carbohydr. Polym. **26**, 11–18

42. Al-Assaf, S., Phillips, G.O., Deeble, D.J., Parsons, B., Starnes, H. and Von Sonntag, C. (1995) Radiat. Phys. Chem. **46**, 207–217

43. Larsen, N.E., Pollack, C.T., Reiner, K., Leshchiner, E. and Balazs, E.A. (1993) J. Biomed. Mater. Res. **27**, 1129–1134

44. Larsen, N.E., Leshchiner, E., Balazs, E.A. and Belmonte, C. (1995) In Polymers in Medicine and Pharmacy: Proceedings of the Materials Research Society, Spring Meeting, 17–21 April 1995, San Francisco, CA (Mikos, A.G., Leong, K.W., Radomsky, M.L., Tamada, J.A. and Yaszemski, M.J., eds.), pp. 149–153, Mater. Res. Soc., Pittsburgh, PA

7

Chemical modification of hyaluronic acid for drug delivery, biomaterials and biochemical probes

Glenn D. Prestwich*†‡¶, Dale M. Marecak*†, James F. Marecek†, Koen P. Vercruysse*† and Michael R. Ziebell*§
*Department of Medicinal Chemistry, 308 Skaggs Hall, College of Pharmacy, The University of Utah, Salt Lake City, Utah 84112, U.S.A., †Department of Chemistry, University at Stony Brook, Stony Brook, New York 11794-3400, U.S.A., ‡Department of Biochemistry & Cell Biology, University at Stony Brook, Stony Brook, New York 11794-3400, U.S.A., and §Department of Physiology & Biophysics, University at Stony Brook, Stony Brook, New York 11794-3400, U.S.A.

Introduction

Hyaluronic acid (HA) is a linear polysaccharide consisting of alternating 1,4-linked units of 1,3-linked glucuronic acid and *N*-acetylglucosamine (Figure 1), and is one of several glycosaminoglycan components of the extracellular matrix (ECM) of connective tissue [1]. Its remarkable viscoelastic properties account for its importance in joint lubrication [2] and its complete lack of immunogenicity makes it an ideal building block for biomaterials needed for tissue engineering [3,4] and drug delivery systems [5,6]. Sodium hyaluronate, the predominant form at physiological pH, and HA are collectively referred to as hyaluronan; in this paper, the chemical modifications always begin with the carboxylic acid form, and we use the abbreviation 'HA'. The three-dimensional structures adopted by HA and HA oligosaccharides in solution [7,8] show extensive intramolecular hydrogen bonding that restricts the conformational flexibility of the polymer chains and induces distinctive secondary (helical) and tertiary (coiled coil) interactions.

In this chapter, we will first summarize background information on (1) key HA–protein interactions important in cell biology, (2) the role of HA in cellular signalling, (3) the biomedical applications of HA, (4) biodegradable polymer scaffolds and (5) chemical modification of HA, leading to the discovery of the hydrazide modification method. Second, we present an overview of our current work on (1) new cross-linkers, (2) optimization of HA hydrogel formation, (3) hydrogel degradation by hyaluronidase (HAse), (4) use of chemically modified hydrogels for tissue engineering, (5) synthesis and applications of functionalized HA probes in cell biology and (6) structural studies of HA–HA binding domain interactions.

¶ To whom correspondence should be addressed (University of Utah).

Figure 1

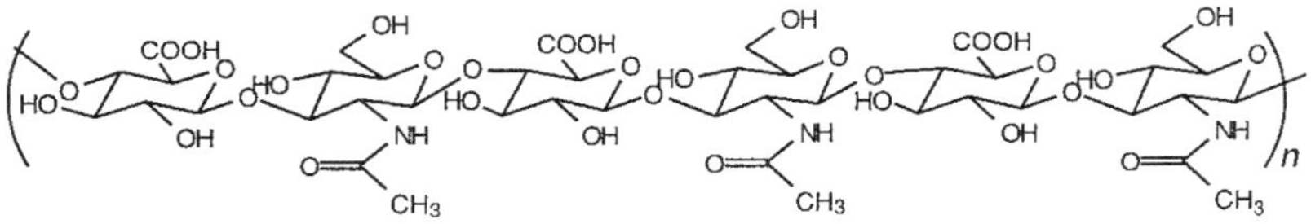

Structure of hyaluronic acid, showing three disaccharide repeat units

HA–protein interactions

Hyaladherins [9], including the link proteins [10], aggrecan, versican and other HA binding proteins, are essential in tissue remodelling processes [11]. HA–protein interactions are important in cell adhesion, growth and migration [9,12,13], stimulation of cell motility and promotion of focal adhesion turnover [14], in cartilage [15], in inflammation and wound healing [16] and in cancer [17,18]. Many of the molecular details of the interactions of HA with HA binding domains (HABDs) of hyaladherins have been elucidated. Indeed, a solution structure of the link module, a hyaluronan-binding domain involved in ECM stability and cell migration, has recently been published [19]. However, HA is not simply 'extracellular glue'; the critical importance of HA receptors in the regulation of intracellular signalling to the cytoskeleton has been recently reviewed [20], and key elements are described below.

Role of HA and HA receptors in metastasis

Extracellular HA acts as a signalling molecule by activating regulatory kinase pathways important for cell cycle progression and movement [21,14,20]. Two types of cellular HA receptors have confirmed roles in signalling: (i) CD44, a family of glycoproteins originally associated with lymphocyte activation, and (ii) RHAMM (receptor for hyaluronan mediated motility), originally identified from transformed fibroblasts. CD44 isoforms found on tumour cells bind HA with higher affinity and promote cell migration [22]. Turley's group found that RHAMM overexpression is itself transforming and regulates events downstream from H-ras [23,14]. Moreover, RHAMM plays a role in the cell cycle, inducing mitotic arrest by suppression of cdc2 and cyclin b1 production [24]. In cancer, HA also has effects on angiogenesis and on an increased HA production in and around tumours [25].

  While the predicted secondary (and tertiary) structures of RHAMM and CD44 are different, both share the HA binding motif common to all known HA binding proteins: BX_7B, where B is His, Arg or Lys [26]. The importance of these two domains for HA binding activity has been determined by deletion mutants and synthetic peptide competition experiments. HA–CD44 complexes mediate lymphocyte–endothelial cell adhesion [27]. A critical Arg^{41} in CD44 was identified by mutagenesis [28]. A series of Lys substitutions in RHAMM confirmed their importance in the mediation of metastatic transformation induced by H-ras [14]. The role of CD44 and RHAMM in determining whether transformed cells become aggressive and metastatic has been extensively investigated [29,30,22,31–37]. The importance of HA–receptor interactions in cancer biology has propelled the development of biophysical and biochemical probes in our laboratories, as well as structural studies of RHAMM polypeptides. This work will be summarized below.

The size of the HA oligosaccharide itself is important. Revascularization during wound healing is inhibited by high-molecular-mass HA but accelerated by shorter HA oligosaccharides [38]. The HA-hexamer blocks HA binding to HABD in plasma membrane and HA-decasaccharide is necessary for inhibition of the interaction of HA with link protein [39]. Indeed, addition of HA or HA oligosaccharides to human articular cartilage explants and monitoring aggrecan and proteoglycan synthesis *in vitro* suggested that chondrocytes may have two HABDs, one for uptake of HA for catabolism and one related to hyaluronectin [40].

Biomedical applications of HA derivatives

Both naturally occurring and chemically modified HA have found use in a broad range of biomedical applications, including ophthalmic surgery [41] and the treatment of arthritis [5,42,43], in particular via the technique known as viscosupplementation [44,45]. A wide variety of uses are described elsewhere in this volume, but several examples are described here to place the present chemical modification work in context. The chemical derivatization of HA by our laboratories and others provides an opportunity to develop materials for new applications that exploit some of the viscoelastic properties of this high-molecular-mass polymers. For example, controlled release of pharmacologically active compounds such as prednisolone [46,47] for use in eye surgery, moisturizing agents, swellable gels, adhesion management aids, cell encapsulation matrices, hydrophilic coatings of plastics [48], mammary implants [49] and certain aspects of wound treatment [41] are some of the applications envisioned for these new materials. Native sodium HA was shown to serve as a template for nerve regeneration [50]. HA is suitable for all of these applications since it is bioerodable and compatible with systemic functions. Insoluble HA derivatives that retain significant bioerodability [51,52] have been used as membranes for the culture of keratinocytes for transfer to human wounds [53]. Other chapters in this volume will provide an expanded repertoire of the specific uses of HA in human medicine. New HA hydrogels can in principle be fashioned into prosthetic implants capable of drug delivery [54], into porous microspheres for culturing cells [55,56] and into gels for bridging gaps in bone or cartilage during cell regrowth [57–59].

Biodegradable polymers

Biodegradable polymer scaffolds based on polyglycolic acid, polylactic acid and related co-polymers have been investigated recently as substrates for tissue engineering [3], and important advances in neocartilage regeneration [60] and joint resurfacing have been described. A high-molecular-mass viscous HA material has also been used alone or in gel-like formulations with decalcified bone matrix [57,58]. Despite promising results, in these cases, the realization of a shape-retaining implant was not achieved. As a result, the need for new biomaterials [4] for tissue engineering and for drug delivery [6] continues to drive the development of new technologies.

Biodegradable hydrogels, extended supermolecules prepared by cross-linking macromolecules into a polymer network, have become the subject of intense study as carriers for controlled release drug delivery systems [61,62]. Hydrogels are generally biocompatible, because of low interfacial tension and advantageous physical properties for cell adhesion and protein adsorption. The major disadvantage is low mechanical strength. New hydrogels based on collagen and other naturally occurring components of the ECM have begun to address this issue [63,64]. Uses of HA and of HA gels for drug delivery have been reviewed [65,66]. Nonetheless, there are very few examples of biocompatible hydrogels with well-characterized functional linkers on to which drugs may be appended [67]. Thus, improvements in our ability to introduce chemical functionality for peptide or drug attachment, in addition to producing a biocompatible, bioerodable hydrogel, should provide access to a unique new set of biomaterials.

Chemical modification of HA

As detailed below, HA can be chemically modified at hydroxy groups of the glucuronic acid or N-acetylglucosamine units, the carboxylic acid of the glucuronate units, or the reducing end of the polysaccharide chain. Since the chemical reactions described herein require the protonated carboxylic acid and are conducted between pH 2.0 and 5.0, we will refer throughout to hyaluronic acid, or simply HA. The resulting materials used in a biological context would be correctly called chemically modified hyaluronan derivatives, referring to a mixture of protonated and sodium and/or potassium salts present in physiological systems. In this chapter, we will focus on controlled modifications of the carboxylic acid under mild aqueous conditions, using water-soluble carbodiimide chemistry.

Figure 2

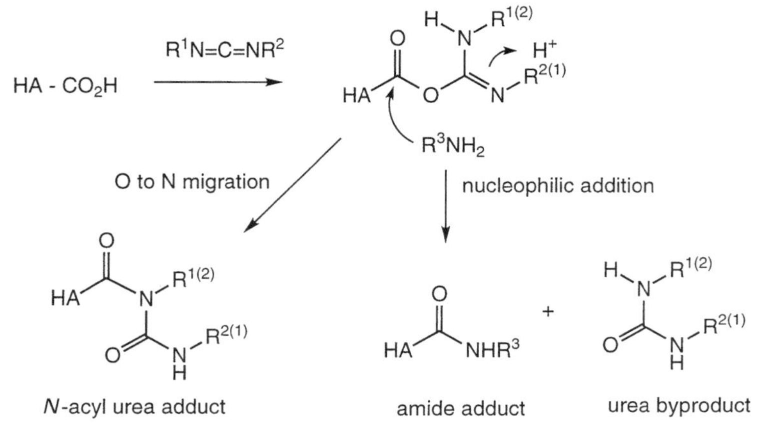

Attempted chemical modification of HA with carbodiimides ($R^1N=C=NR^2$) and primary amines (R^3NH_2) leads to formation of N-acyl ureas by rearrangement of the intermediate O-acyl ureas

The amide and urea by-product are only minor products.

Designed monofunctional (top) and bisfunctional (bottom) carbodiimides

The compounds were synthesized and used to prepare N-acyl urea derivatives and N-acyl urea cross-linked derivatives of HA.

Carbodiimide modification of glycosaminoglycans has three decades of history; unfortunately, for most materials the chemical structures were not well-characterized. We initiated our studies of the modification of HA with ethyl(*N,N*-dimethylaminopropyl)carbodiimide (EDCI) and a variety of 'designer' carbodiimides in 1986. The production of glucuronamides requires the activation of the carboxylic group, which can be accomplished using a water-soluble carbodiimide such as EDCI as the condensing agent. The first report of an HA-glycine methyl ester derivative [68] could not be repeated in our laboratories [69]. We found that the *O*-acyl urea activated complex formed between EDCI and high-molecular-mass HA (2 MDa) at pH 4.75 did not give the expected intermolecular coupling with added diamines or other primary amine-containing nucleophiles (Figure 2). Instead, the *O*-acyl ureas preferentially rearranged to *N*-acyl ureas rather than undergoing coupling to added amine-containing reagents [69]. This was readily seen by a characteristic quartet and triplet pattern for the *N*-ethyl urea moiety in the NMR spectrum of these derivatives. Partially degraded HA (60 kDa) also followed this reaction pathway. We nonetheless took advantage of this observation to prepare lipophilic, aromatic and functionally reactive derivatives of HA based on the robust covalent *N*-acyl urea linkage formed with customized carbodiimides (Figure 3, top) [69]. Biscarbodiimides offered the possibility of cross-linking HA to form stable gels, and both aliphatic and aromatic biscarbodiimides were prepared (Figure 3, bottom) and shown to give swellable, cross-linked products [69,70].

The unambiguous characterization of the HA *N*-acyl ureas, as well as elucidation of the mechanism of formation of these products, was accomplished using isotopically labelled EDCI in conjunction with cross-polarization magic-angle spinning (CP-MAS) ^{13}C- and ^{15}N-NMR [71]. Thus, the [2-^{13}C] and [1-^{15}N]EDCI isotopomers were prepared separately from [2-^{13}C]ethyl isothio-cyanate and ^{15}N-ethylamine, respectively. Using 2 MDa HA at 4 mg/ml with sufficient labelled EDCI to obtain a theoretical maximum of 25% modification of the available glucuronic acid residues, the ^{13}C-CP-MAS of the product showed a new peak at 156 p.p.m. for the *N*-acyl urea with the expected intensity for 25% modification. The chemical shift did not distinguish between two isomeric products that could arise from the rearrangement of the EDCI-HA adduct. The ^{15}N-CP-MAS showed a minor peak at 133 p.p.m. for an unprotonated ^{15}N and a major peak at 83 p.p.m. for a protonated nitrogen. These data suggested that the *O*-acyl urea to *N*-acyl urea conversion was intramolecularly catalysed, as shown in the proposed mechanism (Figure 4) [71].

Meanwhile, many of the other chemical derivatives of HA that involve reaction of the hydroxy- and carboxy-functionalities have produced cross-linked materials. The preparation of these materials is described in detail by P. Band in Chapter 6. For example, HA can be derivatized through the hydroxy group using divinyl sulphone cross-linking on the primary hydroxy groups of HA at high pH [72]. The hylans [73,74], formaldehyde-cross-linked HA derivatives [75], show enhanced rheological properties and longer residence times in tissue relative to native HA, but the processing precludes attachment of chemically reactive linkers for drug attachment. Alternatively, HA can be partially oxidized with periodate and then reductively coupled to Type I collagen [76] to give an artificial ECM, as

Figure 4

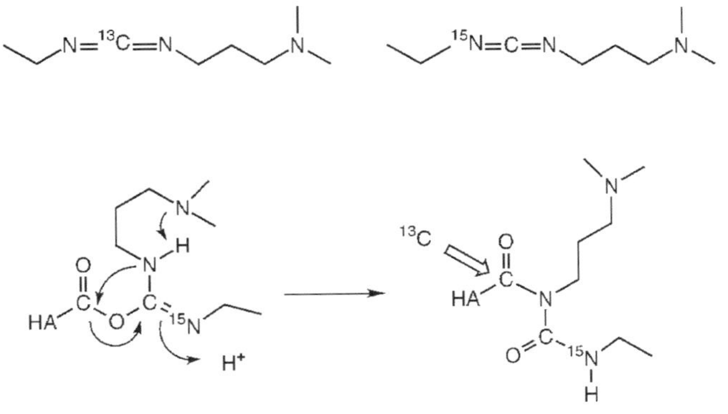

Top, structures of ^{13}C-EDCI and ^{15}N-EDCI. Bottom, intramolecular catalysis of *O*- to *N*-acyl migration during reaction of HA with isotopically labelled EDCI

The regioisomer shown corresponds to an acyl migration of the glucuronate to the more substituted nitrogen. The preferred product has a protonated ^{15}N-ethyl and a ^{13}C-labelled glucuronate (arrow), and was proposed to arise by intramolecular catalysis.

described above [63]. Non-cross-linked ester derivatives have also been developed. Fidia Inc. (Italy) employs a method to convert virtually every carboxylate group of the glucuronic acid moieties of the HA molecule into an ester (e.g. ethyl, benzyl), using the tetra(n-butylammonium) salts of HA dissolved in dimethylformamide [77]. This modification makes the HA molecule lipophilic [78], and the products have been exploited in gauze mats and microspheres with or without ester-linkage tethered drugs [52,51,72,79].

Insoluble HA derivatives, such as the N-acyl urea cross-linked materials [70], have proven useful in prevention of surgical adhesions. Similarly, Genzyme has received patents (U.S. No. 4937270 and U.S. No. 5017229) for modifications of HA for use in prevention of surgical adhesions [80] and as a drug delivery system. Genzyme's modification is to treat HA (or HA-carboxymethylcellulose mixtures) with EDCI in the presence of an amino acid to give a water-insoluble biocompatible material; the spectroscopic characterization of these products has not been presented. Genzyme's Seprafilm®, a carboxymethylcellulose-HA bioresorbable membrane used to prevent surgical adhesions, has received FDA approval and is now being marketed. Clinical or preclinical trials from both industrial and academic research laboratories with other modified HAs include HA-liposomes [81], HA-anti-cancer drugs [82] and HA-anti-inflammatories [83,84].

Recently, the problem of obtaining well-characterized, functionalized derivatives of HA oligosaccharides was advanced significantly by the discovery that dihydrazides could give hydrazide-modified HA for further introduction of drug molecules or intermolecular cross-links to produce hydrogels [67,85]. The hydrazide-derivatized HA technology exhibited a number of important advantages. First, preparation of HA derivatives under mild, aqueous conditions preserves the structural integrity and molecular size range of HA. Second, homogeneous, modified HA materials can be obtained by removal of all side products and unreacted reagents using dialysis, precipitation and/or ultrafiltration. Third, all new HA derivatives based on the bishydrazide technology are chemically well-characterized. Fourth, the mechanical and physicochemical properties can be engineered to fit a variety of applications, including modification of the molecular mass and the extent of intra- and interchain interactions through controlled cross-linking. Fifth, a wide selection of mono-, bis- and polyhydrazide linkers can be used for gel production. These reagents may be either hydrophobic or hydrophilic and may be chemically robust or readily cleavable *in vivo*. Sixth, the ability to *covalently* attach proteins, short peptides, anti-cancer agents, anti-inflammatory or anti-infective drugs, or cell adhesion and growth-promoting ligands to the HA hydrogel provides the potential for drug delivery. Finally, HA hydrogels can be incorporated into other biopolymer matrices, including the preparation of co-cross-linked biomaterials. The remainder of the chapter now deals with the initial development of the hydrazide modification strategy and its application to a variety of new, medically useful derivatives and biochemical probes.

Results and discussion

Modifications with hydrazides

Carbodiimide-mediated coupling of HA to primary amines failed because the acidic pH (<4.75) required to induce HA-O-acyl urea formation also caused protonation of all amine functions. Thus, amines with pK_a values (for the conjugate acids) in the range of 9–13 were >99.99% protonated and thus insufficiently nucleophilic. The realization that hydrazides possessed sufficiently low pK_a amines (pK_a values of 2–3 for the conjugate acids), and that efficient reaction with EDCI-activated HA could take place in water at pH 4.75, was first demonstrated with adipic dihydrazide (ADH). With excess ADH, this reaction gave chemically and thermally stable, colourless and spectroscopically well-defined materials for both oligosaccharides and high-molecular-mass HA (Figure 5) [85]. This derivative, called HA-ADH, was the major product, with the cross-linked HA-ADH-HA appearing as a minor product. Initial work showed that ADH (C_6) was superior to both succinic dihydrazide (C_4) and suberic dihydrazide (C_8). Characterization of the HA oligosaccharide adducts by high-field solution phase ^{1}H-NMR confirmed that, with ADH, >50% of available carboxylates could be modified. The NMR spectra also clearly demonstrated that the HA-ADH oligosaccharide adducts retained a reactive hydrazide for further derivatization. Indeed, HA-ADH was coupled at pH 8.5 to the N-hydroxysuccinimide esters of ibuprofen or hydrocortisone hemisuccinate to produce the

Figure 5

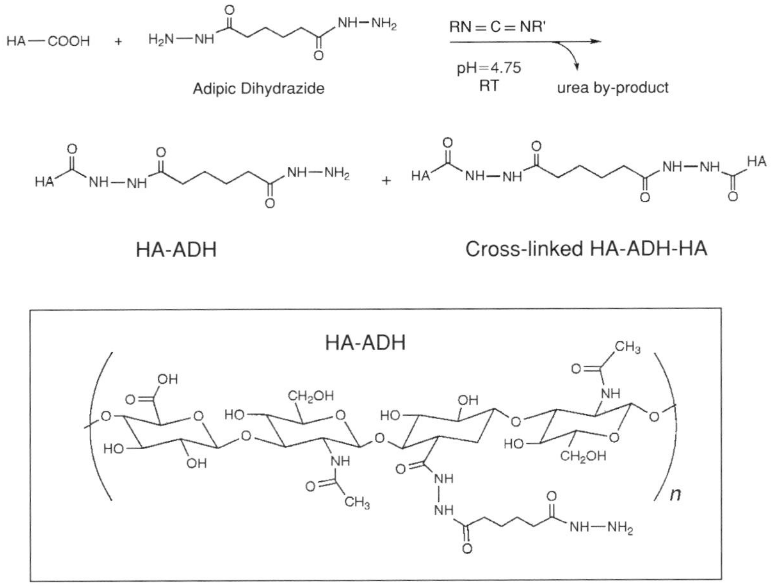

Chemical derivatization of HA with adipic dihydrazide (ADH) using a carbodiimide coupling

The inset shows the molecular details of a region of HA-ADH containing two disaccharide repeat units.

Figure 6

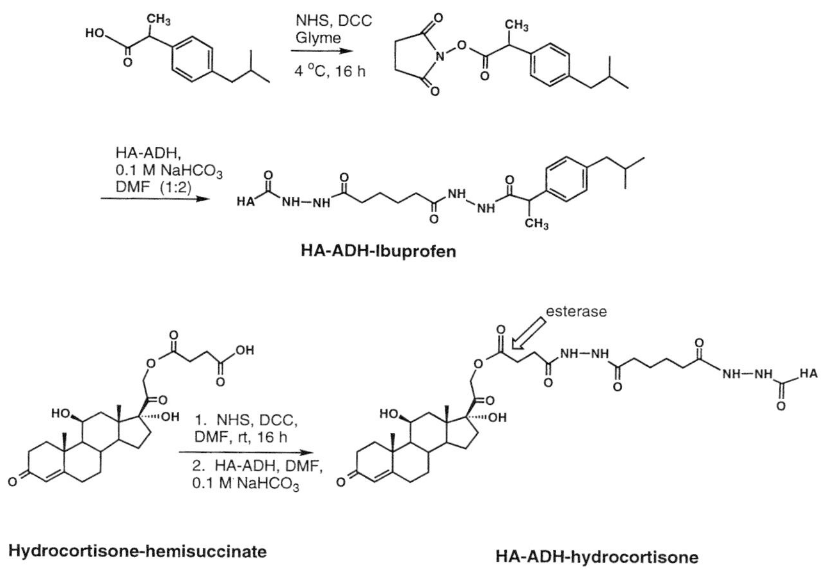

HA-ADH-Ibuprofen

Hydrocortisone-hemisuccinate

HA-ADH-hydrocortisone

Synthesis of two anti-inflammatory drugs attached to HA-ADH

Top, HA-ADH-ibuprofen; bottom, HA-ADH-hydrocortisone hemisuccinate. The open arrow indicates the preferred site of esterase cleavage that would release free hydrocortisone.

corresponding HA-tethered drugs (Figure 6). Finally, exploratory work with four commercially available homobifunctional cross-linkers led to the production of soluble, cross-linked HA-ADH derivatives, as evidenced by broadened signals for the ADH and cross-linker methylenes in the solution NMR [85].

The solution phase work was then extended to HA with a molecular mass of 1.5 MDa. Solid-state ^{13}C-NMR using CP-MAS of lyophilized native HA and HA-ADH revealed that both retained largely solution-like structures in the solid-state [67], although subtle changes in acetamido orientation and glycosidic dihedral angles have been previously discerned [86]. With HA-ADH possessing approx. 20–25% modification of the glucuronates, four homobifunctional cross-linkers were used at a molar ratio of 1:1.4 (HA:cross-linker) to create swellable hydrogels (Figure 7). The CP-MAS spectra gave maximum sensitivity for the determination of the chemical shifts of the carbon atoms of the covalently attached bishydrazide tether and cross-linker. Moreover, subtraction of a native HA spectrum from the derivatized HA spectrum allowed visualization of a residual intensity corresponding to linker groups; calculated chemical shifts confirmed the nature of the linkage. Earlier work had indeed shown broadened resonances for cross-linkers, indicating disorder. In contrast, the HA resonances did not broaden, suggesting a highly ordered, rigid structure. Nonetheless, upfield chemical shifts were observed in one of the HA resonances, suggesting an increase in steric crowding resulting from a change in sugar pucker.

Scanning electron microscopy of lyophilized hydrogels revealed macroporous sponge-like materials with varying surface morphologies, but all

Figure 7

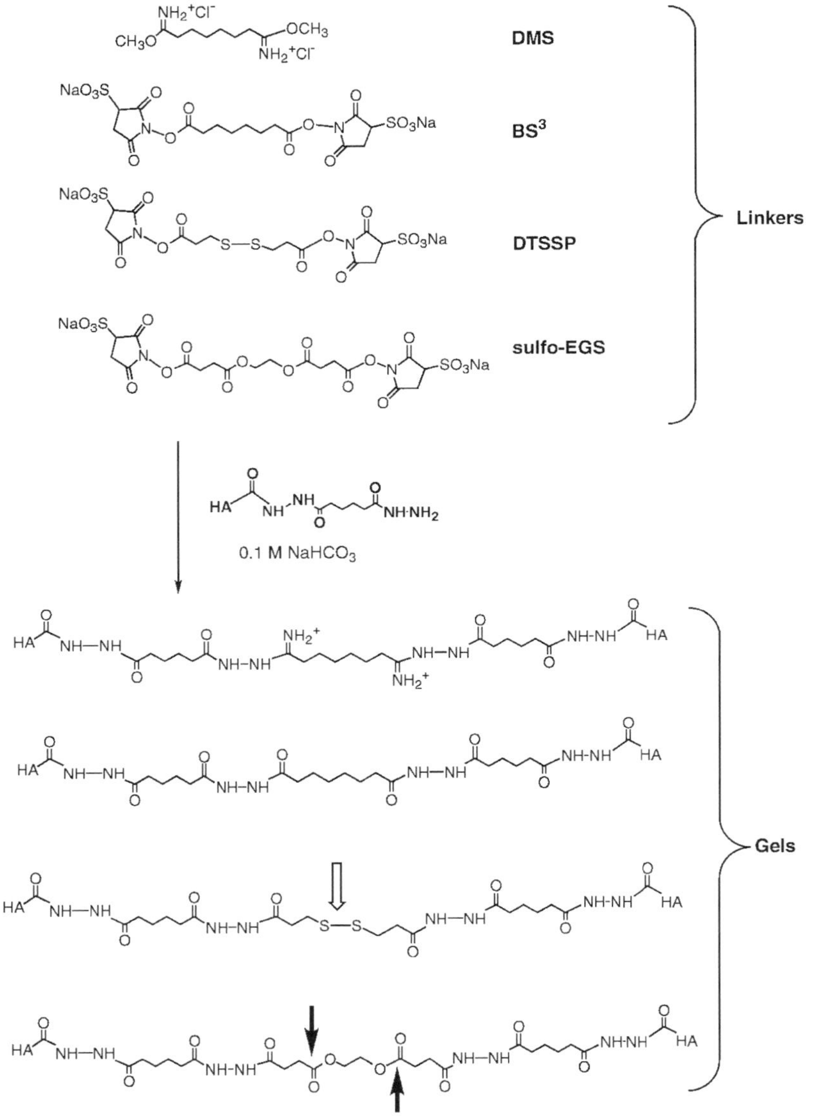

Strategy for cross-linking of HA-ADH to give hydrogels with four different homobifunctional cross-linkers

The open arrow indicates a site for reductive cleavage of the disulphide gel. Similarly, the dark arrows indicate hydrolysis of the diester backbone of the HA hydrogel to release HA-bound gel fragments.

showing average pore diameters of 20 to 50 μm [67]. For example, the HA-ADH-DTSSP-ADH-HA (Figure 7) hydrogel had continuous hexagonal-type pores that were not interconnected, while HA-ADH-BS³-ADH-HA showed elongated pores in a sheet-like network. In preliminary tests for biocompatibility, HA-ADH and its cross-linked derivatives have been examined using a proprietary *in vitro* assay of inflammatory potential (Collaborative Laboratories Inc.). Test substances are incubated with human neutrophils on a model radiolabelled ECM, and neutrophil activation (the first step in inflammatory response) is detected by measuring radiolabelled ECM degradation products resulting from enzymes released by activated neutrophils. In this system, HA-ADH was found to decrease the level of neutrophil activation, i.e. HA-ADH was intrinsically anti-inflammatory (D.C. Watkins, J.A. Hayward and B. Costello, personal communication).

Figure 8

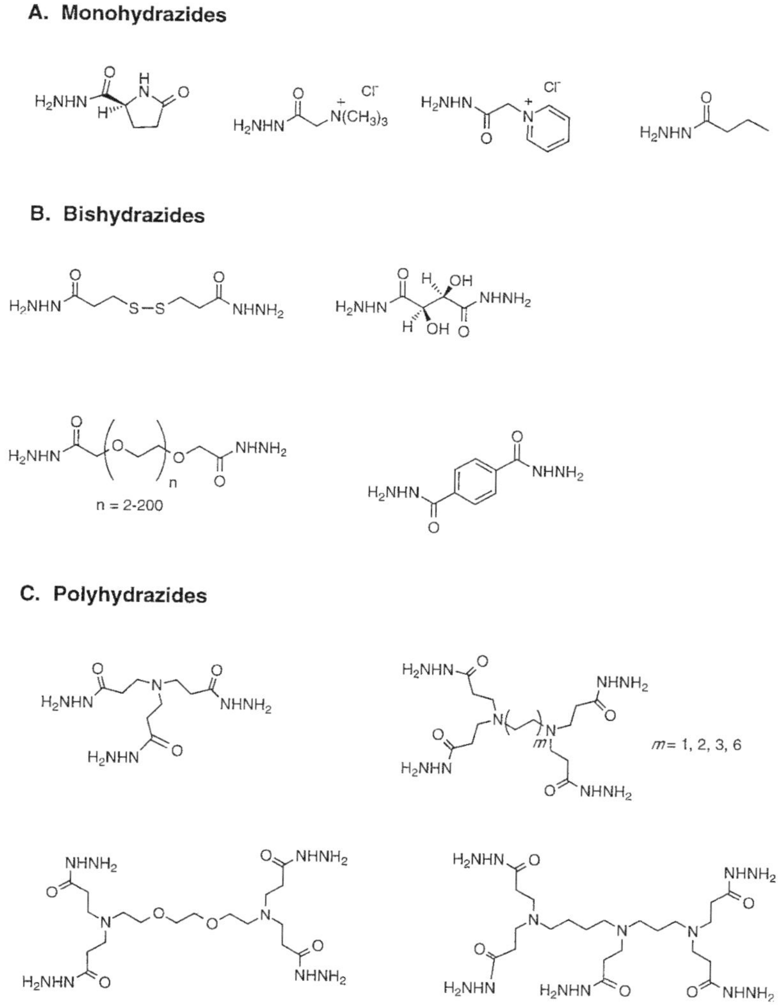

Structures of selected hydrazide reagents used for modification of HA

A, monofunctional hydrazides; B, bisfunctional hydrazides; C, polyfunctional hydrazide cross-linkers.

Figure 9

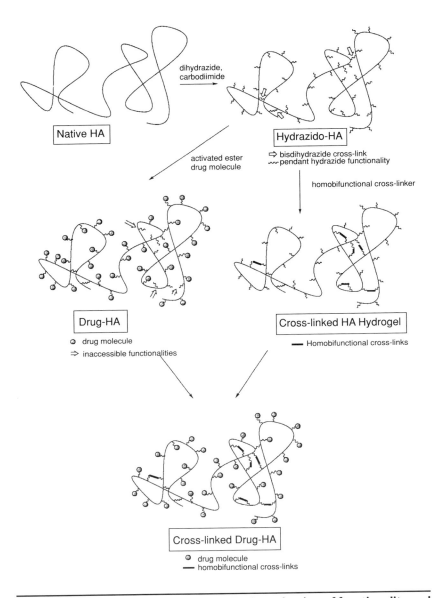

Schematic representation of the stepwise introduction of functionality and cross-links to native HA

A few cross-links (open arrows) occur spontaneously during the original functionalization, while the majority of ADH groups (wavy lines) are available for either attachment of reporter or biologically active groups (spheres) or cross-linkers (filled bars).

We have now extended the scope of the hydrazide technology to include branched and heteroatom-containing bishydrazides, as well as a number of monohydrazides. The structures of these reagents are illustrated in Figure 8. Monohydrazides (Figure 8A) allow the replacement of a defined proportion of

Figure 10

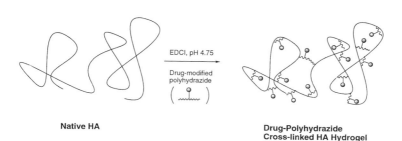

Native HA

Drug-Polyhydrazide
Cross-linked HA Hydrogel

The polyfunctional cross-linker strategy allows simultaneous introduction of cross-linking groups and a pendant chemical functionality for further modification

Alternatively, the bioeffector molecule may be attached to a polyvalent linker after the cross-linking reaction. Fewer pendant groups can be incorporated using this strategy.

---

the glucuronate carboxylic acids with sulphydryl, aliphatic, aromatic, fluorophoric, cationic or bioactive residues. The resulting products have, respectively, modified chemical attachment properties, additional hydrophobic patches, novel reporter functions, inverted charge properties or bioreleasable payloads [87]. The bishydrazides (Figure 8B) were selected or synthesized to enable direct cross-linking of HA in a single step using tethers with distinctive chemistries. Figure 9 illustrates the overall strategy for activation of HA, attachment of a biologically active ligand and cross-linking to form a bioerodable hydrogel that offers slow release of the active agent.

Polyhydrazides (Figure 8C) have also been synthesized based on modifications of dendrimer chemistry using 3–4 mol-equiv. methyl acrylate per mole to mono-, di- and triamines [88], followed by hydrazinolysis (J.F. Marecek, D.M. Marecak and G.D. Prestwich, unpublished work) [87]. These polyfunctional cross-linkers allow cross-linking of HA while retaining a pendant hydrazide functionality for subsequent (or prior) attachment of a bioeffector group possessing an active ester functionality. This modified strategy is illustrated in Figure 10.

Optimization of hydrogel synthesis

The creation of stable hydrogels from HA is challenging. Conditions can be selected that result in hydrogels with different physical properties and rates of gelation. The important parameters include pH, concentration and nature of metal ions present, ratios of HA to EDCI to linker, chemical nature of linker, nature and concentration of buffer, and molecular size range and concentration of HA. Table 1 presents an overview of the reaction parameters involved and their effects on the properties of the resulting HA hydrogels (K.P. Vercruysse, D.M. Marecak, J.F. Marecek and G.D. Prestwich, unpublished work). Softer, more pourable gels are favoured over solid gels by a number of selections: (a) lower concentrations of HA (2.5 versus 8.0 mg/ml), (b) a higher pH (approx. 4.75 versus 3.5), (c) a lower concentration of Bis-Tris buffer (100 versus 500 mM), (d) fewer equiv. of cross-

Table 1

Reagent	Soft-pourable gel	Solid gel
HA	2.5 mg/ml	8.0 mg/ml
0.1 M HCl	pH = 4.7	pH = 3.5
Bis-Tris HCl	100 mM	500 mM
Cross-linker	0.1 mol-equiv.	1.5 mol-equiv.
EDCI	0.5 mol-equiv.	1 mol-equiv.
Buffer	Tris (pK_a = 8.1)	Bis-Tris (pK_a = 6.5)

Reaction parameters and their effects on the physicochemical properties of HA hydrogels

linker (0.1 per glucuronate versus 1.5 mol-equiv.), (e) reduced amounts of EDCI (0.5 versus 1 mol-equiv.) and (f) the type of buffer (Tris versus Bis-Tris). The production of a porous three-dimensional network that is biocompatible, bioerodable, localizable, injectable and capable of furnishing viscosupplementary physical effects would be the ideal target material.

In addition to the parameters in the table, higher temperatures (50–75 °C) are often needed to dissolve higher, more hydrophobic bishydrazides in water. However, while coupling may improve at higher temperatures, cross-linking may be reduced. Co-solvents such as acetone, DMSO, DMF, dioxane and alcohols have been used in amounts that do not affect the integrity of the HA molecular size range or cause precipitation. While it is preferable to avoid non-aqueous solvents in potential medical products, low percentages may assist in obtaining suitable hydrogels. The mode and speed of mixing can influence gelation. It is difficult to obtain a homogeneous, fast-setting gel with simple magnetic stirring. Varying the ratios of HA:bishydrazide:carbodiimide, as shown in the table, is one of the most effective routes to altering gel properties. Conditions may use from 0.01 to 20 equiv. of bishydrazide per HA glucuronate moiety, and from 0.1 to 2 equiv. of carbodiimide per equiv. of bishydrazide.

Finally, the effects of metal ions were examined on the gelation of HA solutions (3.0 and 4.5 mg/ml) with 0.1 mol-equiv. of the hexamethylenediamine tetrahydrazide cross-linker (Figure 8) and 0.4 equiv. of EDCI (D.M. Marecak, unpublished work). Gel formation was attempted in the presence of 200 mM $MnCl_2$, $CaCl_2$, $BeCl_2$, $CoCl_2$, $FeCl_2$, or $CuCl_2$. At 4.5 mg/ml HA, gelation was rapid in the presence of Mn^{2+}, Ca^{2+} or Be^{2+} and in the absence of any salts. In the presence of Co^{2+} the gelation proceeded much more slowly, and in the presence of Fe^{2+} or Cu^{2+} no gels were obtained. With 3.0 mg/ml HA, no gelation occurred in the absence of salts; however, in the presence of Mn^{2+} or Ca^{2+}, gelations were again observed. The firmness of the gels obtained at 4.5 mg/ml decreased in the order $Ca^{2+} > Mn^{2+} > Be^{2+}$ > no salt.

Hydrogel degradation by HAse

In the body, HA is degraded by HAse, which is present in virtually every cell and in serum [89,90]. The kinetics of the degradation of HA in solution is best accomplished using viscometry [91], and a variety of spectrophotometric assays are also commonly used [92]. However, these assays were not suitable for measuring HA hydrogel stability. The biodegradation of hydrogels *in vitro* has been studied by monitoring the release of microspheres from the matrix [93], by

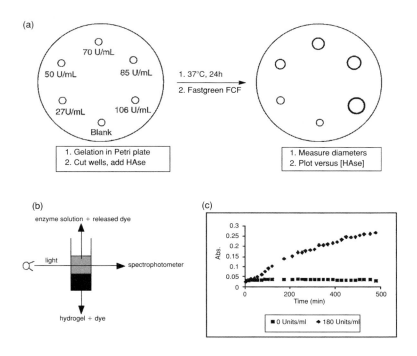

Figure 11

Two assays for HAse degradation of HA hydrogels

(a) Petri plate assay for single time point, multiple concentration screening. (b) Spectrophotometric assay for single concentration, multiple time point determinations. (c) Time course data for HAse degradation in spectrophotometric assay.

monitoring the release of radiolabelled compounds [94], by monitoring the loss of weight or swelling properties [95] or by visual inspection [96]. The biodegradation of HA derivatives by HAse can vary significantly, as determined with testicular HAse under standard assay conditions for a series of HA esters [97]. The latter results showed that fully esterified HA resisted degradation for 7–14 days but 25% esterified HA was partially degraded within minutes. Hydrogels based on glycidol ether bisepoxide cross-linking were examined, and limited access of the enzyme to correct oligosaccharide sites in a rigid lattice resulted in negligible HAse activity [93]. Similar assays [48] to evaluate HA coatings were developed by Biocoat Inc., and a microplate HAse assay was developed for HA substrates in agarose gels [98]. However, none of the assays were suitable for evaluation of the stability of the novel HA hydrogels produced with hydrazide modifications. We therefore developed two assays (Figure 11), the plate assay and the spectophotometric assay, and have used these in evaluation of hydrogels formed with ten different cross-linkers (K.P. Vercruysse, D.M. Marecak, J.F. Marecek and G.D. Prestwich, unpublished work).

In the plate assay, all the reagents in the gelling solution were poured into a Petri dish and allowed to set at room temperature. After gelation, the gel was washed with water, six holes were cut out in a circle and filled with 25 µl of buffer (30 mM citric acid/15 mM Na_2HPO_4/150 mM NaCl; pH = 6.3) and five

Figure 12

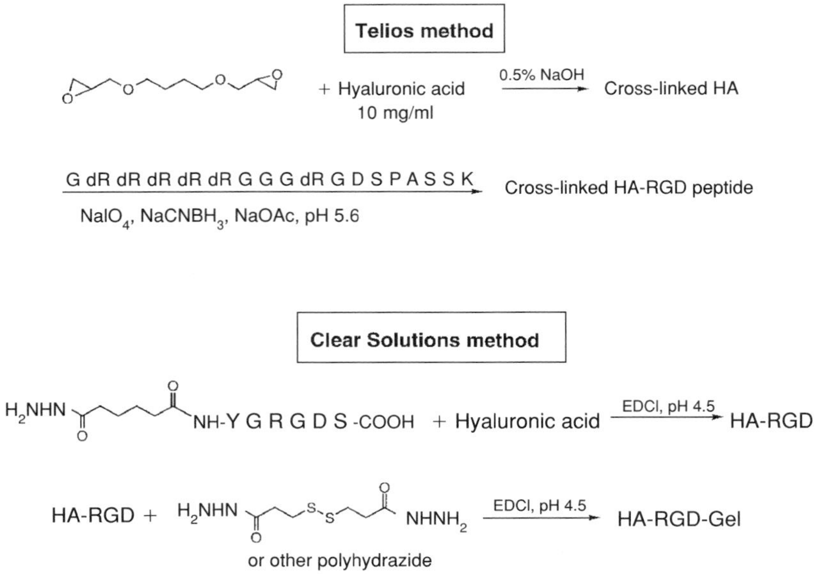

Telios method

+ Hyaluronic acid $\xrightarrow{\text{0.5\% NaOH}}$ Cross-linked HA
  10 mg/ml

G dR dR dR dR dR G G G dR G D S P A S S K   Cross-linked HA-RGD peptide
$\overline{\text{NaIO}_4, \text{NaCNBH}_3, \text{NaOAc, pH 5.6}}$

Clear Solutions method

H_2NHN ⋯ NH-Y G R G D S -COOH   + Hyaluronic acid $\xrightarrow{\text{EDCI, pH 4.5}}$ HA-RGD

HA-RGD + H_2NHN ⋯ NHNH$_2$ $\xrightarrow{\text{EDCI, pH 4.5}}$ HA-RGD-Gel
or other polyhydrazide

Two methods for the preparation of RGD peptide modified HA hydrogels
Top, the Telios Inc. methodology; bottom, the Clear Solutions Biotechnology Inc. hydrazide technology.

HAse samples were dissolved in the same buffer. The plate was stored at 37 °C and, after 24 h, the remaining undigested gel body was stained with a solution of fastgreen FCF. Clear rings of digested gel could be observed against a green background. The assay was performed in triplicate and diameters were plotted graphically as a function of the concentration of HAse.

The spectrophotometric assay allowed measurement of a time course of gel degradation. Thus, HA was cross-linked in the presence of 40 μM Coomassie Brilliant Blue R250. After mixing all the reagents, 1 ml of the gelling mixture was placed in the bottom of a plastic cuvette. This volume of gel does not block the path of the light of the spectrophotometer. When gelation was complete, the gel was washed and then 1 ml of enzyme solution was added on top of the gel, such that the light passed through this solution. Dye release, arising from gel degradation by HAse, was measured as absorbance at 590 nm as a function of the reaction time.

Chemically modified hydrogels for tissue engineering

To facilitate cell adhesion to hydrogels, HA derivatives with pendant fibronectin recognition elements have been prepared. These materials incorporate peptides containing the RGD (Arg-Gly-Asp) domain [99]. Figure 12 compares the two divergent approaches taken by the former Telios Inc. group [100,101] and our methodology (J.F. Marecek, D.M. Marecak and G.D. Prestwich, unpublished work) for preparing RGD-modified HA. The Telios process requires many more steps, conditions that seriously degrade and compromise the molecular size of

Figure 13

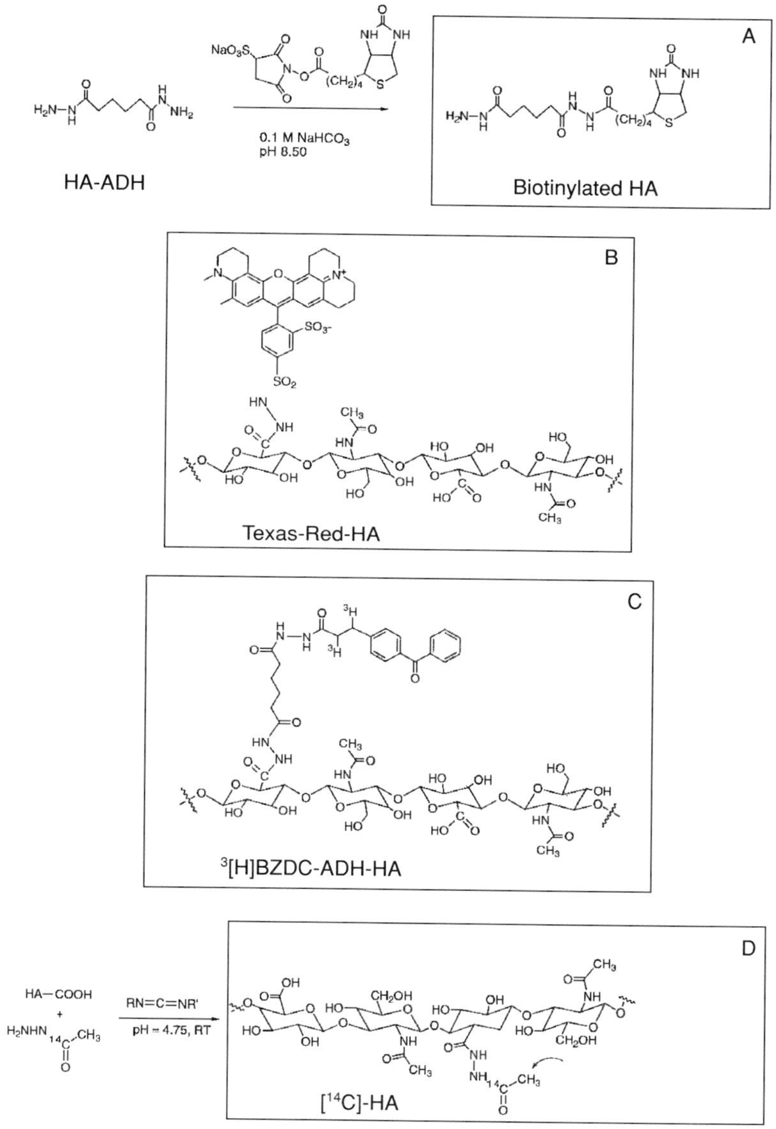

Biochemical probes prepared using hydrazide methodology

Top, (A) biotinylated HA-ADH; (B) Texas-Red-HA, prepared directly from HA and Texas Red sulphonylhydrazide; (C) photoactivatable [³H]BZDC-ADH-HA; and (D) [¹⁴C]acetohydrazide HA.

HA, and reagents that are poorly biocompatible. In contrast, our route requires two sequential, mild aqueous steps in a one-pot process. Similarly, a polypeptide containing the laminin fragment YIGSR has been covalently attached to HA. This mirrors a photochemical attachment mode recently used to pattern neurite growth in two dimensions [102].

Biochemical probes and their applications in cell biology

HA-ADH offers an ideal means to prepare probes for measuring and detecting HA–HABD interactions. Several HA-coupled probes are illustrated in Figure 13. First, high M_r HA and HA oligosaccharides have been converted into biotinylated probes using this methodology [103]. Biotinylated HA has now been employed for the detection of HA-binding proteins in a variety of tissues [104–107]. The biotinyl groups are randomly distributed throughout the HA molecule, usually with a net 1–5% modification, corresponding to 1–5 biotin per 100 glucuronates. This gives maximal sensitivity while preserving the minimum dodecasaccharide recognition unit of HA for HABDs. These probes offer advantages over reducing end-modified HA derivatives [108,109] in that multiple probe moieties can be included per HA molecule, and the probes are positioned in the internal milieu of HA and can be expected to bind in HA binding regions rather than at the periphery of these HABDs.

Second, photoaffinity labelling also offers an approach to identify and map the HA binding site of HA-binding proteins [110,111]. A heterobifunctional reagent, [³H]BZDC-NHS ester, can be used for introduction of a photoactivatable benzophenone group and tritium labels on to ligands bearing amino functionalities [112]. The benzophenone photolabel is stable in ambient light, chemically robust, readily activated at 360 nm and preferentially forms covalent linkages to hydrophobic groups on polypeptide side chains [113,114]. The synthesis of [³H]BZDC-HA-ADH from 3.5–10 kDa HA with approx. 5% modification has been recently accomplished (M.R. Ziebell, unpublished work), and this probe is being validated with RHAMM constructs as described below.

Third, HA has been coupled to Texas Red sulphonylhydrazide, thereby demonstrating that the sulphonylhydrazides as well as carbonylhydrazides are sufficiently nucleophilic to react with HA-EDCI O-acyl urea intermediates. The Texas-Red-HA has been employed in monitoring internalization of HA by transformed fibroblasts and its rapid transit through the cytosol and into the nucleus (E.A. Turley, J. Entwistle and M.R. Ziebell, unpublished work).

Fourth, a novel ¹⁴C-labelled derivative of HA was prepared. [*carbonyl-*¹⁴C]Ethyl acetate was converted into acetic hydrazide with a specific activity of 7 mCi/mmol, which was then condensed with partially degraded (60 and 6 kDa) HA and EDCI at pH 4.5 for three days. The resulting ¹⁴C-HA possessed a maximum of 5% modification (one in 20 glucuronates modified).

Finally, we coupled HA-ADH to the NHS-ester activated resin Affigel-10 to generate an affinity matrix (not pictured) used in our laboratories for the purification of HABDs. An earlier HA affinity matrix was developed by coupling HA to aminohexyl Sepharose [115].

Structural studies of HA–HABD interactions

The location and structure of the HA–protein interactions on HABDs has not been fully resolved, although Drs. Day and Parker report on the solution structure of an HABD of a link protein module [19] in Chapter 15 of this volume.

Figure 14

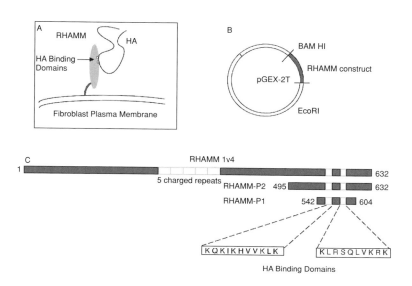

Determining the molecular basis of RHAMM–HA interactions

A, depiction of cellular location and function of RHAMM. B, plasmids containing the GST-fusion constructs of RHAMM polypeptides. C, diagram of the full-length RHAMM 1v4 variant cDNA and two constructs containing the HABD.

Computationally, one can dock HA onto a basic amino-acid-rich region by optimizing electrostatic interactions with negatively charged carboxylates of glucuronic acid moieties. Lerner and co-workers used two-dimensional NOE NMR methods to study short (up to 18-mer) synthetic peptides with HA [116]. They were unable to observe HA–peptide interactions or an induced structure of peptides in the presence of HA. Similar results were found by Turley and co-workers, suggesting that the minimum HABD for structural studies with HA octasaccharides (or higher) would be over 35 residues [26,117].

Recently, two RHAMM polypeptide constructs — 62 ('P1') and 157 amino acids ('P2'), respectively — have been prepared from the full-length RHAMM cDNA provided by E.A. Turley (M.R. Ziebell, unpublished work). As outlined in Figure 14, both constructs were expressed as glutathione transferase fusion proteins in *Escherichia coli* cells, harvested by sonication and pressure-induced cell lysis, and purified on immobilized glutathione columns. The free RHAMM domains were obtained by cleavage with thrombin and repurified on an HA affinity column prepared from Affigel-10 and HA-ADH. Western blots with detection by biotinylated HA confirm the HA binding properties of these recombinant domains. Circular dichroism studies established that both RHAMM-P1 and RHAMM-P2 are predominantly α-helical, in contrast to the predominantly β-barrel nature of the link module [19].

Conclusion

The modification of HA with hydrazides offers a mild and versatile approach that allows both cross-linking to form hydrogels and chemical functionalization to produce drug-delivery systems and biochemical probes. Numerous opportunities exist for the development of new biomaterials from this technology. Moreover, improved understanding of the roles of HA in cell adhesion and cell signalling will be attainable using an increasingly diverse repertoire of HA-coupled biophysical and biochemical reporter groups.

I am grateful to Dr. D.A. Swann (MedChem Inc., Woburn, MA) for introducing me to the importance and challenges of HA modifications, and for support of the initial work of Dr. J-w. Kuo on this project. The hydrazide derivatization was discovered by Dr. T. Pouyani with financial support by the Center for Biotechnology, a New York State Center of Advanced Technology operated through the New York State Science and Technology Foundation, which continued to support research at the University at Stony Brook, in partnership with funding from Collaborative Laboratories Inc. (CLI) and Clear Solutions Biotechnology Inc. (CSBI). Practical applications of this work have also been supported in part with Public Health Service SBIR and STTR awards, and current support includes the Center for Biopolymers at Interfaces (The University of Utah) and the Huntsman Cancer Institute (The University of Utah). We heartily thank Drs. J.A. Hayward, D.C. Watkins, J.R. Martinez, J. Maioriello, B. Costello, and D.T. Aust (all CLI and CSBI) for advice and support, and Dr. E.A. Turley (University of Manitoba; now University of Toronto) for stimulating discussions of the role of HA in cell biology.

References

1. Scott, J.E. (1995) J. Anat. **187**, 259–269
2. Ghosh, P. (1994) Clin. Exp. Rheumatol. **12**, 75–82
3. Freed, L.E., Vunjak-Novakovic, G., Biron, R.J., Eagles, D., Lesnoy, D., Barlow, S. and Langer, R. (1994) Biotechnology **12**, 689–693
4. Peppas, N. and Langer, R. (1994) Science **263**, 1715–1718
5. Swann, D. and Kuo, J-w. (1991) in Biomaterials, Novel Materials from Biological Sources (Byrom, D., ed.) pp. 287–305, Stockton Press, New York
6. Langer, R. (1993) Acc. Chem. Res. **26**, 537–542
7. Holmbeck, S.M.A., Petillo, P.A. and Lerner, L.E. (1994) Biochemistry **33**, 14246–14255
8. Scott, J.E. (1989) in The Biology of Hyaluronan (Evered, D. and Whelan J., eds.), Ciba Foundation Symp. No. 143, pp. 6–20, Wiley, Chichester
9. Knudson, C.B. and Knudson, W. (1993) FASEB J. **7**, 1233–1241
10. Neame, P.J. and Barry, F.P. (1993) Experientia **49**, 393–402
11. Sherman, L., Sleeman, J., Herrlich, P. and Ponta, H. (1994) Curr. Opin. Cell Biol. **6**, 726–733
12. Underhill, C. (1992) J. Cell Sci. **103**, 293–298
13. Turley, E.A. (1989) in The Biology of Hyaluronan (Evered, D. and Whelan J., eds.), Ciba Foundation Symp. No. 143, pp. 121–137, Wiley, Chichester
14. Hall, C.L., Yang, B.H., Yang, X.W., Zhang, S.W., Turley, M., Samuel, S., Lange, L.A., Wang, C., Curpen, G.D., Savani, R.C., Greenberg, A.H. and Turley, E.A. (1995) Cell **82**, 19–28
15. Mason, R.M., Crossman, M.V. and Sweeney, C. (1989) in The Biology of Hyaluronan (Evered, D. and Whelan J., eds.), Ciba Foundation Symp. No. 143, pp. 107–120, Wiley, Chichester
16. Weigel, P.H., Frost, S.J., LeBoeuf, R.D. and McGary, C.T. (1989) in The Biology of Hyaluronan (Evered, D. and Whelan J., eds.), Ciba Foundation Symp. No. 143, pp. 248–264, Wiley, Chichester
17. Banerjee, S.D. and Toole, B.P. (1992) J. Cell Biol. **119**, 643–652

18. Asplund, T. and Heldin, P. (1994) Cancer Res. **54**, 4516–4523
19. Kohda, D., Morton, C.J., Parkar, A.A., Hatanaka, H., Inagaki, F.M., Campbell, I.D. and Day, A.J. (1996) Cell **86**, 767–775
20. Entwistle, J., Hall, C.L. and Turley, E.A. (1996) J. Cell. Biochem. **61**, 569–577
21. Nelson, R.M., Venot, A., Bevilacqua, M.P., Linhardt, R.J. and Stamenkovic, I. (1995) Annu. Rev. Cell Dev. Biol. **11**, 601–631
22. Tanabe, K.K. and Saya, H. (1994) Crit. Rev. Oncog. **5**, 201–212
23. Hall, C.L., Wang, C., Lange, L.A. and Turley, E.A. (1994) J. Cell Biol. **126**, 575–588
24. Mohapatra, S., Yang, X.W., Wright, J.A., Turley, E.A. and Greenberg, A.H. (1996) J. Exp. Med. **183**, 1663–1668
25. Rooney, P., Kumar, S., Ponting, J. and Wang, M. (1995) Int. J. Cancer **60**, 632–636
26. Yang, B.H., Zang, L.Y. and Turley, E.A. (1993) J. Biol. Chem. **268**, 8617–8623
27. Degrendele, H.C., Estess, P., Picker, L.J. and Siegelman, M.H. (1996) J. Exp. Med. **183**, 1119–1130
28. Peach, R.J., Hollenbaugh, D., Stamenkovic, I. and Aruffo, A. (1993) J. Cell. Biol. **122**, 257–264
29. Knudson, W. (1996) Am. J. Pathol. **148**, 1721–1726
30. Bartolazzi, A., Peach, R., Aruffo, A. and Stamenkovic, I. (1994) J. Exp. Med. **180**, 53–66
31. Sleeman, J., Moll, J., Sherman, L., Dall, P., Pals, S.T., Ponta, H. and Herrlich, P. (1995) in Cell Adhesion and Human Disease (Marsh, J. and Goode, J.A., eds.), pp. 142–156, Wiley, Chichester
32. Ponta, H., Sleeman, J., Dall, P., Moll, J., Sherman, L. and Herrlich, P. (1994) Invasion Metastasis **14**, 82–86
33. Günthert, U., Stauder, R., Mayer, B., Terpe, H.J., Finke, L. and Friedrichs, K. (1995) Cancer Surv. **24**, 19–42
34. Culty, M., Shizari, M., Thompson, E.W. and Underhill, C.B. (1994) J. Cell. Physiol., 275–286
35. Mathew, J., Hines, J.E., Obafunwa, J.O., Burr, A.W., Toole, K. and Burt, A.D. (1996) J. Pathol. **179**, 74–79
36. Yeo, T.K., Nagy, J.A., Yeo, K.T., Dvorak, H.F. and Toole, B.P. (1996) Am. J. Pathol. **148**, 1733–1740
37. Zhang, L.R., Underhill, C.B. and Chen, L.P. (1995) Cancer Res. **55**, 428–433
38. Lees, V.C., Fan, T.P.D. and West, D.C. (1995) Lab. Invest. **73**, 259–266
39. Bignami, A., Hosley, M. and Dahl, D. (1993) Anat. Embryol. **188**, 419–433
40. Ng, C.K., Handley, C.J., Preston, B.N., Robinson, H.C., Bolis, S. and Parker, G. (1995) Arch. Biochem. Biophys. **316**, 596–606
41. Goa, K.L. and Benfield, P. (1994) Drugs **47**, 536–566
42. Balazs, E.A. and Denlinger, J.L. (1989) in The Biology of Hyaluronan (Evered, D. and Whelan J., eds.), pp. 265–280, Ciba Foundation Symp. No. 143, Wiley, Chichester
43. Saari, H., Santavirta, S., Nordstrom, D., Paavolainen, P. and Konttinen, Y.T. (1993) J. Rheumatol. **20**, 87–90
44. Pelletier, J.P. and Martel-Pelletier, J. (1993) J. Rheumatol. **20**, 19–24
45. Balazs, E.A. and Denlinger, J.L. (1993) J. Rheumatol. **20**, 3–9
46. Sanzgiri, Y.D., Maschi, S., Crescenzi, V., Callegaro, L., Topp, E.M. and Stella, V.J. (1993) J. Control. Release **26**, 195–201
47. Hume, L.R., Lee, H.K., Benedetti, L., Sanzgiri, Y.D., Topp, E.M. and Stella, V.J. (1994) Int. J. Pharm. **111**, 295–298
48. Lowry, K.M. and Beavers, E.M. (1994) J. Biomed. Mater. Res. **28**, 861–864
49. Lin, K., Bartlett, S.P., Matsuo, K., Livolsi, V.A., Parry, C., Hass, B. and Whitaker, L.A. (1994) Plast. Reconstr. Surg. **94**, 306–315
50. Tona, A. and Perides, G. (1993) Restor. Neurol. Neurosci. **5**, 151–154
51. Benedetti, L.M., Topp, E.M. and Stella, V.J. (1990) J. Control. Release **13**, 33–41
52. Benedetti, L., Cortivo, R., Berti, T., Berti, A., Pea, F., Mazzo, M., Moras, M. and Abatangelo, G. (1993) Biomaterials **14**, 1154–1160
53. Andreassi, L., Casini, L., Trabucchi, E., Diamantini, S., Rastrelli, A. and Donati, L. (1991) Wounds **3**, 116–126
54. Harvey, R.A. (1994) in Implantation Biology, The Host Response and Biomedical Devices (Greco, R.S., ed.), pp. 329–345, CRC Press Inc., Boca Raton, FL
55. Trachtenberg, J.D. and Ryan, U.S. (1994) in Implantation Biology, The Host Response and Biomedical Devices (Greco, R.S., ed.), pp. 81–112, CRC Press Inc., Boca Raton, FL
56. Berthiaume, F., Toner, M., Tompkins, R.G. and Yarmush, M.L. (1994) in Implantation Biology, The Host Response and Biomedical Devices (Greco, R.S., ed.), pp. 363–400, CRC Press Inc., Boca Raton, FL
57. Nevo, Z., Robinson, D. and Halperin, N. (1992) in Bone, Fracture, Repair, and Regeneration (Hall, B.K., ed.), pp. 123–152, CRC Press Inc., Boca Raton, FL
58. Robinson, D., Halperin, N. and Nevo, Z. (1990) Calcif. Tissue Int. **46**, 246–253

59. Dunn, M.G. and Maxian, S.H. (1994) in Implantation Biology, The Host Response and Biomedical Devices (Greco, R.S., ed.), pp. 229–252, CRC Press Inc., Boca Raton, FL
60. Freed, L.E., Grande, D.A., Lingbin, Z., Emmanual, J., Marquis, J.C. and Langer, R. (1994) J. Biomed. Mater. Res. **28**, 891–899
61. Park, K., Shalaby, W.S.W. and Park, H. (1993) Biodegradable Hydrogels for Drug Delivery, Technomic Publishing Co., Lancaster, PA
62. Kamath, K.R. and Park, K. (1993) Adv. Drug Deliv. Rev. **11**, 59–84
63. Yannas, I.V. (1990) Angew. Chem., Int. Ed. Engl. **29**, 20–35
64. Yannas, I.V. (1994) J. Cell. Biochem. **56**, 188–191
65. Shah, C.B. and Barnett, S.M. (1991) ACS Symp. Ser. **480**, 116–308
66. Drobnik, J. (1991) Adv. Drug Deliv. Res. **7**, 295–308
67. Pouyani, T., Harbison, G.S. and Prestwich, G.D. (1994) J. Am. Chem. Soc. **116**, 7515–7522
68. Danishevsky, I. and Siskovic, E. (1971) Carbohydr. Res. **16**, 199–201
69. Kuo, J-w., Swann, D.A. and Prestwich, G.D. (1991) Bioconj. Chem. **2**, 232–241
70. Kuo, J-w., Swann, D.A. and Prestwich, G.D. (1994) U.S. Pat. No. 5 356 883
71. Pouyani, T., Kuo, J-w., Harbison, G.S. and Prestwich, G.D. (1992) J. Am. Chem. Soc. **114**, 5972–5976
72. Balazs, E.A. and Leshchiner, A. (1987) U.S. Pat. No. 4 636 524
73. Takigami, S., Takigami, M. and Phillips, G.O. (1993) Carbohydr. Polym. **22**, 153–160
74. Adams, M.E. (1993) J. Rheumatol. **20**, 16–18
75. Mälson, T. and Lindquist, B. (1986) PCT Int. Appl. WO 8 600 912
76. Gorham, S.D. (1991) in Biomaterials, Novel materials from biological sources, pp. 57–122, Stockton Press, New York
77. Della Valle, F. and Romeo, A. (1988) EP Pat. No. 265 116
78. Sanzgiri, Y.D., Topp, E.M., Benedetti, L. and Stella, V.J. (1994) Int. J. Pharm. **107**, 91–97
79. Goei, L., Topp, E., Stella, V., Benedetti, L., Bivano, F. and Callegaro, L. (1992) in Polymers in Medicine, Biomedical and Pharmaceutical Applications (Ottenbrite, R.M. and Chiellini, E., eds.), pp. 93–113, Technomic Publishing Co., Lancaster, PA
80. Burns, J.M., Skinner, K., Colt, J., Sheidlin, A., Bronson, R., Yaacobi, Y. and Goldberg, E.P. (1995) J. Surg. Res. **59**, 644–652
81. Yerushalmi, N., Arad, A. and Margalit, R. (1994) Arch. Biochem. Biophys. **313**, 267–273
82. Akima, K., Ito, H., Iwata, Y., Matsuo, K., Watari, N., Yanagi, M., Hagi, H., Oshima, K., Yagita, A., Atomi, Y. and Tatekawa, I. (1996) J. Drug Targeting **4**, 1
83. Roth, S.H. (1995) Int. J. Tissue React. **17**, 129–132
84. Moore, A.R. and Willoughby, D.A. (1995) Int. J. Tissue React. **17**, 153–156
85. Pouyani, T. and Prestwich, G.D. (1994) Bioconj. Chem. **5**, 339–347
86. Feder-Davis, J., Hittner, D.M. and Cowman, M.K., ed. (1991) in Water-Soluble Polymers (Shalaby, S.W., McCormick, C.L. and Butler, G.B., eds.), ACS Symp. Ser. vol. 467, pp. 493–501, Am. Chem. Soc., Washington, D.C.
87. Prestwich, G.D. and Marecak, D.M. (1996) Chemical modifications of hyaluronic acid. Pat. pending
88. Tomalia, D.A. (1986) Macromolecules **19**, 2466–2468
89. Kreil, G. (1995) Protein Sci. **4**, 1666–1669
90. Afify, A.M., Stern, M., Guntenhoner, M. and Stern, R. (1993) Arch. Biochem. Biophys. **305**, 434–441
91. Vercruysse, K.P., Lauwers, A.R. and Demeester, J.M. (1995) Biochem. J. **306**, 153–160
92. Vercruysse, K.P., Lauwers, A.R. and Demeester, J.M. (1995) Biochem. J. **310**, 55–59
93. Yui, N., Okano, T. and Sakurai, Y. (1992) J. Control. Release **22**, 105–116
94. Cartlidge, S.A., Duncan, R., Lloyd, J.B., Rejmanova, P. and Kopecek, J. (1986) J. Control. Release **3**, 55–66
95. Shalaby, W.S.W., Blevins, W.E. and Park, K. (1991) in Water-Soluble Polymers. ACS Symp. Ser. vol. 467 (Shalaby, S.W., McCormick, C.L. and Butler, G.B., eds.), pp. 484–492, Am. Chem. Soc., Washington, D.C.
96. Hennink, W.E., Franssen, O., Overbeek, A.Y., Steenbergen, M.J.V. and Talsma, H. (1995) Proc. Int. Symp. Control. Rel. Bioact. Mater. **22**, 23–24
97. Zhong, S.P., Campoccia, D., Doherty, P.J., Williams, R.L., Benedetti, L. and Williams, D.F. (1994) Biomaterials **15**, 359–365
98. Tung, J.S., Mark, G.E. and Hollis, G.F. (1994) Anal. Biochem. **223**, 149–152
99. Hubbell, J.A. (1995) Biotechnology **13**, 565–576
100. Glass, J.R., Dickerson, K.T., Stecker, K. and Polarek, J.W. (1996) Biomaterials **17**, 1101–1108
101. Pierschbacher, M.D., Polarek, J.W., Craig, W.S., Tschopp, J.F., Sipes, N.J. and Harper, J.R. (1994) J. Cell. Biochem. **56**, 150–154
102. Clémence, J-F., Ranieri, J., Aebisher, P. and Sigrist, H. (1995) Bioconj. Chem. **6**, 411–417
103. Pouyani, T. and Prestwich, G.D. (1994) Bioconjugate Chem. **5**, 370–372.

104. Hoare, K., Savani, R.C., Wang, C., Yang, B. and Turley, E.A. (1993) Connect. Tissue Res. **30**, 117–126
105. Yu, Q. and Toole, B.P. (1995) Biotechniques **19**, 122
106. Yang, B.H., Yang, B.L. and Goetinck, P.F. (1995) Anal. Biochem. **228**, 299–306
107. Melrose, J., Numata, Y. and Ghosh, P. (1996) Electrophoresis **17**, 205–212
108. Raja, R.H., McGary, C.T. and Weigel, P.H. (1988) J. Biol. Chem. **263**, 16661–16668
109. Yannariello-Brown, J., Zhou, B., Ritchie, D., Oka, J.A. and Weigel, P.H. (1996) Biochem. Biophys. Res. Commun. **218**, 314–319
110. Yannariello-Brown, J., Frost, S.J. and Weigel, P.H. (1992) J. Biol. Chem. **267**, 20451–20456
111. Yannariello-Brown, J. (1996) Glycobiology **6**, 111–119
112. Olszewski, J.D., Dormán, G., Elliott, J.T., Hong, Y., Ahern, D.G. and Prestwich, G.D. (1995) Bioconj. Chem. **6**, 395–400
113. Dormán, G. and Prestwich, G.D. (1994) Biochemistry **33**, 5661–5673
114. Prestwich, G.D., Dormán, G., Elliott, J.T., Marecak, D.M. and Chaudhary, A. (1997) Photochem. Photobiol. **65**, in the press
115. Tengblad, A. (1979) Biochem. Biophys. Acta **578**, 281–290
116. Horita, D.A., Hajduk, P.J., Goetinck, P.F. and Lerner, L.E. (1994) J. Biol. Chem. **269**, 1699–1704
117. Yang, B.H., Yang, B.L., Savani, R.C. and Turley, E.A. (1994) EMBO J. **13**, 286–296

Metabolism of hyaluronan

Vincent C. Hascall*¶, Csaba Fulop*, Antonia Salustri†, Nicola Goodstone*, Anthony Calabro*, Michael Hogg*, Raija Tammi‡, Markku Tammi‡ and Donald MacCallum§

*Department of Biomedical Engineering, Wb3 Research Institute, Cleveland Clinic Foundation, Cleveland, OH 44195, U.S.A., †Department of Public Health and Cell Biology, University of Rome 'Tor Vergata', 00173 Rome, Italy, ‡Department of Anatomy, University of Kuopio, 70211 Kuopio, Finland, and §Department of Anatomy, University of Michigan, Ann Arbor, MI 48109, U.S.A.

Introduction

Hyaluronan (HA) is an ubiquitous glycosaminoglycan found in almost all tissues. HA was originally isolated by Meyer and Palmer [1] from the vitreus of the eye and shown to contain a hexuronic acid, an amino sugar and acetyl groups with no sulphoester content. Two years later, they isolated HA from umbilical cord and identified glucuronic acid and glucosamine as the sugar constituents [2]. The actual linkages for the repeat disaccharide motif [(-β-1,4-glucuronic acid-β-1,3-N-acetylglucosamine-)$_n$] took longer to determine, culminating in a paper by Weissman and Meyer in 1954 [3].

The number of repeat disaccharides, n, in an HA molecule can reach 30000 (a molecular mass of more than 10 MDa). While this early work defined the somewhat monotonous chemical structure of HA, many other properties of this enigmatic polymer remain elusive to this day, including a clear definition of its metabolism. Thus, mechanisms for both its synthesis and its degradation, and for cellular regulation of both processes remain obscure. Recent advances in a number of laboratories, summarized in this chapter, provide new insights into the metabolism of HA, however, and suggest that many of these problems may well be resolved in the near future.

Current model for biosynthesis of HA

In 1983, Prehm [4] reported the results of a series of experiments that led him to propose a novel mechanism for HA synthesis that was distinctly different from that for other glycosaminoglycans, such as chondroitin sulphate and heparan sulphate. The latter were known to be elongated on core proteins by transfer of an appropriate sugar residue from either UDP-glucuronic acid or an appropriate UDP-N-acetylhexosamine on to the non-reducing terminus of a growing chain (see [5] for review). However, Prehm's study showed that: (1) the particulate membrane fraction of teratocarcinoma cell lysates synthesized HA when provided with UDP-glucuronic acid and UDP-N-acetylglucosamine; (2) small

HA oligosaccharides, i.e. with potential non-reducing terminal acceptors, did not compete for HA elongation; and (3) HA chains, elongated first in the presence of UDP-[^{14}C]glucuronic acid, chased with unlabelled UDP-glucuronic acid and then treated with a combination of the non-reducing terminal exoglycosidases, β-glucuronidase and hexosaminidase, released more radioactivity at early times of digestion than at later times. These data led Prehm to propose that HA synthesis occurs at the reducing terminus of a growing HA chain. In this mechanism, the reducing end sugar (either *N*-acetylglucosamine or glucuronic acid) would remain covalently bound to a terminal UDP, and the alternative substrate for the next sugar to be added (either UDP-glucuronic acid or UDP-*N*-acetylglucosamine) would be transferred onto the reducing end sugar with displacement of this terminal UDP (Figure 1).

In subsequent work, Prehm [6] proposed that the enzyme, or enzyme complex, that synthesizes HA resides at the inner surface of the plasma membrane and that the growing HA chain is extruded through the membrane into the extracellular space. Evidence for this included: (1) the endogenously labelled HA, synthesized by a B6 cell line in the presence of [^{3}H]glucosamine, that remained associated with the cells, was of large molecular size; (2) a large proportion of this cell-associated ^{3}H-HA was digested by treatment of intact cells with *Streptomyces* hyaluronidase; (3) frozen–thawed cells, unlike intact cells, were able to incorporate radioactivity from UDP[^{14}C]glucuronic acid into HA; and (4) this endogenously labelled ^{14}C-HA was of small molecular weight if the intact cells were first treated with the hyaluronidase, but high molecular weight otherwise.

This model for HA synthesis has several attractive features that are supported by other investigations: (1) Mason et al. [7] metabolically labelled proteins in chondrocytes and isolated HA using a combined dissociative and detergent extraction procedure followed by rigorous purification methods. The purified HA contained no covalently bound protein, suggesting that it is synthesized without the necessity of a core protein, in marked contrast to synthesis of glycosaminoglycans on proteoglycans. (2) Philipson and Schwartz [8] used rate zonal sedimentation to partition vesicles formed after homogenization of mouse oligodendroglioma cells and showed that HA synthetic activity correlated most closely with plasma membrane markers. (3) Cote and Robyt [9] showed that the synthesis of a dextran with alternating 1,3 and 1,6 glucose linkages occurred by addition of glucose in the alternate linkages at the reducing end of the polymer with displacement of fructose from the substrate sucrose, a

Figure 1

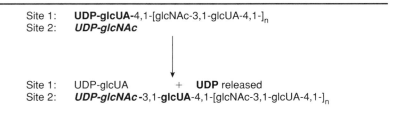

Site 1:   **UDP-glcUA**-4,1-[glcNAc-3,1-glcUA-4,1-]$_n$
Site 2:   ***UDP-glcNAc***

Site 1:   UDP-glcUA   +   **UDP** released
Site 2:   ***UDP-glcNAc*-3,1-glcUA**-4,1-[glcNAc-3,1-glcUA-4,1-]$_n$

Example of proposed HA elongation step on HA synthase protein

very similar general mechanism to that proposed for HA. (4) The extrusion of the growing chain into the extracellular space would also allow unconstrained growth, thereby achieving the large size of HA, whereas confinement of synthesis within a Golgi or *trans*-Golgi compartment could well limit the overall length of the polymers.

HA-synthases

Several studies have attempted to solubilize, identify and purify HA-synthase (HAS) from strains of streptococci that make a capsular coat of HA [10–12] as well as from eukaryotic cells [13]. Prehm and Mausolf [10] used periodate-oxidized UDP-glucuronic acid or periodate-oxidized UDP-*N*-acetylglucosamine, known inhibitors of HA synthesis, to affinity label a protein of ~52 kDa that co-purified with HA. Triscott and van de Rijn [11] were able to solubilize HAS from streptococcal membranes in an active form with digitonin. They also reported that CHAPSO increased HA synthesis in the membrane fraction ~3-fold, even though this zwitterionic detergent did not solubilize any HA synthetic activity. They suggested that this was the result of opening up membrane vesicles, thereby allowing greater access of the UDP-sugar substrates to the site of synthesis. van de Rijn and Drake [12] later selectively tagged three streptococcal membrane proteins of 42, 33 and 27 kDa with a photoactivatable, radiolabelled azido derivative of UDP-glucuronic acid. The authors suggested that the 33 kDa protein is HAS since labelling of this protein with the probe occurred without UV irradiation, was blocked by unlabelled UDP-glucuronic acid and increased in the presence of UDP-*N*-acetylglucosamine.

Dougherty and van de Rijn [14] prepared an acapsular streptococcal mutant from a strain which normally makes an HA capsule by transposon inactivation of the locus required for HA synthesis. This locus consisted of an operon with at least three genes required for HA synthesis, referred to as *hasA*, *hasB*, and *hasC*. The *hasB* gene was identified first and shown to be a UDP-glucose dehydrogenase (required to convert UDP-glucose into UDP-glucuronic acid, one of the substrates for HAS) [14]. Transfection of a combination of *hasA* and *hasB* into either an acapsular streptococcus or a strain of *E. coli* conferred them with the ability to synthesize HA and form a capsule [15]. This provided strong evidence that *hasA* codes for an HAS. Subsequently, the *hasA* gene was sequenced [16] and confirmed [17]. The *hasC* gene was later identified as a UDP-glucose pyrophosphorylase (required to convert glucose-1-phosphate plus UTP to UDP-glucose) [18].

DeAngelis and Weigel [19] provided a major advance by preparing monospecific polyclonal antisera against synthetic peptides from the deduced sequence of *hasA*. These were used to identify a protein of 42 kDa present in wild type and transfected *E. coli* capable of HA synthesis, but absent in strains without HA synthetic activity. Recombinant hasA, immobilized by the antibody, was shown to synthesize HA when supplied with both UDP-sugar substrates. This provided the first proof that a single protein can utilize both substrates and elongate an HA chain.

Itano and Kimata [20] identified the first mammalian *HAS*, now referred to as *HAS1*. They used a mutagen to produce subclones defective in HA synthesis from a murine mammary carcinoma cell line. Three separate classes of subclones were found to be complementary for HA synthesis in somatic cell fusion experiments, suggesting that at least three proteins are required for HA synthetic activity. Two of these classes maintained some HA synthetic activity whereas one showed no activity. The latter was used in transient transfection experiments with cDNA prepared from the parental cells to identify a single protein that restored HA synthetic activity. Sequence analyses revealed a deduced primary structure for a protein of ~65 kDa. The protein contained several stretches of hydrophobic residues similar to the multiple membrane-spanning regions observed in bacterial hasA. The protein, now referred to as murine HAS1 (muHAS1), also showed 57% identity in a 557 amino acid overlap with *Xenopus* DG42 (differentially expressed during gastrulation), a protein known to have significant homology with bacterial HAS. DG42 was originally identified because of its selective expression during embryonic development [21], and recent work has shown that this protein has HA synthetic activity [22,23].

The homology between bacterial HAS and *Xenopus* DG42 was utilized by Spicer et al. [24] to design degenerate primers for RT-PCR amplification from a mouse embryo cDNA library. A product was identified that coded for a distinct protein (now referred to as muHAS2) with homology (~55% identity) to muHAS1. Transfection of *muHAS2* mRNA into COS cells directed *de novo* production of an HA cell coat detected by particle exclusion assay, providing strong evidence that the HAS2 protein can synthesize HA. Fulop et al. [25] independently used a similar strategy to identify *muHAS2* in RNA isolated from ovarian cumulus cells actively synthesizing HA, a critical process for normal cumulus oophorus expansion. Cumulus cell–oocyte complexes were isolated from mice immediately after administering an ovulatory dose of gonadotropins, before HA synthesis begins and 3 and 4 h later when HA synthesis is just beginning (3 h) and already apparent (4 h). RT-PCR showed that mRNA for *HAS2* was absent initially, but expressed at high levels 3 and 4 h later.

Remarkably, two additional studies [26,27], published with that of Spicer et al. [24] in the same issue of the *Journal of Biological Chemistry*, identified the human transcripts for *HAS1* (*huHAS1*) and *HAS2* (*huHAS2*). Watanabe and Yamaguchi [27] also used degenerate primers based on the bacterial HAS and DG42 and identified *huHAS2* in a human brain stem cDNA library. Thus, all three studies [24,25,27] that used this approach identified the HAS2 form of the enzyme. A different strategy was used by Shyjan et al. [26]. They expressed proteins from a human mesenteric lymph node library in murine mucosal homing lymphocyte T-cells and screened for the ability of transfectants to adhere in a rosette assay. Adhesion of one transfect was inhibited by antisera to CD44, a known cell surface HA-binding protein, and was abrogated directly by pretreatment with an HA-specific hyaluronidase. Thus, the rosetting required synthesis of HA by the transfectant. Cloning and sequencing of the responsible mRNA identified *huHAS1*. Itano and Kimata also described *huHAS1* by screening a human fetal brain library with probes based upon the *muHAS1* sequence [28].

Regulation of HA synthesis

Expansion of the mouse cumulus cell–oocyte complex (COC) *in vitro* and *in vivo* requires synthesis and organization of an extracellular matrix in which HA is a necessary and dominant component [29–32]. Immediately after administering an ovulatory dose of gonadotropins, ~1000 cumulus cells are closely adherent to the oocyte forming the compact COC. When compact COCs are isolated and cultured in the presence of either FSH (follicle-stimulating hormone) or EGF (epidermal growth factor) at physiological concentrations (~5 ng/ml and ~50 pg/ml, respectively), HA synthesis is first detected at ~3 h, rises rapidly to a maximum level that is sustained between 6–12 h and declines to zero by ~15 h [29]. However, cumulus cells cultured in the presence of either of these factors, but in the absence of the oocyte, synthesize only ~10% as much HA, which indicates that the oocyte produces a factor required to achieve maximum rates of synthesis [30]. Recent work [32] has shown that either the FSH or EGF need be present only during the first 2 h, before HA synthesis actually begins, for subsequent synthesis to be maximal. Conversely, the oocyte (or oocyte factor) must be continuously present to achieve maximum synthesis of HA. If oocytes are removed at 6 h after initiating the process by adding FSH or EGF to compact COCs, HA synthesis rapidly declines, only producing an amount equivalent to that which would be achieved if the maximal rate were sustained for 2–3 h after removing the oocytes. Further, addition of actinomycin D to COCs at 6 h, when HA synthesis is maximal, also shuts down HA synthesis to the same extent.

These results and those described above for up-regulation of mRNA for *HAS2* [25] suggest that regulation of HA synthesis in cumulus cells is primarily controlled at the level of transcription, and they are consistent with the following model: (1) the cumulus cells have both FSH and EGF receptors which, when activated, initiate intracellular signalling pathways that eventually converge on a promoter region for *HAS2* (as well as other proteins that are required for production and organization of the extracellular matrix). (2) FSH operates in part through a cAMP-mediated signalling pathway and in part through a tyrosine kinase pathway, while EGF operates only through tyrosine kinase pathways [32]. (3) The oocyte factor, operating through a separate receptor and signalling pathway, promotes the continuous, high level transcription of the *HAS2* gene (and other relevant genes), leading to maximum production of the HAS2 protein, and subsequently of HA. (4) Removal of the oocyte factor rapidly down-regulates transcription of *HAS2*, and actinomycin D rapidly blocks its transcription. In both cases, HA synthesis, indicative of active HAS2, continues for a period of time (2–3 h) during which synthesis of new HAS2 protein stops, as *HAS2* mRNA is depleted, and subsequent delivery of HAS2 protein to the site of HA synthesis ceases. This model predicts that both *HAS2* mRNA and HAS2 protein have short half-lives during the COC expansion process.

Is HA synthesis compartmentalized?

In contrast, under certain experimental conditions, HA synthesis is remarkably stable. Previously, we studied the effects of brefeldin A on glycosaminoglycan synthesis by rat chondrosarcoma chondrocytes [33]. This fungal metabolite rapidly and reversibly interferes with antegrade vesicular transport. This disassembles the Golgi complex, with redistribution to the endoplasmic reticulum, and isolates the *trans*-Golgi cisternae from the *trans*-Golgi network. Chondroitin sulphate synthesis decreased to less than 1% of control within 15 min after adding the reagent to the cells, whereas HA synthesis continued at or above control levels for at least 8 h. If HAS2 protein is normally turning over in these cells, this stability of HA synthesis suggests that brefeldin A isolates the HA synthesis compartment by blocking both entry and exit of HAS2.

Even more surprising results were obtained when these cells were treated with staphylococcal α-hemolysin. This protein forms heptamers in the plasma membranes of cells, creating pores that allow small molecules of molecular mass ~2 kDa or less to diffuse freely in or out of the cytoplasm [34]. Chondrosarcoma cultures permeabilized with α-hemolysin lose intracellular ATP (to less than 10% of control), and decrease synthesis of chondroitin sulphate (to a few % of control), as measured with [³H]glucosamine as a metabolic precursor, within 1–2 h after permeabilization (N. Goodstone, V.C. Hascall and A. Calabro, unpublished work). Nevertheless, HA synthesis was sustained at a constant level (~70% of control) for at least 24 h. The loss of ATP could interfere with vesicular transport, which requires metabolic energy, and isolate the site of HA synthesis by a similar mechanism as for brefeldin A treatment. However, in the case of the α-hemolysin, much more stringent constraints would be put on the properties of the compartment. Elongation of HA by HAS requires UDP-glucuronic acid and UDP-*N*-acetylglucosamine as substrates. Thus, the precursor used, [³H]glucosamine, must be phosphorylated, acetylated and reacted with UTP to form UDP-*N*-acetyl[³H]glucosamine, and all these enzymic steps require metabolic energy. Additionally, glucose must be phosphorylated, reacted with UTP, and oxidized to form the other substrate, UDP-glucuronic acid. The fact that HA synthesis continues in the face of decreased ATP suggests that the site of HA synthesis must utilize the residual ATP very efficiently or have access to a source of metabolic energy. The unusual results with the α-hemolysin could be explained if the compartment in which HA synthesis occurs is isolated from the cytoplasm and contains the enzymes necessary to carry out these processes. The concentrations of the sugars, glucose and glucosamine, in this compartment could equilibrate with cytoplasmic concentrations by diffusion. This model would be distinct from the antiport process required to transport UDP-sugars into the *trans*-Golgi network to support chondroitin sulphate synthesis (see [5,35] for reviews). The antiport process requires metabolic energy in the cytoplasmic compartment and may be compromised rapidly by the loss of ATP after permeabilization.

Steady state metabolism of HA

While the COC model can provide insight into regulation of HA synthesis, the matrix that is formed is not catabolized to any significant extent by the cumulus cells. In most tissues, however, the cells not only synthesize HA but also actively catabolize HA. When these two processes are balanced, the metabolism of HA is in steady state. Studies with two tissues, hyaline cartilage and epidermis, have measured metabolic parameters for HA in steady state conditions and illustrate the wide range of rates that can occur.

Hyaline cartilage

The concentration of HA in the extracellular matrix of hyaline cartilages can reach 1–2 mg/g wet weight, approximately the same concentration found in the expanded COC matrix (see [36] for review). In cartilages, however, HA provides a scaffold for binding the much higher concentration of aggrecan (typically 20–50 mg/g wet weight) in the matrix, thereby providing these tissues with their characteristic capacity to bear weight with minimal deformation.

Explant cultures, prepared from the metacarpophalangeal cartilages of young bovines, have been shown to exhibit steady state parameters for proteoglycan metabolism when the culture medium is supplemented with serum or appropriate growth factors (see [37] for review). Typically, a newly synthesized aggrecan molecule has a half-life in the matrix of ~20 days before being catabolized. While the fate of aggrecan can be readily followed using [^{35}S]sulphate as a metabolic precursor, which is essentially specific for proteoglycans, it is much more difficult to study the metabolism of HA in this tissue. Two studies have utilized either [^{3}H]glucosamine [38] or [^{3}H]acetate [39] as a metabolic precursor to pulse label glycosaminoglycans and glycoproteins in explant cultures. The fate of the labelled HA with continued time of culture was monitored by using isolation procedures and enzymes that specifically degrade HA. In the former study [38], a stringent protocol was used to purify native aggregates, i.e. aggrecan and link protein bound to central strands of HA. Pulse labelled cultures, chased 3–20 days, were digested with highly purified collagenase to allow extraction of native proteoglycan aggregates in high yield with an associative solvent. Both isopycnic and velocity gradient ultracentrifugation procedures were then used to purify native proteoglycan aggregates with a final yield of ~40% of the total in the tissues. Analyses of the purified aggregates revealed that: (1) HA constituted ~3% of the total mass of glycosaminoglycan; (2) ^{3}H-labelled HA likewise constituted ~3% of the total label in glycosaminoglycan independent of chase time; (3) the half-life of label in both HA and aggrecan was therefore identical and determined to be ~20 days; and (4) while ~95% of the labelled aggrecan was recovered in the culture medium in a proteolytically degraded form, none of the labelled HA lost from the tissue appeared in the medium compartment. The latter study [39] obtained the same results and revealed additional points: (1) a proportion of the pulse labelled HA (~25%) escaped into the medium during the first day of chase with no further release at later chase times; and (2) the average molecular size of the labelled HA in the tissue was larger than that of the resident, unlabelled HA at early chase times, but decreased in size toward that of the resident HA distribution, with a half-life of 1–2 weeks.

Results from the above experiments indicate that: (1) most of the HA synthesized by chondrocytes in hyaline cartilage from young animals becomes incorporated into aggregates in the matrix; (2) the rates of synthesis of HA and aggrecan are proportional to their mass ratio; (3) the catabolic process removes both from the matrix at the same time; and (4) while most of the aggrecan is proteolytically cleaved and released from the matrix, the HA is completely degraded by the cells. Thus, the catabolic unit, situated on or near the surface of the chondrocyte, would probably require three activities: (1) a receptor for binding HA; (2) a protease (the so-called aggrecanase); and (3) an endo-active hyaluronidase. A model consistent with the results would involve: (1) binding the catabolic unit to an exposed region of HA in an aggregate; (2) cleaving the HA at that point; (3) cleaving any accessible aggrecan at the specific aggrecanase site; and (4) progressively internalizing one of the cleaved strands of HA while repeating the proteolytic steps. The initial cleavage of HA would on average be in the middle of the aggregate, and the polarity of the process would leave half of the HA in the matrix. This is consistent with the observation that the newly synthesized HA is larger than the equilibrium distribution of sizes represented by the resident HA.

Additional evidence provides indirect support for such a catabolic mechanism. The cell surface HA-binding protein CD44 is present on chondrocytes, and in cell culture has been implicated in catabolism of exogenously added HA [40]. The dominant cleavage on aggrecan during normal and cytokine or retinoic acid accelerated catabolism occurs at a specific glutamate–alanine bond located between the G1 (HA-binding) and G2 globular domains (the aggrecanase site) [41–43]. The specificity of cleavage is distinct from known proteases, such as stromelysin (MMP-3), which have been considered to have roles in aggrecan catabolism, indicating that the enzyme is closely regulated, perhaps by intercalation in the cell membrane within the catabolic unit. Finally, there is precedent for cell surface hyaluronidase activity [44,45]. A glycosyl-phosphatidyl inositol-anchored hyaluronidase, tethered on the surface of the sperm head, has been shown to be required for successful penetration into the HA matrix surrounding the oocyte prior to fertilization.

Epidermis

Suprisingly, HA is highly concentrated (~2 mg/ml) in the limited extracellular space surrounding keratinocytes in the epidermis [46], a tissue derived from ectoderm, unlike the mesenchymal origin of most connective tissues such as cartilages. Explant cultures of human skin have been pulsed with [³H]glucosamine and the epidermis separated from dermis at different chase times [47]. Labelled HA in the epidermis had a short half-life, ~1 day, indicating that metabolism of HA in this tissue is exceptionally active. The epidermis is separated from the underlying dermis by a basement membrane, which most likely presents a barrier to diffusion of large macromolecules such as HA from the tissue. Thus, catabolism of HA must occur to a large extent within the tissue, and steady state metabolism would require that catabolism be as active as synthesis. Epidermal keratinocytes synthesize a variety of splice variant CD44 molecules, which are highly glycosylated, frequently with glycosaminoglycans (see [48] for review). CD44 localizes on the surfaces of the cells in close contact with the HA in the surrounding matrix [49], and it likely facilitates organization of the HA at the cell surface and participates in catabolic processes.

We have recently studied metabolism of HA in a rat epidermal keratinocyte cell line that undergoes organotypic differentiation when cultured on a reconstituted, native collagen substrate at an air/medium interface (R.Tammi, M. Tammi, M. Hogg, D.K. MacCallum and V.C. Hascall, unpublished work). Fully differentiated cultures contain a basal layer on the substrate, a spinous cell layer, a granular cell layer with keratohyaline granules and a surface stratum corneum. On collagen substrate alone, the basal layer is porous, and most of the labelled macromolecular HA synthesized by the cells diffuses into the collagen substrate and the underlying medium compartment. A series of experiments tested whether the presence of an intact basement membrane would alter HA metabolism. Canine kidney tubule epithelial cells (the MDCK cell line) were first cultured on the collagen substrate where they deposit an intact basement membrane. These cells were then removed and the keratinocytes cultured on the modified substrate, i.e. on native collagen with an intervening basement membrane. Fully differentiated cultures were then metabolically labelled with [³H]glucosamine for times up to 24 h. Label in HA reached steady state (i.e. equal rates of synthesis and catabolism of labelled HA) by ~10 h when the basement membrane was present. Further, unlike cultures without an intervening basement membrane, most (~70%) of the labelled HA was retained in the cell layer, and only smaller molecular weight HA fragments escaped into the substrate and medium compartments. These results indicate that the half-life of HA in this culture model is only a few hours, dramatically different from the value for hyaline cartilage, and that epidermal keratinocytes have highly active processes for catabolism of HA.

Concluding remarks

This chapter has focused on novel models for biosynthesis and catabolism of HA that are consistent with previous studies and with novel, often surprising, results of studies in progress. Future research will determine which aspects of these speculations have merit and which need to be modified or discarded. It is clear that identification of the eukaryotic HAS variants at the mRNA level has set the stage for rapid advances in unravelling regulatory mechanisms for transcription at the gene level and for probing details of the synthetic mechanism through expression of modified recombinant HAS proteins. Almost certainly, as has been a hallmark of studies in the past, the results of future work will take many unusual turns before HA reveals its many remaining hidden charms.

References
1.   Meyer, K. and Palmer, J.W. (1934) J. Biol. Chem. **107**, 629–634
2.   Meyer, K. and Palmer, J.W. (1936) J. Biol. Chem. **114**, 689–703
3.   Weissman, B. and Meyer, K. (1954) J. Am. Chem. Soc. **76**, 1753–1757
4.   Prehm, P (1983) Biochem. J. **211**, 181–189

5. Hascall, V.C., Heinegard, D.K. and Wight, T.N. (1991) in Cell Biology of Extracellular Matrix, 2nd edn. (Hay, E.D., ed.), pp. 149–175, Plenum Press, New York
6. Prehm, P. (1984) Biochem. J. **220**, 597–600
7. Mason, R.M., D'Arville, C., Kimura, J.H. and Hascall, V.C. (1982) Biochem. J. **207**, 445–457
8. Philipson, L.H. and Schwartz, N.B. (1984) J. Biol. Chem. **259**, 5017–5023
9. Cote, G.L. and Robyt, J.F. (1982) Carbohydr. Res. **101**, 57–74
10. Prehm, P. and Mausolf, A. (1986) Biochem. J. **235**, 887–889
11. Triscott, M.X. and van de Rijn, I. (1986) J. Biol. Chem. **261**, 6004–6009
12. van de Rijn, I. and Drake, R.R. (1992) J. Biol. Chem. **267**, 24302–24306
13. Ng, K.F. and Schwartz, N.B. (1989) J. Biol. Chem. **264**, 11776–11783
14. Dougherty, B.A. and van de Rijn, I. (1993) J. Biol. Chem. **268**, 7118–7124
15. DeAngelis, P.L., Papaconstantinou, J. and Weigel, P.H. (1993), J. Biol. Chem. **268**, 14568–14571
16. DeAngelis, P.L., Papaconstantinou, J. and Weigel, P.H. (1993) J. Biol. Chem. **268**, 19181–19184
17. Dougherty, B.A. and van de Rijn, I. (1994) J. Biol. Chem. **269**, 169–175
18. Crater, D.L., Dougherty, B.A. and van de Rijn, I. (1995) J. Biol. Chem. 270, 28676–28680
19. DeAngelis, P.L. and Weigel, P.H. (1994) Biochemistry **33**, 9033–9039
20. Itano, N. and Kimata, K. (1996) J. Biol. Chem. **271**, 9875–9878
21. Rosa, F., Sargent, T.D., Rebbert, M.L., Michaels, G.S., Jamrich, M., Grunz, H., Jonas, E., Winkles, J.A. and Dawid, I.B. (1988) Dev. Biol. **129**, 114–123
22. Meyer, M.F. and Kreil, G. (1996) Proc. Natl. Acad. Sci. U.S.A. **93**, 4543–4547
23. DeAngelis, P.L., and Achyuthan, A.M. (1996) J. Biol. Chem. **271**, 23657–23660
24. Spicer, A.P., Augustine, M.L. and McDonald, J.A. (1996) J. Biol. Chem. **271**, 23400–23406
25. Fulop, C., Salustri, A. and Hascall, V.C. (1997) Arch. Biochem. Biophys. **337**, 261–266
26. Shyjan, A.M., Heldin, P., Butcher, E.C., Yoshino, T. and Briskin, M.J. (1996) J. Biol. Chem. **271**, 23395–23399
27. Watanabe, K. and Yamaguchi, Y. (1996) J. Biol. Chem. **271**, 22945–22948
28. Itano, N. and Kimata, K. (1996) Biochem. Biophys. Res. Commun. **222**, 816–820
29. Salustri, A., Yanagishita, M. and Hascall, V.C (1989) J. Biol. Chem. **264**, 13840–13847
30. Salustri, A., Yanagishita, M. and Hascall, V.C. (1990) Dev. Biol. **138**, 26–32
31. Salustri, A., Yanagishita, M., Underhill, C.B., Laurent, T.C. and Hascall, V.C. (1992) Dev. Biol. **151**, 541–551
32. Tirone, E., D'Alessandris, C., Hascall, V.C., Siracusa, G. and Salustri, A. (1997) J. Biol. Chem. **272**, 4787–4794
33. Calabro, A. and Hascall, V.C. (1994) J. Biol. Chem. **269**, 22764–22770
34. Song, L., Hobaugh, M.R., Shustak, C., Cheley, S., Bayley, H. and Gouaux, J.E. (1996) Science **274**, 1859–1866
35. Hirschberg, C.B. and Snider, M.D. (1987) Annu. Rev. Biochem. **56**, 63–87
36. Wight, T.N., Heinegard, D.K. and Hascall, V.C. (1991) in Cell Biology of Extracellular Matrix, 2nd edn. (Hay, E.D., ed.), pp. 45–78, Plenum Press, New York
37. Hascall, V.C., Luyten, F.P., Plaas, A.H.K. and Sandy, J.D. (1990) in Methods in Cartilage Research (Maroudas, A. and Kuettner, K., eds.), pp. 108–112, Academic Press, New York
38. Morales, T.I. and Hascall, V.C. (1988) J. Biol. Chem. **263**, 3632–3628
39. Ng, C.K., Handley, C.J., Preston, B.P. and Robinson, H.C. (1992) Arch. Biochem. Biophys. **298**, 70–79
40. Hua, Q., Knudson, C.B. and Knudson, W. (1993) J. Cell Sci. **106**, 365–375
41. Sandy, J.D., Neame, P.J., Boynton, R.E. and Flannery, C.R. (1991) J. Biol. Chem. **266**, 8683–8685
42. Ilic, M.Z., Handley, C.J., Robinson, H.C. and Mok, M.T. (1992) Arch. Biochem. Biophys. **294**, 115–122
43. Lark, M.W., Gordy, J.T., Weidner, J.R., Ayala, J., Kimura, J.H., Williams, H.R., Mumford, R.A., Flannery, C.F., Carlson, S.S., Iwata, M. and Sandy, J.D. (1995) J. Biol. Chem. 270, 2550–2556
44. Gmachl, M., Sagan, S., Ketter, S. and Kreil, G. (1995) FEBS **336**, 545–548
45. Lin, Y., Mahan, K., Lathrop, W.F., Myles, D.G. and Primakoff, P. (1994) J. Cell Biol. **125**, 1157–1163
46. Tammi, R., Ripellino, J.A., Margolis, R.U., Maibach, H.I. and Tammi, M. (1989) J. Invest. Dermatol. **92**, 326–332
47. Wang, C., Tammi, M. and Tammi, R. (1992) Histochemistry **98**, 105–112
48. Lesley, J., Hyman, R., and Kincade P.-W. (1993) Adv. Immunol. **54**, 271–335
49. Tammi, R., Saamanen, A-M., Maibach, H.I. and Tammi, M. (1991) J. Invest. Dermatol. **97**, 126–130

9

Structure and regulation of mammalian hyaluronan synthase

Paraskevi Heldin

Department of Medical and Physiological Chemistry, Biomedical Center, Uppsala University, Box 575, S-751 23 Uppsala, Sweden

Introduction

Hyaluronan is a naturally occurring glycosaminoglycan (GAG) composed of alternating glucuronic acid and *N*-acetylglucosamine residues. It is found in all mammalian species and also, for example, in the extracellular capsule of group A and C streptococci [1]. Hyaluronan is, through its interactions with specific hyaluronan binding proteins, implicated in a variety of cell-biological processes such as cell migration and proliferation. Furthermore, hyaluronan has important structural functions in the extracellular matrix of all tissues [2–5].

The biosynthetic pathway of hyaluronan differs from those of the other GAGs, which are synthesized in processes involving elongation of the sugar chains bound to core proteins and modifications in the Golgi apparatus, followed by vesicular transport to the cell surface [6]. In contrast, hyaluronan synthesis occurs at or close to the cell membrane both in bacteria and mammalian cells [7,8]. In streptococci the key enzyme in the biosynthetic pathway, hyaluronan synthase, polymerizes hyaluronan by alternate transfer of monosaccharides from UDP-sugar nucleotide precursors [1,9,10]. Even though bacterial and mammalian hyaluronan is structurally similar, knowledge of the vertebrate hyaluronan synthetic system is still incomplete. This is due in part to the difficulty, despite the efforts of many researchers, to solubilize and purify the enzyme(s) in active form. Ng and Schwartz [11] solubilized a functional hyaluronan synthase complex from mouse oligodendroglioma cells using digitonin, but the components in the complex were not further purified and characterized. To understand the molecular mechanisms behind the mammalian hyaluronan synthesizing system, the exact location of hyaluronan synthase and the characterization of other proteins involved in hyaluronan biosynthesis, as well as the direction of polymer synthesis and the role, if any, of primers for hyaluronan synthesis need to be explored.

Regulation of hyaluronan biosynthesis in mammalian cells

The mechanism of hyaluronan biosynthesis has been studied extensively in streptococci, because of the simplicity of this organism. Moreover, streptococci do not synthesize other GAGs and have the capacity to produce large amounts of hyaluronan [1,10]. In eukaryotes, characterization of the regulation of hyaluronan biosynthesis has been more difficult. However, recently the molecular mechanisms of the regulation of hyaluronan synthesis have been partially unravelled. Hyaluronan synthesis is stimulated by various growth factors, such as platelet-derived growth factor (PDGF), transforming growth factor-β (TGF-β), and insulin-like growth factor-I (IGF-I) [12,13]. Bansal and Mason [14] reported a rapid metabolic turnover of hyaluronan synthase that correlated with the rate of cellular multiplication; an accumulation of hyaluronan was seen during mitosis, leading to cell detachment that is necessary for mitosis [15]. Hyaluronan synthase up-regulation was also seen after pre-treatment of cells with hyaluronidase [16], possibly due to the relief of a feedback control mechanism since high-molecular-mass hyaluronan can inhibit hyaluronan production [17]. A unique mechanism for hyaluronan synthase regulation has been suggested both in bacteria and in mammalian cells; the hyaluronan synthase is shed from streptococci or mammalian cells together with the hyaluronan chain resulting in its inactivation [18,19].

The elevation of hyaluronan in the early phase of embryonal development, tissue repair and other inflammatory conditions as well as in the stroma of certain cancer forms [2,20], suggests the involvement of growth factors and other inflammatory mediators that stimulate hyaluronan synthesis in autocrine or paracrine manners. Previous reports suggested two different ways through which tumour cells or macrophages in culture stimulate hyaluronan synthesis in mesenchymal cells. One way is through interactions between tumour cells and connective tissue cells [21,22], and the other by secreted hyaluronan stimulatory factors [23–25].

Studies on the signal transduction pathways that mediate the hyaluronan stimulatory effects of PDGF-BB and TGF-β revealed that protein kinase C (PKC) plays a crucial role in the transduction of the effects. Direct or indirect activation of existing hyaluronan synthase molecules as well as induction of new enzyme molecules are involved [26,27] (Figure 1). This mode of action seems to be valid for cells of mesenchymal origin such as human foreskin fibroblasts and mesothelial cells. Interestingly, the hyaluronan stimulatory effects of PDGF-BB and TGF-β on human fibroblast cultures were not correlated to their mitogenic effects [12]. Additional studies revealed the involvement of another serine/threonine kinase, protein kinase A (PKA), as well as the involvement of a tyrosine kinase [26,13,28,29]. Studies of embryonic tissues have also suggested that growth factors such as TGF-β and bFGF (basic fibroblast growth factor) control hyaluronan synthesis [30]. Furthermore, osteoblast-like cell cultures exhibited a dramatic increase in hyaluronan synthesis in response to parathyroid hormone [31]. The above-mentioned studies have been performed in intact cells which in response to PDGF-BB, fetal calf serum, prostaglandin E_2 or parathyroid hormone exhibited a transient increase in hyaluronan synthase activity, reaching a peak of activity 4–7 h after stimulation. Interestingly, when foreskin fibroblasts were stimulated with TGF-β the peak of activity was seen later, at 24 h. This delay in hyaluronan stimulatory effect of TGF-β on human fibroblasts did not involve induction of PDGF, but may involve the induction of other factors that in turn stimulate hyaluronan production [26,27].

Figure 1

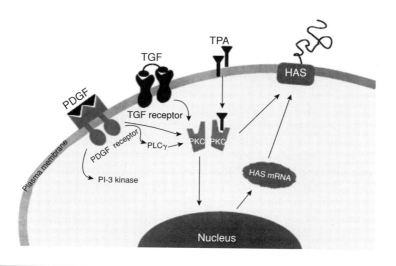

Regulation of hyaluronan synthase (HAS) activity

PDGF-BB, TGF-β and 12-O-tetradecanoylphorbol-13-acetate (TPA) mediate their hyaluronan stimulatory effects partly through activation of PKC. The stimulatory effects of PDGF-BB and TGF-β are partly dependent on new synthesis of enzyme molecules whereas those of TPA are not. PLC-γ (phospholipase C-γ) is involved in the transduction of PDGF-BB-mediated hyaluronan biosynthesis but phosphatidylinositol (PI)-3 kinase is not.

The high amounts of hyaluronan detected at the tumour–stromal interface of invasive tumours *in vivo* are not only due to the production of hyaluronan by host connective tissue cells in response to factors released by the tumour cells, but also to high hyaluronan synthesizing activity of tumour cells themselves [32,33]. Metastatic human colon carcinoma cells and metastatic breast carcinoma cells exhibited both high ability to synthesize hyaluronan and to bind hyaluronan. These cell types may provide useful models to study the interplay between synthase activity and hyaluronan receptors and how the synthesized hyaluronan and its interactions with hyaladherins affect the metastatic behaviour of these cells.

Increased knowledge about the signal transduction pathways that mediate the effects of various extracellular stimuli on hyaluronan biosynthesis may be the basis for the development of specific drugs to regulate hyaluronan synthesis.

Synthesis and assembly of hyaluronan-containing pericellular matrix

A pericellular matrix or cell coat surrounds many cells *in vitro* and is primarily composed of hyaluronan-aggrecan (chondrocytes) or hyaluronan–versican (mesenchymal cells) complexes. It is invisible with phase contrast microscopy but it can be visualized by the exclusion of formalin-fixed red blood cells [34] and video-enhanced light microscopy [35]. The function of the cell coat is not well-defined but it appears to be involved in the assembly of extracellular matrix and to contribute to the ability of tumour cells to evade cellular immune attack [36,37].

Normal human mesothelial (NHM) cells and their transformed counterparts, mesothelioma cells, provide a good model to study the mechanisms of biosynthesis of hyaluronan and its interaction with specific cell surface receptors, hyaluronan chains and other extracellular matrix components (Figure 2). In patients with mesothelioma, large amounts of hyaluronan are accumulated in pleura fluid and tumour tissue. However, the hyaluronan is not synthesized by mesothelioma cells; rather, mesothelioma cells produce factors that stimulate hyaluronan synthesis in mesothelial cells and fibroblasts. Part of the stimulatory activity is due to PDGF and bFGF but other factors also seem to be involved [23]. Furthermore, NHM cells but not mesothelioma cells are surrounded by cell coats [38] (Figure 2). The coat around NHM cells is composed of newly synthesized hyaluronan and is stabilized by interactions between hyaluronan molecules or between hyaluronan and proteins, which involves an epitope of hyaluronan that is larger than six monosaccharide units. It is induced *in vitro* by PDGF, which was also detected in pleura fluid from mesothelioma patients and thus may also induce hyaluronan synthesis *in vivo*. TGF-β has been shown to stimulate coat formation in chick-embryo limb mesodermal cells, illustrating that the hyaluronan content in pericellular matrices can be regulated by different factors in various cell types [39]. In the mesothelial/mesothelioma cell model, transformation of NHM cells leads to inhibition of hyaluronan synthesis and thereby to loss of hyaluronan-containing coats as well as to expression of specific hyaluronan receptors [23,25].

Although the exact function of the cell coat remains to be elucidated it is possible that it has protective roles against, for example, bacteria and other infectious micro-organisms. It is possible that, under certain pathological conditions, transformed mesothelioma cells both secrete factors that stimulate hyaluronan synthesis by neighbouring cells, and release matrix components such as chondroitin sulphate proteoglycan, which stabilize hyaluronan-containing

Figure 2

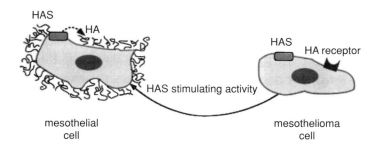

A model to study hyaluronan biosynthesis and interactions

Human mesothelial cells synthesize and release into culture media large amounts of hyaluronan and are surrounded by hyaluronan-containing coats. Mesothelioma cells possess very low hyaluronan synthesizing activity, release factors with hyaluronan stimulatory activity and express highly specific hyaluronan receptors. HA, hyaluronan; HAS, hyaluronan synthase.

coats around mesothelial cells [23,40]. Moreover, it has been shown that hyaluronan in the culture media, through its interaction with the hyaluronan receptor CD44, stimulate interleukin-1β, tumour necrosis factor-α and IGF-1 mRNA expression in bone-marrow-derived macrophages [41]. It is thus possible that the pericellular matrix composes a microenvironment between the plasma membrane and the extracellular matrix which plays regulatory roles in the interaction of soluble hyaluronan with its receptor.

Cloning of vertebrate hyaluronan synthase genes

To understand the regulation of the biosynthesis of hyaluronan in mammalian cells, the structural characterization of hyaluronan synthase and molecular cloning of genes encoding enzymes that take part in hyaluronan biosynthesis are necessary.

The first efforts to identify genes that are involved in hyaluronan biosynthesis were performed in streptococci. Using transposon mutagenesis, stable acapsular mutants occurred. Subsequent identification of the DNA flanking the inserted transposon revealed the gene *hasA*, which encodes a protein of molecular mass 42 kDa and possesses hyaluronan synthase activity [42,43]. Database searches revealed that *hasA* shows similarity to *Rhizobium nodC* gene, to yeast chitin synthases and to DG42, which was identified in *Xenopus*. In streptococci, the genes that participate in hyaluronan synthesis are encoded in the *has* operon, which contains three genes, *hasA*, *hasB*, and *hasC* [44,10].

Recently, mouse mammary carcinoma cells with high hyaluronan synthesizing capacity were mutagenized with *N*-methyl-*N'*-nitro-*N*-nitro soguanidine, resulting in mutant cell lines deficient in hyaluronan synthase activity. Then, using a transient expression system a clone that could direct the expression of hyaluronan synthase was isolated, and termed *HAS* [45]. Another mouse cDNA for hyaluronan synthase, *Has2*, was cloned using degenerate primers designed on the amino acid sequence homology between hasA, NodC and DG42 proteins [46]. *HAS* and *Has2* encode plasma membrane proteins of molecular masses 65 500 Da and 63 000 Da, respectively, and display about 51% similarity.

A human cDNA for hyaluronan synthase (*HuHAS1*) was cloned using a functional cloning approach; CHO cells transfected with a cDNA encoding synthase activity conferred increased hyaluronan production in cell cultures and mediated hyaluronidase-sensitive adhesion to the lymphocyte hyaluronan receptor CD44 [47]. Using the sequence similarity in the intracellular region of bacterial, fungal and vertebrate DG42 proteins, two other human cDNAs for hyaluronan synthase were cloned [48,49]. Comparison of the amino acid sequences of the three cloned human hyaluronan synthase cDNAs, reveal that two of three are almost identical and may be derived from the same gene [47,49], whereas the third shows about 52% similarity [48]. It is thus possible that mammalian cells have more than one hyaluronan synthase gene. More recently, expression of the cDNA for DG42 in *Saccharomyces cerevisiae*, which does not produce either hyaluronan or UDP-GlcA, resulted in hyaluronan synthase activity in membrane preparations of those cells, suggesting that the DG42 protein possesses both D-glucuronosyl-(GlcA) and *N*-acetyl-D-glucosaminyl-(GlcNAc) transferase activities [50]. This characteristic of DG42 is not unique. A 70 kDa protein that is involved in heparin/heparan sulphate synthesis possesses both glycosyltransferase activities [51].

A hydrophilic region is present in the cloned hyaluronan synthase cDNAs between an amino-terminal hydrophobic stretch and carboxy-terminal hydrophobic stretches. It is highly conserved between the bacterial and vertebrate enzymes and may contain the enzymic activity. This intracellular region also possesses consensus sequences for phosphorylation by PKC, PKA and tyrosine kinases. The significance of these sites for the regulation of hyaluronan synthase activity through phosphorylation/dephosphorylation remains to be determined. Klewes and Prehm [29] reported a decrease in the membrane-associated enzymic activity upon phosphatase treatment. In addition to the putative phosphorylation sites, hyaluronan binding motifs are also present. It is possible that these sites bind hyaluronan and align it to the synthase during its polymerization.

The vertebrates have a much more complicated biosynthetic machinery than bacteria; the long hyaluronan chain is extruded through the plasma membrane while it is synthesized. It is therefore possible that interactions with other proteins are important for hyaluronan synthase function. To get insight in the biosynthesis of hyaluronan, further studies on the recombinant hyaluronan synthase, and the solubilization of active synthase followed by identification of components important for the enzymic activity, will be necessary. Increased knowledge of hyaluronan biosynthesis will contribute to our understanding of the biological roles of hyaluronan in tissue structure during development, inflammation and cancer.

I would like to thank Prof. Torvard Laurent and Drs. Håkan Pertoft, Tomas Asplund, Priit Teder, Masanobu Suzuki and Michael Briskin for fruitful collaboration. This project was supported in part by grants from The Swedish Cancer Foundation, The Swedish Medical Research Council (03X-4), Gustaf V:s 80-års fond and The Swedish Heart Lung Foundation.

References

1. Sugahara, K., Schwartz, N.B. and Dorfman, A. (1979) J. Biol. Chem. **254**, 6252–6261
2. Laurent, T.C. and Fraser, J.R.E. (1992) FASEB J. **6**, 2397–2404
3. Turley, E.A. (1992) Cancer Metastasis Rev. **11**, 21–30
4. Knudson, C.B. and Knudson, W. (1993) FASEB J. **7**, 1233–1241
5. Laurent, T.C., Laurent, U.B.G. and Fraser, J.R. (1995) Ann. Rheum. Dis. **54**, 429–432
6. Kjellen, L. and Lindahl, U. (1991) Annu. Rev. Biochem. **60**, 443–475
7. Prehm, P. (1984) Biochem. J. **220**, 597–600
8. Philipson, L.H. and Schwartz, N.B. (1984) J. Biol. Chem. **259**, 5017–5023
9. Prehm, P. (1983) Biochem. J. **211**, 191–198
10. DeAngelis, P.L. and Weigel, P.H. (1995) in Genetics of streptococci, enterococci and lactococci (Ferretti J.J., Gilmore, M.S., Klaenhammer T.R. and Brown F., eds.), **85**, pp. 225–229, Dev. Biol. Stand., Karger, Basel
11. Ng, K.F. and Schwartz, N.B. (1989) **264**, 11776–11783
12. Heldin, P., Laurent, T.C. and Heldin, C-H. (1989) Biochem. J. **258**, 919–922
13. Honda, A., Noguchi, N., Takehara, H., Ohashi, Y., Asuwa, N. and Mori, Y. (1991) J. Cell Sci. **98**, 91–98
14. Bansal, M.K. and Mason, R.M. (1986) Biochem. J. **236**, 515–519
15. Brecht, M., Mayer, U., Schlosser, E. and Prehm, P. (1986) Biochem. J. **239**, 445–450

16. Larnier, C., Kerneur, C., Robert, L. and Moczar, M. (1989) Biochim. Biophys. Acta **1014**, 145–152
17. Smith, M.M. and Gosh, P. (1987) Rheumatol. Int. **7**, 113–122
18. Mausolf, A., Jungmann, J., Robenek, H. and Prehm, P. (1990) Biochem. J. **267**, 191–196
19. Kitchen, J.R. and Cysyk, R.L. (1995) Biochem. J. **309**, 649–656
20. Fraser, J.R.E. and Laurent, T.C. (1996) in Extracellular Matrix (W. D. Comper, ed.), vol. 2, pp. 141–199, Harwood Academic, Amsterdam
21. Knudson, W., Biswas, C. and Toole, B.P. (1984) Proc. Natl. Acad. Sci. U.S.A. **81**, 6767–6771
22. Knudson, W. and Toole, B.P. (1988) J. Cell. Biochem. **38**, 165–177
23. Asplund, T., Versnel, M.A., Laurent, T.C. and Heldin, P. (1993) Cancer Res. **53**, 388–392
24. Teder, P., Nettelbladt, O. and Heldin, P. (1995) Am. J. Respir. Cell Mol. Biol. **12**, 181–189
25. Teder, P., Versnel, M.A. and Heldin, P. (1996) Int. J. Cancer **67**, 393–398
26. Heldin, P., Asplund, T., Ytterberg, D., Thelin, S. and Laurent, T.C. (1992) Biochem. J. **283**, 165–170
27. Suzuki, M., Asplund, T., Yamashita, H., Heldin, C-H. and Heldin, P. (1995) Biochem. J. **307**, 817–821
28. Honda, A., Sekiguchi, Y. and Mori, Y. (1993) Biochem. J. **292** (Pt 2), 497–502
29. Klewes, L. and Prehm, P. (1994) J. Cell. Physiol. **160**, 539–544
30. Toole, B.P., Munaim, S.I., Welles, S. and Knudson, C.B. (1989) in The Biology of Hyaluronan (Evered, D. and Whelan, J., eds.), Ciba Foundation Symp. No. 143, pp. 138–149, Wiley, Chichester
31. Midura, R.J., Evanko, S.P. and Hascall, V.C. (1994) J. Biol. Chem. **269**, 13200–13206
32. Mitchell, B.S., Whitehouse, A., Prehm, P., Delpech, B. and Schumacher, U. (1996) Clin. Exp. Metastasis **14**, 107–114
33. Heldin, P., de la Torre, M., Ytterberg, D. and Bergh, J. (1996) Oncol. Rep. **3**, 1011–1016
34. Clarris, B.J. and Fraser, J.R.E. (1968) Exp. Cell Res. **49**, 181–193
35. Lee, G.M., Johnstone, B., Jacobson, K. and Caterson, B. (1993) J. Cell Biol. **123**, 1899–1907
36. Toole, B.P. (1991) in Cell Biology of Extracellular Matrix (Hay, E.D., ed.), pp. 305–339 , Plenum Press, New York
37. Gately, C.L., Muul, L.M., Greenwood, M.A., Papazoglou, S., Dick, S.J., Kornblith, P.L., Smith, B.H. and Gately, M.K. (1984) J. Immunol. **133**, 3387–3395
38. Heldin, P. and Pertoft, H. (1993) Exp. Cell Res. **208**, 422–429
39. Munaim, S.I., Klagsbrun, M. and Toole, B.P. (1991) Dev. Biol. **143**, 297–302
40. Heldin, P., Suzuki, M., Teder, P. and Pertoft, H. (1995) J. Cell Physiol. **165**, 54–61
41. Noble, P.W., Lake, F.R., Henson, P.M. and Riches, D.W.H. (1993) J. Clin. Invest. **91**, 2368–2377
42. DeAngelis, P.L., Papaconstantinou, J. and Weigel, P.H. (1993) J. Biol. Chem. 19181–19184
43. Dougherty, B.A. and van de Rijn, I. (1994) J. Biol. Chem. **269**, 169–175
44. Crater, D.L. and van de Rijn, I. (1995) J. Biol. Chem. **270**, 18452–18458
45. Itano, N. and Kimata, K. (1996) J. Biol. Chem. **271**, 9875–9878
46. Spicer, A.P., Augustine, M.L. and McDonald, J.A. (1996) J. Biol. Chem. **271**, 23400–23406
47. Shyjan, A.M., Heldin, P., Butcher, E.C., Yoshino, T. and Briskin, M.J. (1996) J. Biol. Chem. **271**, 23395–23399
48. Watanabe, K. and Yamaguchi, Y. (1996) J. Biol. Chem. **271**, 22945–22948
49. Itano, N. and Kimata, K. (1996) Biochem. Biophys. Res. Commun. **222**, 816–820
50. DeAngelis, P.L. and Achyuthan, A.M. (1996) J. Biol. Chem. **271**, 23657–23660
51. Lind, T., Lindahl, U. and Lidholt, K. (1993) J. Biol. Chem. **268**, 20705–20708

10

Catabolism of hyaluronan

J.R.E. Fraser*‡, T.J. Brown* and T.C. Laurent†
*Department of Biochemistry and Molecular Biology, Monash University, Clayton, Victoria, Australia, and †Department of Medical and Physiological Chemistry, University of Uppsala, Box 575, S-751 23, Uppsala, Sweden.

Hyaluronan (hyaluronic acid, hyaluronate) is distributed throughout the extracellular matrix, whether in tiny amounts or in more obvious abundance in the soft connective tissues and viscous body fluids. With the burgeoning interest in the part played by hyaluronan in physiology and disease, its turnover in the body has assumed a corresponding importance.

The amounts and concentrations of hyaluronan in the tissues are necessarily determined by the balance between its rates of synthesis and of elimination. The negative phase of its turnover depends almost entirely on elimination through catabolism rather than excretion. This remains true even though much of it takes place somewhat remote from its original sources. Further discussion of its transport and other aspects of its turnover can be found elsewhere [1,2]. It will suffice to observe here that the daily turnover of hyaluronan is in the order of one-third of the total body content [3], though the regional turnover does vary greatly from place to place in both fractional and absolute terms. In the following we shall briefly review its catabolism in terms of the enzymic mechanisms, the tissues and the kinds of cell involved, and give an account of these processes as they are manifest in intact tissues within the living body.

Enzymic mechanisms

The substrate

The structure of hyaluronan is well established as a uniform unbranched linear polymer of D-glucuronic acid (GlcUA) and N-acetyl-D-glucosamine (GlcNAc) in $\beta1\rightarrow3$, $\beta1\rightarrow4$ linkage. It is polydisperse as recovered from tissues but with a few exceptions its mean molecular weight is usually very high, ranging from $1-2 \times 10^6$ Da in skin to $7-8 \times 10^6$ Da in synovial fluid. This feature, together with its expanded molecular configuration, is relevant to the cellular intake that precedes its catabolism. When synthesized from labelled substrates by cultured cells in optimum conditions, we have found the mean molecular weight to be in the high range with less polydispersity. Evidence of partial postsynthetic reduction in polymer size has been found in tissue cultures of skin [4] and of cartilage [5]. This might be explained by enzymic activity, free radicals or possibly by mechanical shearing (in joints or bloodstream, for example), although, as noted below, clear evidence of extracellular hyaluronidase activity is lacking.

The enzymes

The enzymic degradation of hyaluronan to its constituent monosaccharides has been exhaustively analysed with crude tissue and cell extracts and with highly

‡To whom correspondence should be addressed, at 131 Manning Road, East Malvern, VIC 3145, Australia.

purified preparations. Shortly after the recognition of hyaluronan, hyaluronidases were identified as the commonest forms of Duran–Reynals' spreading factors. Their sources included mammalian tissues, notably testicular, various reptilian and insect venoms, and leeches. The participation of monosaccharidases in complete degradation was then quickly established to define three participating enzymes in mammalian physiology. Their essential characteristics have been progressively reviewed since 1947 [6,7].

Hyaluronidase

The mammalian forms are endohexosaminidases that randomly reduce the polymers through a range of oligosaccharides to finally yield tetrasaccharide residues with GlcUA at the non-reducing end. They fall into two classes according to the ionic conditions that determine their activity. Consistent with its extracellular site of action, testicular hyaluronidase has a pH optimum of about 5 to 6.2 but remains very active at pH 7.5. No other hyaluronidase from normal mature tissues has been shown to be active in extracellular fluids, though a neutrally active form has been found in embryogenesis [8] and might therefore be inducible in other conditions. The other class is typically lysosomal with a lower pH optimum, and with rising pH becomes inactive at a level still below the lowest pH that can develop in extracellular pH. The enzymes found in extracellular fluids (for example, plasma [9] and synovial [10]) fall within this category and therefore have no functional significance.

Exoglycosidases: β-D-glucuronidase and β-N-acetyl-D-hexosaminidase

These two enzymes hydrolyse the appropriate linkages of hyaluronan oligosaccharides sequentially from the non-reducing end. Both occur in lysosomes but also in other intracellular sites, and show subunit or other structural variations that might underlie differences in their behaviour in laboratory conditions.

Outcome of concerted digestion

Concurrently with or following hyaluronidase, the exoglycosidases reduce hyaluronan to its monosaccharides *in vitro* but with a 10–15% residue of the disaccharide, GlcUA($\beta1\rightarrow3$)GlcNAc (*N*-acetylhyalobiuronic acid), which is resistant to further digestion [11]. A similar result follows incubation with isolated liver lysosomes [12]. The exoglycosidases will also hydrolyse polymeric hyaluronan without hyaluronidase, but leave a high residual fraction of small oligosaccharides in altered monosaccharide sequence, indicating transglycosylation [13]. Degradation by the exoglycosidases *in vitro* appears to be greatly facilitated by a trace of hyaluronidase [13]. Transglycosylation can also occur with hyaluronidase, but may be an artefact of ionic conditions [14].

Although some cells contain the exoglycosidases without hyaluronidase, the latter enzyme occurs in numerous organs and tissues [15,16], notably in the main sites of hyaluronan catabolism, and the former have other substrates. From the foregoing kind of study, the most likely sequence in normal catabolism is preliminary digestion by hyaluronidase to the level of large oligosaccharides with subsequent hydrolysis by the exoglycosidases [7,17]. The question of the residues will be taken up again in the light of catabolic studies *in vivo*.

Hyaluronan catabolism in physiological conditions

Anatomical location

It has long been believed that hyaluronan is degraded within its tissues of origin, a conclusion supported in many instances by the localization of hyaluronidase, and consistent with the resistance to displacement that might be expected from the physical character of concentrated hyaluronan itself.

The discovery that large amounts of hyaluronan enter the bloodstream through lymphatic channels [18], and the observation that its fractional turnover rate in plasma ranges as high as 70% per min [19,20], have since led to the discovery of significant hyaluronan uptake at two levels remote from its origins. The lymph nodes can extract 50–90% from peripheral lymph [21,22], and may be responsible for the largest fraction of hyaluronan catabolism. In the bloodstream, the liver and kidneys extract, respectively, >80% and approx. 10% [23]. The spleen has a high affinity for hyaluronan in some animals [19,24] but its contribution is moderated by species variation in its circulation and relative size [25].

Cellular aspects
Classes of cell involved

Uptake of hyaluronan from blood and lymph is most prominent in the endothelial cells of the liver sinusoids, medullary sinuses of the spleen, and the sinuses in the lymphatic channels of the nodes. Kupffer cells may play a minor part in liver, and a macrophage-like cell closely associated with endothelial cells in spleen and lymph node sinuses. The best characterized are the hepatic sinusoidal endothelial cells, whose uptake is associated with a particular phenotype. In the tissues, resident macrophages, epidermal cells, chondrocytes, fibroblasts and other cell types may take part. In tissues without lymph drainage, they must be entirely responsible. In others, their activity may be small unless specifically enhanced by as yet undefined physiological signals, inflammation or other disease processes.

Modes of cellular intake

At the most active levels of normal turnover, catabolism is facilitated by receptor-mediated uptake; notably in the liver, where it achieves high extraction rates from very low concentrations. To date, catabolism has been associated with two kinds of receptor. One consists of variants of CD44, which has been identified with uptake and catabolism in virus-transformed fibroblasts and alveolar macrophages [26]. Expression of these variants can be induced or up-regulated at several levels of control and might thus become more active in disease. Another distinct kind of receptor has been identified with hyaluronan catabolism in liver endothelial cells [27] but not yet associated with any particular class of cell-surface adhesion molecule. It shares its specificity with chondroitin sulphates. The presumed receptor in lymphatic sinus lining cells is unidentified but the pathway is also shared with chondroitin sulphates. In the high hyaluronan concentrations in tissues, non-specific endocytosis or low-affinity receptors may play a larger part in cellular intake of hyaluronan.

Catabolism in the living body

Isotopic tracing has been used specifically to study the catabolism of hyaluronan *in vivo* in two ways: by incorporation of ^3H or ^{14}C in its natural structure, or of ^{125}I-labelled substituents that modify its structure [28]. The latter are particularly useful to identify the sites of catabolism by their persistent intracellular residues, though these by their nature must nevertheless modify the final stages of catabolism. In the following we shall focus on the former, since they have been most helpful in identifying the catabolic steps predicted from enzymic studies *in vitro*, and at the same time in providing data on the kinetics of turnover and catabolism. Labelling with ^3H is preferred in the acetyl group since it is stable in the alkyl configuration and its catabolic end-product is ^3H_2O, which enters a less labile body pool than $^{14}CO_2$. However, ^{14}C is preferable for labelling the saccharide since ^3H is too freely exchangeable on hydroxy groups. [^{14}C]Saccharide labelling must be done by biosynthesis, whereas [^3H]acetate can also be substituted chemically, depending on other experimental requirements.

Labelled hyaluronan has been introduced by injection into the bloodstream, lymph vessels, eye, skin and joints, and recently by absorption through the intact skin of hairless mice [29]. Small amounts have been used in some studies to allow estimates of basal turnover rates and saturation doses or sustained infusions for other purposes and to seek particular metabolic products. The findings relevant to its catabolism are as follows.

First steps
Extracellular modification
As noted, the hyaluronidases in extracellular fluids are normally inactive but post-secretory reduction in polymer size occurs in skin and cartilage. The mean molecular weight of residual hyaluronan diminishes in the blood stream during absorption [30] and also after passage through lymph nodes [21,22]. In unpublished studies with R.N.P. Cahill, we found a 50% reduction of molecular weight in the residue remaining in a sheep's hock joint 48 h after injection in the synovial cavity. In the lymph node, greater receptor affinity for larger polymers [27] might be responsible since the effect is nullified by concurrent infusion of excess chondroitin sulphate [31]. Mechanical shearing is the more likely reason in joints, since the turnover of smaller polymers is faster. Shearing may also operate in the bloodstream in addition to selective absorption. We have not found extracellular oligosaccharides or monosaccharides in these studies.

Intracellular
We have injected doubly labelled hyaluronan (^3H-acetyl, ^{14}C-glycosyl) intravenously in large amounts to achieve zero-order elimination kinetics and saturate the catabolic pathways. The $^3H:^{14}C$ ratio in plasma did not change during absorption of labelled hyaluronan, thus excluding extracellular de-acetylation. The only metabolic products recovered from the liver 12 min after injection were GlcNAc, with an increased $^3H:^{14}C$ ratio compared with hyaluronan, and, in stoichiometric yield, another monosaccharide fraction containing ^{14}C only. This consisted of GlcUA, but already had a high content of gulonic acid, the first step in its established catabolic pathway.

With [³H]acetyl -labelled hyaluronan, a 22-saccharide fragment is the smallest oligosaccharide we have recovered from liver or skin. No disaccharide has been detected in any of our studies. We cannot completely exclude its occurrence *in vivo*, but it would necessarily be in much smaller proportions than found in the quoted enzymic studies [11–13]. We have not detected intermediate oligosaccharide residues that would correspond with the transglycosylation products of digestion with isolated enzymes.

Intermediate stages

The later products of hyaluronan catabolism *in vivo* have been shown to include *N*-acetylglucosamine phosphates. These appeared to be GlcNAc-1P in fetal sheep liver [20] and GlcNAc-6P in adult sheep lymph nodes [21], but the two were not sharply discriminated in those studies. Using an alternative chromatographic method, we have since clearly identified both concurrently in the same liver samples after intravenous injection of [³H]acetyl-labelled hyaluronan in rabbits. After phosphorylation of GlcNAc at C6, a ready exchange must occur between the C6 and C1 phosphates. This will provide a low-energy source of UDP-GlcNAc for the synthetic requirements of the cell [7] but, except for a little lipid, very little of the residual ³H label is recovered as anything but ³H$_2$O. Cultured liver endothelial cells generate [³H]acetate from labelled hyaluronan [32], showing that their catabolic pathway extends to GlcNAc-6P, a necessary step for deacetylation of GlcNAc. Hepatocytes contain very little deacetylase and are a major site of synthesis, whereas deacetylase is abundant in the hepatic sinusoidal lining cells [33]. This strongly suggests that the exchange also occurs in the latter, and the GlcNAc-1P pool functions there mainly as a subsidiary reservoir for GlcNAc-6P. We have not traced the further catabolism of Glc-6P, which is well known [7].

End results

The only catabolic products recovered from extracellular fluids have been [³H]acetate, which appears especially after large loads of labelled hyaluronan and is soon accompanied and then replaced by ³H$_2$O. Evidence from the recovery of ³H$_2$O after introduction of labelled hyaluronan into eye, joints and skin indicates that more than 90% of the GlcNAc moiety is metabolized to this degree [34]. A similar fate can be assumed for GlcUA, since 24 h after intravenous injection of [¹⁴C]glucose-labelled hyaluronan, most of it was excreted by respiration as ¹⁴CO$_2$ [35]. There is a minor conservation of its acetyl group in lipid [19,21]. Some labelled hyaluronidase-resistant macromolecule has been recovered in the liver from [¹⁴C]glucose-labelled hyaluronan [24,35] and from [³H]acetyl-labelled material [19].

The pathways of hyaluronan catabolism are set out diagrammatically in Figure 1.

Kinetic aspects

[³H]Acetate and ³H$_2$O have been detected within 9 to 20 min after intravenous injection of large-polymer HYA [19,20,36]. The following steps must precede these endpoints: (1) binding to receptor; (2) internalization as endosome; (3) fusion of endosome with primary lysosomes; (4) intralysosomal digestion to

Figure 1

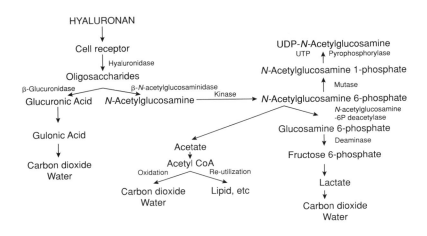

Pathways of hyaluronan catabolism in mammalian cells

monosaccharides by integrated action of hyaluronidase and exoglycosidase; (5) active transport of GlcNAc through lysosomal membrane to cell sap; (6) its phosphorylation; and (7), deacetylation.

Summary

In the normal body, the disposal of hyaluronan is almost entirely effected by complete catabolism rather than excretion. This takes place at three levels; in the tissues where it is synthesized, in the lymph nodes after displacement from tissue fluids, and after discharge of lymph to the bloodstream, mainly by uptake in the liver and to a lesser extent in kidneys with a minor participation by spleen and bone marrow [24]. In blood and probably in lymph, cellular uptake is facilitated by hyaluronan receptors.

Enzymic studies indicate that the most likely sequence in its degradation to monosaccharides is initial digestion by hyaluronidase to large oligosaccharides, which are then hydrolysed to single sugars by the sequential action of β-D-glucuronidase and β-N-acetyl-D-hexosaminidase at their non-reducing ends. This leaves a resistant residue of disaccharide. In certain circumstances slightly larger products of altered sugar sequence persist, indicating transglycosylation.

Studies with isotopically labelled hyaluronan in the living body have shown no evidence of significant extracellular enzymic degradation. The first labelled catabolic products consist of large oligosaccharides (of about 22 sugars), glucuronic acid and N-acetylglucosamine, but no intermediate oligosaccharides. Gulonic acid, the next step in glucuronic acid catabolism, is detected almost immediately. The further catabolism of N-acetylglucosamine has been confirmed *in vivo* by detection of its 6-carbon phosphate, an essential step for its deacetylation. This product also establishes equilibrium with N-acetylglucosamine-1P, the entry point for re-utilization in synthesis, but the ^{3}H-labelled acetyl group is almost completely accounted for as ^3H_2O, indicating that the synthetic pathway is not a significant route of disposal. These findings conform with enzymic studies *in vitro*, except that no evidence of residual disaccharide or transglycosylation products has been found.

The intracellular mechanisms involved in hyaluronan catabolism at its main sites of physiological turnover are remarkably rapid, integrated and complete.

References

1. Laurent, U.B.G. and Reed, R.K. (1991) Adv. Drug Deliv. Rev. **7**, 237–256
2. Fraser, J.R.E. and Laurent, U.B.G. (1998) in Connective Tissue Biology: Integration and Reductionism (Reed, R.K. and Rubin, K., eds.), pp. 49–69, Portland Press, London
3. Laurent, T.C. and Fraser, J.R.E. (1991) in Degradation of Bioactive Substances: Physiology and Pathophysiology (Henriksen, J.H., ed.), pp. 249–265, CRC Press, Boca Raton, FL
4. Tammi, R., Ågren, U.M., Tuhkanen, A-L. and Tammi, M. (1994) Prog. Histochem. Cytochem. **29**(2), 1–81
5. Holmes, M.W.A., Bayliss, M.T. and Muir, H. (1988) Biochem. J. **250**, 435–441
6. Meyer, K. (1947) Physiol. Rev. **27**, 335–358
7. Rodén, L., Campbell, P., Fraser, J.R.E., Laurent, T.C., Pertoft, H. and Thompson, J.N. (1989) in The Biology of Hyaluronan (Evered, D. and Whelan, J., eds.), Ciba Foundation Symposium No. 143, pp. 60–86, Wiley, Chichester
8. Bernfield, M.,Banerjee, S.D., Koda, J.E. and Rapraeger, A.C. (1984) in The Role of Extracellular Matrix in Development (Trelstad, R.L., ed.), pp. 545–572, Adam R. Liss, New York
9. De Salegui, M. and Pigman, W. (1967) Arch. Biochem. Biophys. **120**, 60–67
10. Stephens, R.W., Ghosh, P. and Taylor, T.K.F. (1975) Biochim. Biophys. Acta **399**, 101–112
11. Linker, A., Meyer, K. and Weissmann, B. (1955) J. Biol. Chem. **213**, 237–248
12. Aronson, Jr., N.N. and De Duve, C. (1968) J. Biol. Chem. **243**, 4564–4573
13. Longas, M.O. and Meyer, K. (1981) Biochem. J. **197**, 275–282
14. Gorham, S.D., Olavesen, A.H. and Dodgson, K.S. (1975) Connect. Tissue Res. **3**, 17–25
15. Bollet, A.J., Bonner, Jr., W.M. and Nance, J.L. (1963) J. Biol. Chem. **238**, 3522–3527
16. Vaes, G. (1965) Biochem. J. **97**, 393–402
17. Weissmann, B., Cashman, D.C. and Santiago, R. (1975) Connect. Tissue Res. **3**, 7–15
18. Laurent, U.B.G. and Laurent, T.C. (1981) Biochem. Int. **2**, 195–199
19. Fraser, J.R.E., Laurent, T.C., Pertoft, H. and Baxter, E. (1981) Biochem. J. **200**, 415–424
20. Fraser, J.R.E., Dahl, L.B., Kimpton, W.G., Cahill, R.N.P., Brown, T.J. and Vakakis, N. (1989) J. Dev. Physiol. **11**, 235–242
21. Fraser, J.R.E., Kimpton, W.G., Laurent, T.C., Cahill, R.N.P. and Vakakis, N. (1988) Biochem. J. **256**, 153–158
22. Fraser, J.R.E., Cahill, R.N.P., Kimpton, W.G. and Laurent, T.C. (1996) in Extracellular Matrix, vol. 1, Tissue Function (Comper, W.D., ed.), pp. 109–130, Harwood Academic, Amsterdam
23. Bentsen, K.D., Henriksen, J.H. and Laurent, T.C. (1986) Clin. Sci. **71**, 161–165
24. Fraser, J.R.E., Appelgren, L-E. and Laurent, T.C. (1983) Cell Tissue Res. **233**, 285–293
25. Fraser, J.R.E., Alcorn, D., Laurent, T.C., Robinson, A.D. and Ryan, G.B.(1985) Cell Tissue Res. **242**, 505–510
26. Culty, M., Nguyen, H.A. and Underhill, C.B. (1992) J. Cell Biol. **116**, 1055–1062
27. Laurent, T.C., Fraser, J.R.E., Pertoft, H. and Smedsrød, B. (1986) Biochem. J. **234**, 653–658
28. Dahl, L.B., Laurent, T.C. and Smedsrød, B. (1988) Anal. Biochem. **175**, 397–407
29. Brown, T.J. and Fraser, J.R.E. (1995) in Third International Workshop on Hyaluronan in Drug Delivery, (Willoughby, D.A., ed.), Round Table Series, vol. 40, pp. 31–37, Royal Society of Medicine Press, London
30. Fraser J.R.E. (1995) in Round Table Series, vol. 36, pp. 8–10, Royal Society of Medicine Press, London
31. Tzaicos, C., Fraser J.R.E., Tsotsis, E. and Kimpton W.G. (1989) Biochem. J. **264**, 823–828
32. Smedsrød, B., Pertoft, H., Eriksson, S., Fraser, J.R.E. and Laurent, T.C. (1984) Biochem. J. **223**, 617–626
33. Campbell, P., Thompson, J.N., Fraser, J.R.E., Laurent, T.C., Pertoft, H. and Rodén, L. (1990) Hepatology **11**, 199–204
34. Brown, T.J., Laurent, U.B.G. and Fraser, J.R.E. (1991) Exp. Physiol. **76**, 125–134

35.   Nimrod, A., Ezra, E., Ezov, N., Nachum, G. and Parisada, B. (1992) J. Ocul. Pharmacol. **8**, 161–172

36.   Fraser, J.R.E., Laurent, T.C., Engström-Laurent, A. and Laurent, U.B.G. (1984) Clin. Exp. Pharmacol. Physiol. **11**, 17–25

11

Degradation of hyaluronan systems by free radicals

Glyn O. Phillips

Research Transfer Ltd., 2 Plymouth Drive, Radyr, Cardiff CF4 8BL, U.K.

Introduction

Hyaluronan can be degraded by chemical methods involving heat [1], alkaline [2] and acidic [3] conditions. Physical stress, as introduced by freeze-drying, high-speed stirring or critical shearing also induce degradation [4,5]. Enzymic degradation pathways have also been identified [6]. Various irradiation conditions can also lead to depolymerization whether by direct excitation using ultrasound [7], ultraviolet light or ionizing radiations [8,9]. Singlet oxygen also reduced the viscosity of hyaluronan [10]. Certain of these agents will promote free-radical participation, which is the subject of this paper.

There are clinical and practical implications associated with free-radical interactions with hyaluronan systems. Two sources of free radicals are considered to be associated with the breakdown of hyaluronan in the connective tissue. One is the effect of ionizing radiations, where excitation and ionization processes produce highly reactive free-radical systems [11]. Another is the inflammation stage of disease, particularly arthritis, which leads to the reduction in the viscosity of synovial fluid, and the depolymerization of the hyaluronan therein [12–15].

It has been demonstrated that it is the OH radical that is the most potent in initiating the degradation of hyaluronan [16–18]. When water is radiolysed, both transient and stable products are formed:

$$H_2O \xrightarrow[\text{radiation}]{\text{ionizing}} {}^{\bullet}OH,\ \bar{e}_{aq},\ {}^{\bullet}H,\ {}^{\bullet}H_2O_2 + H_2$$

For di- and oligosaccharides, a direct consequence of OH radical attack is scission of the glycosidic linkage. For hyaluronan, the radiation chemical yields of strand breaks have been measured by steady-state radiolysis, both in the presence and absence of oxygen [18]. To determine the kinetics of the process, fast reaction techniques were employed.

Using pulse radiolysis, combined with monitoring transient product formation with conductometry and laser-light scattering, the kinetics of hydroxyl-strand breakage of hyaluronan were determined [16]. At neutral pH, hyaluronan ($pK_a = 2.9$) [19] is a single-stranded polyelectrolyte, and hence the formation of strand breaks can be measured conductometrically. The high electric field of polyelectrolyte anions causes the condensation of cationic counterions around the polyelectrolyte and only a fraction of the counterions remain free.

Therefore, when a strand break occurs the electric potential at the newly formed ends is lowered, and the counterions are released, leading to a conductivity change [16]. A more direct method of detecting strand breakage is by means of low-angle laser-light scattering (LALLS). By utilizing a low angle of scatter (4.5–5.5°), the amount of light scattering is affected only by the molecular weight of the hyaluronan and is unaffected by the conformational changes [16].

To produce an almost pure source of OH radicals, saturation of the solution with N_2O has been used. The solvated electrons are converted into OH radicals:

$$e_{aq}^- + N_2O \longrightarrow {}^{\bullet}OH + N_2 + OH^-$$

Under these conditions, the yield of OH radicals, expressed as the G value (yield in molecules per 100 eV energy input) is 5.6, whereas $G(\bullet H)$ is only 0.6. Both radicals react with hyaluronan by abstracting carbon-bound hydrogen atoms:

$$^{\bullet}OH(^{\bullet}H) + RH \longrightarrow H_2O(H_2) + R^{\bullet}$$

The OH radical is generally not very selective, and the available carbon-bound hydrogens are most likely abstracted at random [20]. In the presence of oxygen (N_2O/O_2 4:1 v/v saturated solutions) the majority of the H atoms will be scavenged by oxygen to give the hydroperoxide radical:

$$^{\bullet}H + O_2 \longrightarrow {}^{\bullet}HO_2 \rightleftharpoons H^+ + O_2^-$$

The superoxide radical ion (O_2^-) is often invoked to account for metabolic and disease-related effects. In oxygen too, the carbohydrate radicals are converted into the corresponding peroxy radicals:

$$R^{\bullet} + O_2 \longrightarrow RO_2^{\bullet}$$

Both $R^{\bullet}$ and $RO_2^{\bullet}$ can lead to strand breakage in hyaluronan.

The production of free radicals has also been considered to occur during the inflammation stage of arthritic disorders, although it must be stressed that the evidence remains completely circumstantial. McCord and Fridovich [21] were the first to propose that O_2^- could be generated biologically. Subsequently, several excellent reviews of this subject have appeared [22–24]. It has often been suggested that the O_2^- radical itself is the direct entity causing tissue damage. Free-radical chemistry, however, has demonstrated unequivocally that O_2^- is a sluggish radical when abstraction reactions are considered as a preliminary to polymer degradation. More likely now is that the O_2^- is converted into more reactive free radicals, particularly the OH radical, which can act effectively in chain-breakage reactions [17]. The most relevant reaction associated with O_2^- alone is the dismutation reaction:

$$O_2^- + O_2^- \xrightarrow{\ \ 2H^+\ \ } H_2O_2 + O_2$$

This can occur spontaneously, but can proceed at a faster rate when catalysed by the intercellular enzyme, superoxide dismutase. Thus processes by which O_2^- can be converted into OH radicals need to be identified. One possibility is the Haber–Weiss reaction:

$$O_2^- + H_2O_2 \longrightarrow O_2 + \ ^\cdot OH + OH^-$$

This reaction, however, is slow unless catalysed by a metal ion [25–27]:

$$Fe\ (chelate)^{3+} + O_2^- \longrightarrow Fe\ (chelate)^{2+} + O_2$$

It is not clear whether the concentration of free metal ion in tissues such as synovial fluid is sufficient to catalyse the classical Fenton's reaction:

$$Fe^{3+} + O_2^- \longrightarrow O_2 + Fe^{2+}$$
$$Fe^{2+} + H_2O_2 \longrightarrow Fe^{3+} + OH^- + \ ^\cdot OH$$

An important question is whether the protein-bound metal ions, e.g. ferritin, transferrin, lactoferrin, haemoglobin, etc., can also reduce ferric to ferrous ions. There are other naturally occurring substances that would convert Fe^{3+} into Fe^{2+}, notably ascorbate, possibly also glutathione, NAD, reduced NADH, etc. 'Site-specific' hydroxy radical formation could also be responsible for strand breakage in a polyanion such as hyaluronan at the actual site of OH radical generation [28]. The binding of iron to its storage proteins is pH dependent. The low pH of ischaemic tissue favours its release [29]. While the pH of the bulk phase of a synovial fluid effusion is not low enough to cause this, the pH of ischaemic micro-environments within inflamed joint tissue could fall to these levels.

There are two hypotheses about the manner in which O_2^- is produced biologically. The two suggested pathways are either via phagocytic cells or via the ischaemia reperfusion concept. Both hypotheses have their devoted proponents. Evidence in support of either mechanism is derived from scavenger studies; using catalase, for example, to remove H_2O_2 and/or chelate systems to remove metal ions. These would appear to reduce the observed effect, thereby taken as evidence for O_2^- participation. The various proposals have been reviewed within the overall context of our present studies [30]. All concerned now implicitly attribute the molecular changes to the intervention of free radicals.

The objective of this chapter is to identify the mechanisms whereby radicals degrade hyaluronan, whether produced via the participation of metal ions or not. We shall demonstrate also that cross-linked or aggregated hyaluronan systems (hylans) provide greater free radical (and radiation) stability. This stability can be related to the molecular aggregate structure, which is also responsible for the effectiveness of such systems in the viscosupplementation of osteoarthritic diseased joints. The relevance of such radical degradation to the connective tissue has been described elsewhere [31].

Molecular changes induced in hyaluronan systems by OH radicals

Hyaluronan

Of all the reactive species produced by radiolysis of water, the most effective in initiating degradation of hyaluronan is the OH radical, as demonstrated by the bimolecular rate constants measured by pulse radiolysis (Blake and co-workers, [32]; where HA is hyaluronan):

$$\cdot OH + HA \quad 9 \times 10^8 \; M^{-1} \cdot s^{-1} \quad \text{(independent of pH in range 3–7)}$$
$$\cdot H + HA \quad 6 \times 10^7 \; M^{-1} \cdot s^{-1} \quad \text{(pH2)}$$
$$e_{\overline{aq}} + HA \quad < 10^7 \; M^{-1} \cdot s^{-1}$$

Whereas, the role of complex redox systems has received a great deal of discussion [33–35], less information is available at the molecular level, which is the subject of this chapter and previous studies [17,18].

Pulse-conductivity measurements on N_2O-saturated (where $\cdot OH$ is the major attacking species) 2.5×10^{-3} M hyaluronan solutions showed that at acid, neutral and alkaline pH there was a conductivity increase following the pulse, and are consistent with counterion (K^+) being released on chain cleavage (Figure 1). The rate of strand breakage is independent of dose in the range 5–50 Gy. From the variation of the overall half-life for the conductivity build-up with pH, the formation of breaks is catalysed by acid and base. Since the overall half-life is independent of the dose rate, the processes involved are kinetically of first order, although this cannot be attributed to one single process. The catalytic effect of base and acid is thus best expressed in terms of the overall variation in half-life (Table 1).

In oxygen, using LALLS in conjunction with pulse radiolysis, the concentration of chain breaks as a function of time after the pulse can be expressed (Figure 2). The kinetics of strand breakage is again complex. However, the concentration of strand breaks can be plotted against dose at various pH values (Figure 3).

Table 1

Dose = 5.35 Gy									
pH	4.8	5.4	6.95*	7.4*	9.1	9.5	9.8	10.0	10.2*
$t_{1/2}$/ms†	0.6	1.0	1.4	1.3	0.5	0.4	0.4	0.18	0.1

*10^{-4} M phosphate.

†Independent of dose.

Overall half-lives for release process of counterions from hyaluronan (0.25 mM)

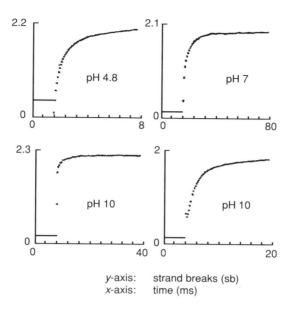

Figure 1

y-axis: strand breaks (sb)
x-axis: time (ms)

Conductivity changes (μS) as a function of time (ms) on pulse irradiating (30 Gy) nitrous oxide-saturated hyaluronan solution

To quantify the effect of OH radicals on the overall molecular weight of hyaluronan, the changes in molecular weight with dose were measured using LALLS and viscometrically [30]. In both instances the yield–dose plot is linear (Figure 4). However, on standing at room temperature, there was a continual further decrease in molecular weight of the irradiated hyaluronan. This slow post-irradiation depolymerization can be brought to completion by heating the

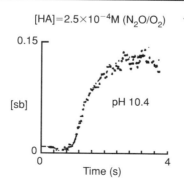

$[HA]=2.5\times10^{-4}M\ (N_2O/O_2)$

Figure 2

The concentration of strand breaks (sb; μM, measured by light scattering) as a function of time (s)

HA, hyaluronan.

Figure 3

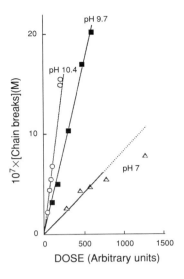

Chain breaks induced in hyaluronan (0.25 mM) (N_2O/O_2)

samples for 1 h at 60 °C. The initial G value for the ·OH induced degradation is 4, and after completion $G = 6$, when the molecular weight is followed viscometrically or by LALLS. Having regard to the fact that $G(·OH) = 5.6$ and $G(·H) = 0.6$, then 6.2 hyaluronan radicals are formed by abstraction of these species at the

Figure 4

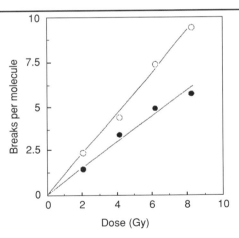

Yield–dose plot showing the number of chain breaks per molecule induced in N_2O-saturated aqueous solution of hyaluronan at pH7 in phosphate buffer, 2×10^{-3} M

(●) *No post-irradiation treatment;* (○) *post-irradiation incubation for 1 h at 60 °C. Weight average molecular weight were determined by LALLS.*

Scheme 1

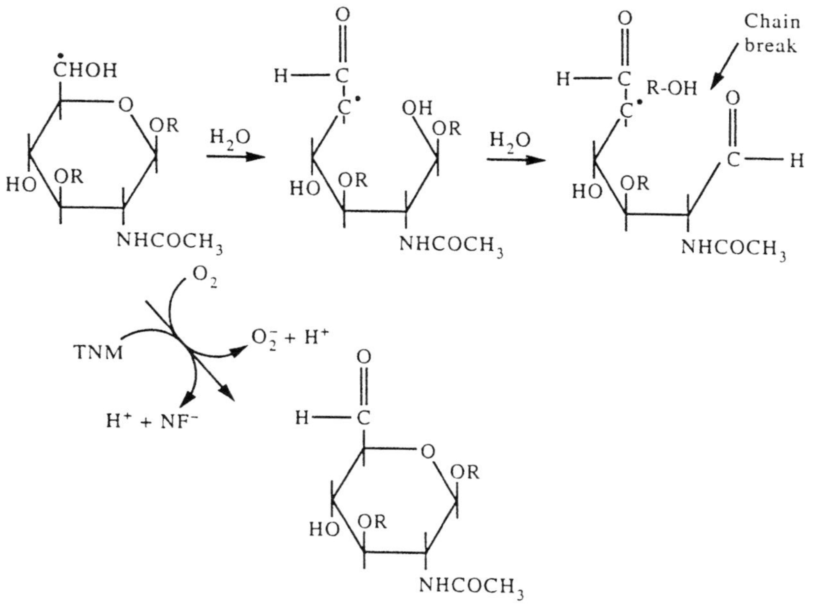

CHOH · O OR HO OR NHCOCH₃

H—C=O ·C OH OR HO OR NHCOCH₃

H—C=O ·C R-OH Chain break O C—H HO OR NHCOCH₃

H_2O H_2O

O_2

TNM

$O_2^- + H^+$

$H^+ + NF^-$

H—C=O O OR HO OR NHCOCH₃

Thermal post-irradiation transformation of radical of C6 at N-acetyl glucosamine moiety leading to oxidation and chain break

carbohydrate skeleton. It would appear that 65% of the radicals lead to immediate (i.e. within approx. 15 min) chain scission, whereas the vast majority of the remaining radicals give rise to thermally labile products that eventually lead to chain scission. If the hyaluronan is made acid (pH approx. 2, H_2SO_4) or alkaline (pH approx. 11, NaOH) immediately after irradiation, the full yield of breaks ($G = \sim 6$) is obtained. This general acid/base catalysis, as noted in the kinetic observations also, is strong evidence for the involvement of a hydrolytic reaction in converting 'latent' breaks into actual breaks. The presence of oxygen (N_2O/O_2) reduces the value of G (degradation) to 3.2. This protective effect of oxygen is well known in carbohydrate radiation chemistry [36] and can be explained as in Scheme 1. Lal [37], also reported that the decrease in viscosity was reduced when hyaluronan was irradiated in oxygen.

We have used a specialized technique to distinguish between two types of attack on the hyaluronan chain, first used by Fujita and Steenken [38] with pyrimidines. A reducing radical is able to reduce the oxidant tetranitromethane (TNM). The nitro-form (NF^-) from TNM can be determined spectrophotometrically [39].

In solutions of D-glucose, OH radical attack is essentially random [40–44]. Hyaluronan would behave similarly and most of the radicals produced would be β-hydroxy or β-alkoxy radicals and these would react with TNM to give NF^- [40]. However, from NF^- yields in irradiated hyaluronan (1×10^{-3} M, N_2O, pH 7) solutions containing TNM, it is evident that not all the hyaluronans are reducing. In contrast, similar experiments using β-cyclodextrin in place of hyaluronan give 100% reducing radicals [41]. The probable 'oxidizing' radical in hyaluronan is that at C-5 of the glucuronic acid residue. One of the main features of the free-radical chemistry of simple carbohydrates is the elimination of water from 1,2-dihydroxy radicals with the formation of the corresponding β-carbonyl radical:

$$-\overset{\cdot}{C}OHCHOR \longrightarrow -CO\overset{\cdot}{C}H + HOR$$

Similarly, 1-hydroxyl-2-alkoxy radicals can eliminate the alcohol, again forming the corresponding β-carbonyl radical. Both reactions are acid and base catalysed [42]. Out of the eleven possible hyaluronan radicals, those at C-2, C-4 and C-6 of the acetamido-glucose moiety and those at the C-2 and C-3 of the glucuronic acid moiety are 1-hydroxy-2-alkoxy radicals. Alkoxy elimination from all these radicals with the exception of the C-6 radical involves strand breakage (the alkoxy group is the adjacent sugar sub-unit). For the C-6 radical, ring opening would occur with the formation of a hemiacetal at C-1. This would give a strand break and would also lead to strand breakage as shown in Scheme 1. The 43% protection afforded by TNM can be rationalized on this basis [30].

Figure 5

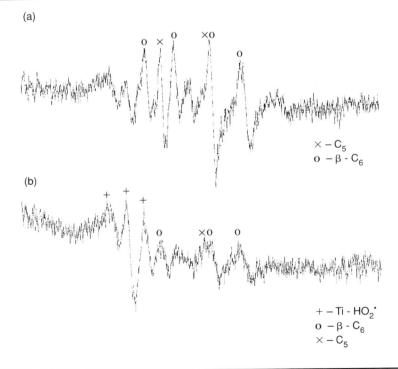

(a)

$\times - C_5$
$o - \beta - C_6$

(b)

$+ - Ti - HO_2^{\cdot}$
$o - \beta - C_6$
$\times - C_5$

(a) Electron spin resonance spectrum of hyaluronan radicals produced by OH radical abstraction; (b) comparable spectrum for hylan

For hylan the spectrum of Ti-HO$_2^{\cdot}$ radical is also evident Ti^{3+}, 7 mM; H$_2$O$_2$, 50 mM; substrate, 5 g dm^{-3}.

Until now it has been impossible to resolve the electron spin resonance (ESR) spectra of hyaluronan free-radicals produced by OH radical abstraction. Their formation could only be inferred from kinetic and pulse radiolysis observation. Now for the first time the actual polymeric free-radicals can be measured and identified (Figure 5a and 5b). These spectra were achieved by using a rapid mixing flow technique (see Acknowledgements, p.109 of this volume) whereby Ti^{3+} (7 mM), H_2O_2 (50 mM) and hyaluronan (or hylan) (5 g dm^{-3}) were flowed simultaneously into an aqueous flat cell and the ESR spectra recorded within 30 ms, using either a Bruker ESP 300 or a Joel RE 1-X X-band ESR spectrometer with 100 kHz modulation. The spectra can be assigned to radicals based at C-5 of the glucuronic acid moiety and the C-6 of the N-acetyl glucosamine moiety, which are confirmed also by computer simulation of the spectra using the assigned hyperfine coupling constants of these radicals. It is, of course, not possible to exclude radicals formed at other positions in the sugar rings that degrade quicker than the timescale of our experiments (30 ms), and indeed this is highly probable from the kinetic data which has been presented. Our detailed ESR observations, however, exclude only the formation of the radical at C-2 of the N-acetyl glucosamine residue, undoubtedly due to the powerful deactivating effect of the carbonyl function.

Hylan

Cross-linked forms of hyaluronan have been prepared by Balazs and co-workers, and were designated hylans [45]. Here we use hylan A, where formaldehyde was used to cross-link, for this material is water soluble. Using multi-angle laser light scattering with a DAWN laser photometer, absolute molecular weights of the hylan samples used in our investigation were measured (Wyatt Technology Corporation) (Table 2).

The LALLS and viscometric methods have similarly been applied to study the degradation by OH radicals of hylan. The results are illustrated using one hylan (Sample 5, Table 2). Figure 6 shows the yield–dose curve for this hylan sample, after the post-irradiation degradation process had been completed (after incubation for 1 h at 60 °C). This procedure was adopted since a similar thermal post-irradiation process was observed with hylan as was described for hyaluronan. This after-effect is not as pronounced as for hyaluronan, but remains significant.

The results in Figure 6 are based on LALLS measurements, and have, thereafter, been used to calibrate the viscosity measurements. A double logarithmic plot of the molecular weights obtained by LALLS against the intrinsic

Hylan	G value (degradation)	M_w	
1	2.0	3.3×10^6	**Table 2**
2	1.8	2.5×10^6	
3	1.9	6×10^6	
4	1.9	6×10^6	
5	2.3	5.6×10^6	

G values for the degradation of various hylan samples

Figure 6

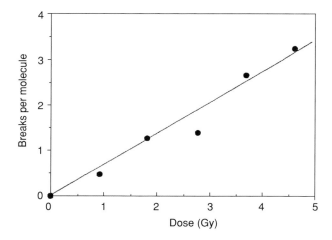

Yield–dose plot showing the number of chain breaks per molecule induced in N₂O saturated aqueous solution of hylan (H44-2D-5) at pH 7 in phosphate buffer 2 × 10⁻³ M

Following irradiation the solutions were incubated for one hour at 60 °C. Weight average molecular weights were determined by LALLS.

viscosities of the same hylan solutions was made [30]. The Mark–Houwink equation, which relates intrinsic viscosity (η) to weight average molecular weight (M_w), for hylan is $[\eta] = KM_w^a$. Hence, $\log \eta = a\log M_w + \log K$. Thus from the slope and y-intercept at $\log M_w = 0$ for this plot it is possible to evaluate K and a values. In this way the Mark–Houwink constants K and a were evaluated. For hylan the $K = 0.033$ and $a = 0.77$, compared with $K = 0.029$ and $a = 0.80$ for hyaluronan. Based on these values the change in intrinsic viscosity with dose could be used to measure the G value of hylan degradation also. Table 2 illustrates these results. Thus for all the hylan samples studied, the structure is three times more stable to degradation by OH radicals than non-cross-linked hyaluronan.

Factors that influence the stability of hylan to free radicals

Molecular aggregation

The size of the reduction in the molecular weight of a polymer by free-radical attack depends on the number of free radicals reacting with each polymer molecule. For radiation-induced free radicals, the number of free radicals produced is directly proportional to the radiation dose. The number of polymer molecules present in solution is directly proportional to the concentration of polymer (g dm⁻³) and inversely proportional to the molecular weight (number average) of the polymer. To directly compare the behaviour of a linear polymer such as hyaluronan with a cross-linked aggregate polymer matrix such as hylan requires further consideration. The G values already measured reflect molecular weight changes with radiation dose, which is directly equivalent to OH radical concentration. Two situations were considered. First, equimolar solutions of hylan and hyaluronan were irradiated. Second, solutions of hylan and hyaluronan of the same initial specific viscosity were compared. In both instances it is evident that hylan solution is more resistant to free-radical attack.

Specific viscosity is a concentration-dependent variable. A much more definitive characteristic of a polymer is its intrinsic viscosity, which is independent of polymer concentration, and is a measure of the size of the polymer molecule. It would be preferable, if possible, to compare the radiation-induced decrease in intrinsic viscosity of the two polymers at equal concentration and polymer molecular weights. The hyaluronan system is now so well-characterized that it is possible to calculate intrinsic viscosity–dose curves for the same hyaluronan concentration and weight average molecular weights as the hylan solutions. These hylan solutions were irradiated and their intrinsic viscosity–dose curves measured along with the corresponding data for hyaluronan. The results again confirm the greater resistance of hylan to free-radical attack. Details of these experiments have been described [30].

The reaction rate of OH radicals and H atoms were measured by pulse radiolysis using a competition method, with tryptophan and 5-bromouracil as the references. The rates that we found were:

$$\cdot OH + hylan = 8.5 \times 10^8 \ mol^{-1} \cdot dm^3 \cdot s^{-1}$$
$$\cdot H + hylan = 9.3 \times 10^7 \ mol^{-1} \cdot dm^3 \cdot s^{-1}$$

These rates are not significantly different from those of hyaluronan, which would be anticipated. The radicals encounter the same structures, and wherever generated have access to N-acetyl glucosamine and glucuronic acid moieties. The extent of radical attack is identical on hylan and hyaluronan. The measured G values are based on molecular weight changes when the hylan structures have a distinctive capacity to resist disintegration to units of lower molecular weight.

A powerful method for studying the structures of such high-molecular-weight aggregate hylan systems is atomic force microscopy [46]. This study has yielded the first such images of hyaluronan structures. When 10 μg ml^{-1} aqueous hylan and hyaluronan are deposited on to mica, highly entangled networks are evident for both polymers. However, on dilution to 1 μg ml^{-1} before deposition on to mica, the hyaluronan network is broken up, and only individual hyaluronan chains are observed. Hylan, on the other hand, when similarly diluted preserves the large aggregates (see Figure 7). From these about 12 chains can be identified as leaving the aggregate, indicating a molecular mass of about 10–12 × 10^6 Da for the aggregate, if it is considered that 6 hyaluronan chains intertwine to make up the aggregate. Full details have been described [47]. The forces holding the aggregate together prevent complete disintegration following individual chain breaks, as occurs for non-cross-linked hyaluronan. It is possible to observe also single chains of cross-linked hylan with molecular weight of approx. 2 × 10^6 Da (Figure 8). The ESR spectrum (Figure 5b) confirms the greater radiation stability of the hylan

Figure 7

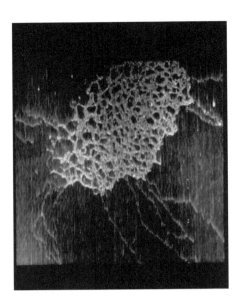

Atomic force microscope image of hylan (scan size 1900 × 1900 nm) illustrating aggregate structure (deposited at 1 μg ml⁻¹)

aggregate structure. The radical yield is considerably lower than for hyaluronan and the spectrum shows greater anisotropy. Both spectra of hyaluronan and hylan radicals indicate the presence of high-molecular-weight radicals that tumble relatively slowly in solution, resulting in the broadening of the spectral lines at the higher field end of the spectrum. This behaviour is more pronounced for hylan, which is a further indication of a larger aggregate structure that tumbles less freely.

It is evident that the method of preparing hylan using formaldehyde to cross-link when extracting from the rooster combs preserves the *in situ* structures

Figure 8

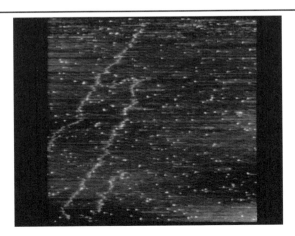

Single strand of cross-linked hylan as shown by atomic force microscopy (scan size 1250 × 1250 nm; deposited at 1 μg ml⁻¹)

in part, and provides a stability to such aggregates of hyaluronan chains that does not extend to processed hyaluronan, which we have examined using the same techniques. Identification of these aggregate structures provides a reasonable explanation for a number of distinctive physical and chemical characteristics that we have found to be different for hylan compared with non-aggregated hyaluronan.

When OH radical attack initiates strand breaks, a structure based on dynamic entanglement, as in concentrated hyaluronan solutions, dramatically loses viscosity when exposed to OH radicals in the entangled region [31]. Strand breakage leads to disintegration of the entangled viscoelastic network. With hylan, however, the network is not dependent on transient forces to achieve its stability, and strand breakage can occur without its overall disintegration. The G value of 2 molecules degradation (per 100 ev energy input) for hylan compared with 6 for hyaluronan, illustrates this threefold greater ability to preserve its network structure.

Rheological characteristics

Extensional viscosity measurements were carried out using a Rheometrics RFX Fluid analyser, and compared with shear flow viscosity measurement using a Carrimed 100 Controlled Stress Rheometer. Only recently has suitable experimentation become available to quantify such extensional flow behaviour [48]. Double logarithmic plots of the extensional and shear-viscosity–strain rate profiles of a 0.2% solution of hylan (M_w 4.4 × 10⁶ Da) and hyaluronan (M_w 2.0 × 10⁶ Da) are shown in Figure 9. The shear viscosities, while diverging somewhat at low shear rates, converge at high shear rates. The extensional viscosities, however, of hylan and hyaluronan are markedly different, with the hylan being consistently greater at all strain rates and also the increment between the behaviour of hylan and hyaluronan becoming even more pronounced. The

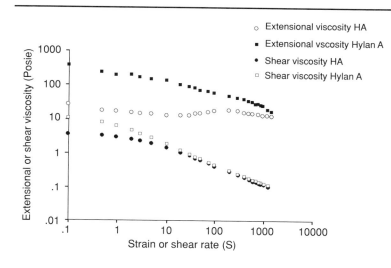

Figure 9

Extensional and shear viscosity profiles of 0.2% aqueous solution of hyaluronan and hylan at pH 7 in 2 × 10⁻³ M phosphate buffer

Figure 10

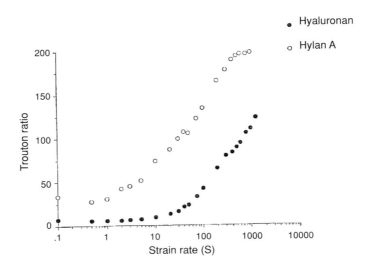

Trouton ratio profiles of 0.2% aqueous solution of hyaluronan and hylan A
at pH 7 in 2×10^{-3} M phosphate buffer

shear thinning, which characterizes the shear behaviour, is not so pronounced in extensional flow and strain thickening is evident here. This point is further illustrated by the changes with strain rate in the Trouton ratio (T_R), which is the ratio of the extensional flow at a particular strain rate to the shear viscosity at an equivalent shear rate (Figure 10). While for hylan and hyaluronan T_R increases with the strain rate, the value of T_R for hylan is significantly greater than for hyaluronan, and the divergence increases with strain rate.

Evidence for greater structure of the hylan in aqueous solution comes also from measurements of G', the dynamic elastic (storage) modulus and G'', the dynamic viscous (loss) modulus, which indicates that hylan exhibits a weak gel character whereas at the same concentration (0.2%) the hyaluronan performs as a viscous solution ([45] and the present study).

As noted for hyaluronan, the network structure is produced purely by entanglement of the polysaccharide chains, is of a dynamic character and is dependent on hydrogen bonds for its stability. Evidence that some other linkage system is operative in aqueous solutions of hylan is indicated by this rheological behaviour. Extensional viscosity and oscillation experiments both point to a more elastoviscous structure of hylan than hyaluronan. At a strain/shear rate of 10, for example, despite comparable shear viscosities, the extensional viscosity of hylan is ~4 times greater than for hyaluronan. It points to greater mechanical and physical resistance. The Trouton ratio quantifies this difference. In shear flow, both polymers exhibit strain thinning, owing to the molecular entanglements. In extensional flow there is evidence of strain thickening of hyaluronan, which can be attributed to strain-induced 'stretching' of the initially compact coil to more extended conformations, thereby resulting in increased drag. This effect does not occur to the same extent for the higher molecular aggregate-structure hylan. The

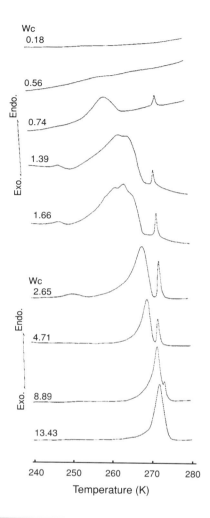

Figure 11

Differential scanning calorimetry heating curves for water–hylan systems (W_c = weight of water in g to 1 g of hylan)

oscillation experiments also point to the greater elasticity of hylan than hyaluronan. Effectively, therefore, hylan is more 'stretchable' than hyaluronan.

Water binding characteristics

The network structure of hylan also provides a means of understanding the remarkable ability of hylan to interact with water [49,50], which was studied using differential scanning calorimetry (DSC). Typical DSC curves are shown in Figure 11. Three types of water can be distinguished: non-freezing, freezing-bound and free. When the water content of the system is increased even up to 10 g water to 1 g hylan, almost all the water remains in the freezing-bound state with ΔH less

Figure 12

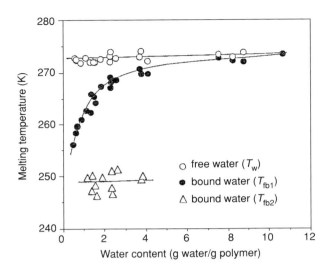

Variation of melting temperature with water content

than free water. Several metastable states could be detected within the structured hylan–water matrix, indicative of defects in the frozen-bound ice structure. The maximum amount of non-freezing water, intimately associated with the hydrophilic groups of hylan, corresponds to 13 water molecules per disaccharide unit of the hylan chain. A comparison is possible with hyaluronan [51] and other polymers by considering the value W_c, which is the highest water content where there is distinctive freezing-bound transition discernible from free water. Figure 12 shows the variation in melting temperature for the two types of bound water

Table 3

Sample	W_c
Poly(8-hydroxystyrene)	0.080
Poly(2-acetoxystyrene)	0.026
Cellulose I	0.50
Cellulose II	0.06
Sodium carboxymethyl cellulose	0.08
Sodium polystyrene sulphonate	1.3
Sodium cellulose sulphonate	1.4
Sodium xanthan	2.0
A. senegal	3–6
Hyaluronan	3.0
Hylan fluid HC and HD	10.0
Hylan gel GFD	12.0
Hylan gel GP	15.75

Weight content (W_c in g/g) of water in various polymers necessary to achieve saturation

Hylan fluid HC was obtained by freeze-drying a concentrated solution, hylan fluid HD by freeze-drying a dilute solution. GFD indicates freeze-dried hylan gel, GP indicates precipitated and freeze-dried hylan gel.

and the free water in hylan–water systems, which enable W_c to be estimated. Hylan in its variously prepared forms is vastly superior to any other polymer (Table 3). The voids within the hylan network structure provide sites where water can interact with the polysaccharide matrix and freeze to give ice in a dislocated form. These results provide further evidence for the manner in which structures are produced by hylan with water, which influence its free-radical stability and rheological properties.

Significance for viscosupplementation in osteoarthritic joints

There are obvious implications for these results in considering the utilization of hylan and hyaluronan in viscosupplementation. The relative permanence of the aggregate structure in hylan, compared with hyaluronan, is reflected in their relative rheological properties. The network structure evident in the atomic force microscopic images provides stable thermodynamic traps for water binding. This in turn leads to the formation of structure with water that has both viscous and elastic properties, more akin to a gel than a viscous solution. The behaviour in extension is considerably better than the transient entangled hyaluronan network. Better extensional viscosity would imply that when called upon to perform a lubricating function in a moving joint it would prove more effective. When influenced by the inflammation condition during osteoarthritis, there is generally degradation of the hyaluronan structure as previously demonstrated. It is after such degradation that the elastoviscous characteristics of the synovial fluid are impaired. Hylan would be more resistant to this degradation by a factor of 3 as reflected by the -G value (degradation per 100 ev energy input) of 2 for hylan compared with 6 for hyaluronan. The aggregate structures observed by atomic force microscopy fully provide an explanation for the rheological characteristics and enhanced free-radical stability of hylan. In seeking viscosupplementation materials, therefore, it would be prudent also to build in the necessary structures that can provide effective water binding, namely good elastoviscous matrices that are stable OH radicals.

Endre A. Balazs first introduced me to hyaluronic acid in 1963, and here I wish to thank him for his continued inspiration and guidance over the subsequent 33 years. We are still collaborating in research, and enjoying every minute of our joint discussions and even the occasional argument! Thank you Bandi especially for your loyal friendship. I also wish to acknowledge the support of my co-workers in these investigations, notably Saphwan Al-Asaaf, Professor Barry Parsons, Professor Peter A. Williams, Professor C. von Sonntag and the late Dr David Deeble.

Electron spin resonance measurements were undertaken by Saphwan Al-Asaaf and Clare L. Hawkins in the University of York, U.K. under the direction of Professor Bruce C. Gilbert and Professor B. Parsons. Full details will be published elsewhere.

References

1. Bothner, H., Waaler, T. and Wik, O. (1988) Int. J. Biol. Macromol. **10**, 289–291
2. Hirano, S., Ishigami, M. and Koga, Y. (1975) Connect. Tissue Res. **3**, 73

3.   Cleland, R.L. (1984) Biopolymers **23**, 647–666
4.   Davies, A., Gormally, J., Wyn-Jones, E., Wedlock, D.J. and Phillips, G.O. (1983) Int. J. Biol. Macromol. **5**, 186–188
5.   McGary, C.T., Raja, R.H. and Weigel, P.H. (1989) Biochem J. **257**, 875–884
6.   Rodén, L., Campbell, P., Fraser, J.R.E., Laurent, T.C., Pertoft, H. and Thomson, J.N. (1989) in The Biology of Hyaluronan (Evered, D. and Whelan, J., eds.), Ciba Foundation Symposium No. 143, pp. 60–86, Wiley, Chichester
7.   Chabrecek, P., Soltes, L. and Orvisky, E. (1991) J. Appl. Polym. Sci. Symp. **48**, 233–241
8.   Balazs, E.A., Laurent, T.C., Howe, A.F. and Varga, L. (1959) Radiat. Res. **11**, 149–164
9.   Balazs, E.A., Davies, J.V., Phillips, G.O. and Young, M.D. (1967) Radiat. Res. **31**, 243–255
10. Andley, U.P. and Chakrabarti, B. (1983) Biochem. Biophys. Res. Commun. **115**, 894–901
11. Phillips, G.O. (1968) in Energetics and Mechanisms in Radiation Biology, pp. 527, Academic Press, London & New York
12. Gutteridge, J.M.C. and Wilkes, S. (1983) Biochim. Biophys. Acta, **759**, 38–41
13. Balazs, E.A. and Denlinger, J.L. (1985) in Osteoarthritis, Current Clinical and Fundamental Problems (Peyron, J.G., ed.), pp. 165–174, Ciba-Geigy, Rueil-Malmaison, France
14. Balazs, E.A. and Denlinger, J.L. (1989) in The Biology of Hyaluronan (Evered, D. and Whelan, J., eds.), Ciba Foundation Symposium, No. 143, pp. 265–280, Wiley, Chichester
15. Balazs, E.A. (1982) in Disorders of the Knee (Helfel, A., ed.), Lippocott, Philadelphia, 63–74
16. Deeble, D.J., Bothe, E., Schuchmann, H-P., Parsons, B. J., Phillips, G.O. and von Sontag, C. (1990) Z. Naturforsch., C Biochem. Biophys. Biol. **45**, 1031–1043
17. Deeble, D.J., Phillips, G.O., Bothe, E., Schuchmann, H-P. and von Sonntag, C. (1991) Radiat. Phys. Chem. **37**, 115–118
18. Myint, P., Deeble, D.J., Beaumont, P.C., Blake, S.M. and Phillips, G.O. (1987) Biochim. Biophys. Acta **925**, 194–202
19. Cleland, R.L., Wang, J.L. and Detweiler, D.M. (1982) Marcomolecules **15**, 386–395
20. Schuchmann, M.N. and von Sonntag, C. (1977) J. Chem. Soc., Perkin Trans. II, 1958–1963
21. McCord, J.M. and Fridovich, I. (1969) J. Biol. Chem. **244**, 6049–6055
22. Halliwell, B. and Gutteridge, J.M.C. (1984), Biochem. J. **219**, 1–14
23. Cohen, M.S., Britigan, B.E. and Hasset, D.J. and Rosen, G.M. (1988) Rev. Infect. Dis. 10, 1088–1096
24. Bellavite, P. (1988) Free Radical Biol. Med. **4**, 225–261
25. van Hemmen, J.J. and Meuling, W.J.A. (1977) Arch. Biochem. Biophys. **182**, 743–748
26. McCord, J.M. and Day, E.M. (1978) FEBS Lett. **86**, 139–142
27. Halliwell, B. (1978) FEBS Lett. **96**, 238–242
28. Czapski, G., Aronovitch, J. and Samieni, A. (1983) in Oxy Radicals and their Scavanger Systems, Molecular Aspects (Cohen, G. and Greenwald, R., eds.), vol. 1, pp. 11–115, Elsevier, Amsterdam
29. Halliwell, B., Gutteridge, J.M.C. and Blake, D.R. (1985) Phil. Trans. R. Soc. London. Biol. **311**, 659–671
30. Al-Asaaf, S., Phillips, G.O., Deeble D.J., Parsons, B., Starnes, H. and von Sonntag, C. (1995) Radiat. Chem. Phys. **46**(2), 207–217
31. Phillips G. O. (1992) in Viscoelasticity of Biomaterials (Glasser, W.G. and Hatakeyama, H., eds.), ACS Symp. Ser., vol. 489, pp. 168–183, American Chemical Society, Washington, D.C.
32. Blake, S., Deeble, D.J., Beaumont, P.C., Parsons, B.J. and Phillips, G.O. (1989) in Free Radicals, Metal Ions and Biopolymers (Beaumont, P.C., Deeble, D.J., Parsons, B.J. and Rice-Evans, C., eds.), pp. 159–183, Richeleau Press, London
33. Moseley, R., Waddington, R., Evans, P., Halliwell, B. and Embery, G. (1995) Biochem. Biophys. Acta, **1244**, 245–252
34. Uchiyama, H., Dobashi, Y., Ohkouchi, K. and Nagasawa, K. (1990) J. Biol.Chem. **265**, 7753–7759
35. Wong, S.F., Halliwell, B., Richmond, R. and Skowroneck, W.R. (1981) J. Inorg. Biochem. **14**, 127–134
36. von Sonntag, C., Dizdarogly, M. and Schultze-Frohlinde, D. (1976) Z. Naturforsch. **31**, 857–864
37. Lal, M. (1985) J. Radioannl. Nucl. Chem. Articles, **92**, 105–112
38. Fujita, S. and Steenken, S. (1981) J. Am. Chem. Soc. **103**, 2540–2545
39. Rabani, J. Mulac, W.A. and Matheson, M.S. (1965) J. Phys. Chem. **69**, 53–70
40. von Sonntag, C. (1980) The Chemical Basis of Radiation Biology, Taylor and Francis, London
41. Myint, P., Deeble, D.J. and Phillips, G.O. (1989) IAEA Pro. Advisory Group Meeting, IAEA-TEC Doc-527, p.105
42. Steenken, S., Davies, M.J. and Gilbert, B.C. (1986) J. Chem. Soc., Perkin Trans. II, 1003–1010
43. Schuchmann, M. N. and von Sonntag, C. (1978) Int. J. Radiat. Biol. **34**, 397–400
44. von Sonntag, C. (1980) Adv. Carbohydr. Chem. Biochem. **37**, 7–77

45. Balazs, E.A. and Leschiner, E.A. (1989) in Cellulosic Utilisation (Inagaki, H. and Phillips, G.O., eds.), pp. 233–241, Elsevier Applied Science, London
46. Gunning, A.P., Kirby, A.R., Morris, V.J., Wells, B. and Brooker, B.E. (1995) Polym. Bull. **34**, 615–619
47. Gunning, A.P., Morris, V.J., Al-Asaaf, S. and Phillips, G.O. (1996) Carbohydr. Polym. **30**, 1–8
48. Al-Asaaf, S., Phillips, G. O., Meadows, J., and Williams, P., (1996) Biorheology **33**, 319–332
49. Takigami, S., Takigami, M. and Phillips, G.O. (1993) Carbohydr. Polym. **22**, 153–160
50. Takigami, S., Takigami, M. and Phillips, G.O. (1995) Carbohydr. Polym. **26**, 11–18
51. Yoshida, H., Hatakayama, T. and Hatakayama, H. (1990) Polymer **31**, 693–698

12

Hyaluronan-binding matrix proteins

Dick Heinegård*‡, Sven Björnsson†, Matthias Mörgelin* and Yngve Sommarin*
*Department of Cell and Molecular Biology Section for Connective Tissue Biology, Lund University, P.O. Box 94, S-22100 Lund Sweden and †Department of Clinical Chemistry, Lund University, Sweden

Hyaluronan is found in all connective tissues. It is an extremely large polymer made up of disaccharide repeats of N-acetylglucosamine and glucuronic acid. The number of such repeats in a given polymer is up to several thousand, resulting in chains of many millions in molecular weight. Hyaluronan is synthesized at the cell surface without a protein core precursor and is then released into the extracellular matrix. Some of the hyaluronan appears to remain attached to the cell surface, where it participates in interactions with surrounding matrix molecules. There is also hyaluronan bound to the cell surface via specific receptors. Mechanisms for the release of the newly synthesized hyaluronan to the matrix are not known.

Although hyaluronan has in itself no protein backbone, it participates in interactions with numerous other matrix molecules. These interactions modify the structure of hyaluronan and of course the properties of the molecule. Another putative feature of such complexes is that the properties and arrangement of the interacting molecules in the matrix are governed by the association. Furthermore, this binding to hyaluronan may represent an important feature in retaining the molecules in the proper localization in a given matrix compartment. However, to date there are only a few cases where we understand at least some of the consequences of these interactions. Below, an account is given of representative interactions that have been clearly established between hyaluronan and molecules residing in the extracellular matrix.

The first described specific interactions between hyaluronan and matrix molecules were those between the large proteoglycan aggrecan, the link protein and hyaluronan. These will be described more extensively as a model for hyaluronan interactions in the matrix.

Aggrecan, link protein–hyaluronan interactions

Aggrecan has a core protein with several functional domains. The most N-terminal one, schematically illustrated in Figure 1, contains two PTR-loops and an Ig-fold structure (see Chapter 15 in this volume by Day and Parkar) and is referred to as the G1- or hyaluronan-binding domain. It has the capacity to specifically interact with a decasaccharide sequence of hyaluronan, thus linking many aggrecan molecules to one hyaluronan molecule in a single aggregate, shown in

‡To whom correspondence should be addressed.

Figure 1

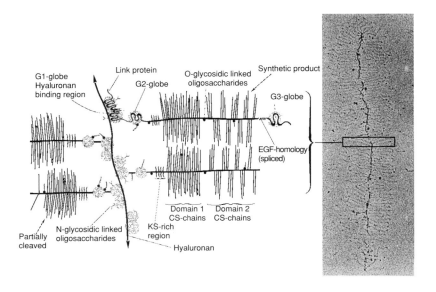

G1-globe
Hyaluronan
binding region

Link protein

G2-globe

O-glycosidic linked
oligosaccharides

Synthetic product

G3-globe

EGF-homology
(spliced)

Domain 1
CS-chains

Domain 2
CS-chains

KS-rich
region

N-glycosidic linked
oligosaccharides

Partially
cleaved

Hyaluronan

Schematic illustration of a proteoglycan aggregate with aggrecan

For comparison, an electron-microscopic picture of a rotatory-shadowed aggregate is shown where each aggrecan appears as a side chain filament, where the globular domains can be observed. The glycosaminoglycan chains cannot be seen in this picture. The central filament represents hyaluronan with the G1-domain of aggrecan and link protein. CS, chrondroitin sulphate; KS, keratin sulphate; EGF, epidermal growth factor.

Figure 1. In addition, this domain can specifically interact with a small protein, the link protein, which has homologous structure and actually also binds with the same specificity to hyaluronan (for refs. see [1]). Hereby a ternary complex involving two hyaluronan decasaccharide sequences, the link protein and the G1-domain of aggrecan, is formed. While the binding constant to hyaluronan of each individual molecule is about $K_D = 10^{-8}$, the ternary complex has a binding several orders of magnitude stronger (for refs, see [2]). The remainder of the aggrecan molecule contains, in sequence from the binding site, the interglobular extended domain, the G2 globular domain with homology to the G1 structure, an extended filament followed by a keratan-sulphate-rich domain and the major chondroitin sulphate carrying region, forming a domain which is extremely highly, negatively charged. At the C-terminal end we find the G3 globule with domains of lectin homology, epidermal growth factor repeats and a complement regulatory protein homology (for refs. see [3]).

Aggregation appears to occur extracellularly as shown using rat chondrosarcoma cell cultures [4]. Subsequently, Kimura and co-workers [5] provided evidence that the link protein shows a molar ratio to proteoglycan of 1 in the stable aggregates. In addition, those aggrecan molecules not complexed in aggregates already appear to contain a link protein molecule. The data were taken to indicate that the link protein could bind to the monomer before aggregation and it has even been suggested that the aggrecan–link complex is formed inside the

Figure 2

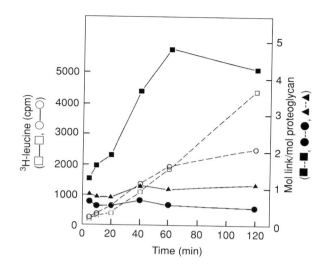

Link protein content in monomeric and aggregate aggrecan with time

Early kinetics of appearance of aggrecan, aggrecan–link-protein as well as aggregate with excess link protein (LP) bound to hyaluronan in medium from chondrocytes cultured in suspension. (□) Total LP in aggregates, (○) total LP in monomeric aggrecan, (■) molar ratio LP to aggrecan in aggregate, (●) molar ratio LP in monomeric aggrecan and (▲) molar ratio LP to aggrecan in total proteoglycan.

cell [6]. Using rapid zonal rate centrifugation to separate aggregates, monomeric aggrecan and free link-protein in medium from chondrocytes grown in suspension, it has been possible to show that indeed aggregation is an extracellular event. At short times, the presence of aggregate in the medium is low, in the order of 10% over the first one hour. After 24 h of incubation, in contrast, 82% of the aggrecan molecules are present in aggregates [7]. Interestingly a closer study of the presence of link protein in those early aggregates versus monomeric aggrecan molecules (Figure 2) showed that the molar ratio of link protein to aggrecan in the monomer was <1 over the first two hours of incubation, ranging from 0.7 to 0.5. Over this time period, the percentage of monomeric aggrecan barely changed, except for the longest time point of two hours, when it increased slightly. Interestingly, the relative content of link protein in aggregates, presumably rich in hyaluronan, increased from a molar ratio just above 1 to one of almost 5 (link protein over aggrecan in aggregates).

In summary it appears that aggregation is an extracellular process and that aggregation can occur either via link protein–aggrecan complex binding to hyaluronan or an aggrecan molecule binding to a link protein–hyaluronan complex. Indeed the data possibly indicate that both pathways occur within the same order of magnitude.

What is the role of the link protein in the formation of aggregates? An early observation was that the aggrecan when first secreted from the cell did not bind tightly to hyaluronan [8]. However, more efficient binding could occur with a large excess of hyaluronan, indicating that binding is occurring, albeit weakly. Pretreatment of the monomer with mild base [9] or mechanical loading of the tissue also allowed the aggrecan molecules to form much tighter complexes with hyaluronan at short times [10]. In addition, we found a higher degree of sensitivity to proteinase digestion of newly formed aggregates in culture (S. Björnsson and D. Heinegård, unpublished work). Interestingly, in a set of experiments aimed at dissociating newly formed aggregates by competition with hyaluronic acid oligosaccharides (10–12 saccharides), hyaluronan-binding region of aggrecan and high-molecular-weight hyaluronan, it was possible to show that newly secreted link protein formed fully stable complexes with hyaluronan and exogenous aggrecan. However, complexes involving newly secreted aggrecan could be dissociated by high-molecular-weight hyaluronan, but not by the hyaluronan binding region or by oligosaccharide approaching the size of the binding domain [7]. We therefore hypothesize that the newly secreted aggrecan contains a binding region that is not completely stable, while the link protein has, upon secretion, already reached its final stable conformation. It is possible that there are disulphide bridges within the hyaluronan binding region of aggrecan that are only formed after its interaction with hyaluronan [11]. A functional role of such modified aggregation could be that the aggrecan molecules do not become irreversibly fixed to one site along the hyaluronan molecule, but could be shuffled around. This hypothesis, however, needs to be further studied.

Binding of aggrecan to hyaluronan at the chondrocyte cell surface

It appears that in chondrocyte suspension culture a portion of the aggrecan molecules are secreted into the medium and then bound back at the cell surface [12]. In experiments to elucidate this binding it was shown that the aggrecan binds to hyaluronan attached at the cell surface. The binding rapidly becomes stabilized such that only a small proportion of the bound aggrecan (less than one third) can be removed by competition with aggrecan, hyaluronan binding region, hyaluronan or hyaluronan oligosaccharides (10 or larger) [12]. Specificity for competition at early times indicated binding via a decasaccharide sequence of hyaluronan. Thus, aggrecan appears to be bound at the cell surface via a link-stabilized interaction to hyaluronan. The complex could be removed by hyaluronidase digestion [12]. Since neither added excess oligosaccharides nor added hyaluronan removed the bound aggrecan, it appears that the binding of the central hyaluronan in this aggregate is not mediated via cell surface receptors such as CD44 or RHAMM (receptor for hyaluronan-mediated activity). It is possible, then, that the major proportion of the hyaluronan is actually bound at the cell surface via the synthetase, as proposed by Prehm [13]. Such an organization at the cell surface, where hyaluronan assembles aggrecan molecules into an aggregate, may represent a precursor pool of extracellular matrix aggregates, at least in the territorial compartment. Furthermore, the organization of aggrecan at the cell surface in an aggregate will create an environment that organizes this matrix and

Figure 3

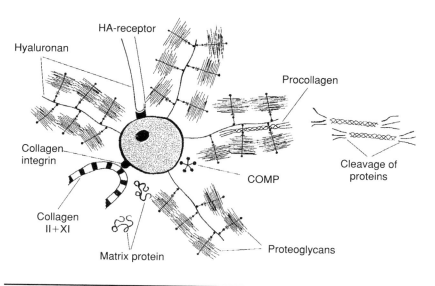

Schematic illustration of a chondrocyte and its pericellular environment

The picture shows a chondrocyte with attached hyaluronan–aggrecan and several other macromolecules. The aggregate provides a matrix that has the potential for regulating diffusion of newly secreted molecules, exemplified by procollagen, such that the assembly of collagen fibres may be directed and potentially delayed. HA, hyaluronan.

regulates the secretion and perhaps the assembly of extracellular matrix molecules, as outlined in Figure 3.

Other features of binding of aggrecan and the link protein to hyaluronan

As is seen in Figure 1, the central filament of the aggrecan, where the G1 hyaluronan-binding region of aggrecan and the link protein are bound to hyaluronan, has a rather compact appearance with no space between the globular domains. Using hyaluronan of defined molecular masses, it could be shown [14] that the length of the hyaluronan filament decreased upon binding of either aggrecan or link protein in a similar manner, while binding of both did not further decrease the length of the filament. Thus, it is clear that the hyaluronan changes its conformation when either ligand is bound, such that it becomes shorter. It is likely that the polysaccharide acquires a more coiled conformation, and some pictures using rotary shadowing can actually be taken to indicate that this occurs [14]. Future studies of the conformation of the binding site, as further detailed in the presentation on another hyaluronan-binding molecule, TSG6 (tumour necrosis factor-stimulated gene-6), in Chapter 15 in this volume by Day and Parkar, will further elaborate the structural requirements for binding.

Interestingly, studies of subsaturation binding of aggrecan–link-protein complexes to hyaluronan indicated that there was no cooperativity between the individual complexes, such that there was no clustering along the hyaluronan filament (for refs, see [1]). It has, however, since then been shown [15] that aggregates can be extracted without dissociation from rat chondrosarcoma tissue, and that in a subpopulation of these aggregates there was clustering of aggrecan–link-protein complexes along the hyaluronan filament, interrupted by regions with no apparent bound ligands to the hyaluronan. At the same time other fractions contained a continuous complex of aggrecan–link protein bound along the hyaluronan molecule. This could point to a cooperativity at the tissue level, or alternatively it may be that some of these complexes had been removed by, for example, proteolytic catabolism. More work on this topic is of fundamental importance to decide what is governing aggregation in the tissue.

Other large hyaluronan-binding proteoglycans

There are three additional well-defined, and one less well-defined, large proteoglycans with hyaluronan-binding capabilities and a C-terminal lectin homology domain. They all have very similar hyaluronan-binding domains and similar lectin domains, while the parts containing the glycosaminoglycan chains vary considerably, such that they usually contain fewer side chains than aggrecan. The proteoglycans are listed in Table 1.

Table 1

Molecule	Features, procedure	Specificity	Tissue
Proteoglycans	PTR repeats, Ig-fold	HA-10	
Aggrecan			Cartilage
Versican			Ubiquitous
Fibroblast PG			Fibrous tissue
Brevican			Brain
Neurocan			Brain
GHAP	Probably versican MMP cleavage?		Brain
Hyaluronectin			
TS-6	PTR-repeat		TNF-α stimulation, suggested interaction with Iαl
BEHAB	PTR-repeats, Ig-fold		Brain, glia development
SHAP	Iαl	Covalent	Synovial fluid
P-32	2 × 34 kDa		Kidney
HABP-102	Not known		Chondrosarcoma
Collagen VI	N-terminal globule		Ubiquitous
Fibronectin	N-terminal part		Ubiquitous
C1q	Isolation of complex		
Hemopexin	Intermediate gel electrophoresis		

Hyaluronan-binding proteins

Iαl, inter-α-trypsin inhibitor; TNF-α, tumour necrosis factor-α; BEHAB, brain-enriched hyaluronan-binding protein.

Versican occurs in several splice forms of different length. Versican is present in most tissues. Brevican and neurocan have a much more limited distribution and are primarily found in the brain. These proteoglycans have, at least in vitro, a capacity to bind to hyaluronan similarly to aggrecan. In the case of versican, it has also been shown that it can form link-stabilized aggregates *in vitro* [16].

Interestingly, there is a fifth member of this family, which has not been clearly defined as yet. This is represented by a large proteoglycan isolated from sclera and tendon. It shares many features with aggrecan, e.g. G1, G2 and G3 domains, although it appears to have fewer glycosaminoglycan side chains. In addition, this proteoglycan can form stable aggregates with link protein and hyaluronan [16].

The role of these different proteoglycans in the matrix is not clear. They may, however, serve roles as cross-linking agents. Thus, versican has been shown to bind to Tenascin R via its C-terminal lectin domain, although this is not a carbohydrate-dependent interaction [17]. This interaction may serve as a model for the other proteoglycans, where they have one functional N-terminal domain binding to hyaluronan and another in the C-terminus binding to other matrix constituents. The domain carrying all the negatively charged chondroitin sulphate chains will of course provide a region with extremely high, fixed charge density, having important consequences for tissue swelling and water retention.

Cell surface proteins binding hyaluronan

Much information on hyaluronan binding has been obtained through studies of two cell surface proteins with capacity for binding hyaluronan. These are CD44, existing in a number of different splice forms, and RHAMM. The extensive work performed on these molecules will be discussed in separate chapters and is therefore not covered here. It is worth noting, however, that the binding specificity is for a shorter stretch of hyaluronan, i.e. a hexasaccharide sequence.

Other hyaluronan-binding proteins in the matrix

Extracellular matrices contain a number of hyaluronan-binding proteins that are much less well-characterized. The exception is TSG6, which is a protein that has been found to contain the PTR-repeat characteristic of the aggrecan family of proteins. It is upregulated by tumour necrosis factor α, which may indicate a role in inflammation. Furthermore, an interaction with IαI (inter-α-trypsin inhibitor) has been suggested. Detailed studies of the structures of this protein have been conducted and indicate details of the hyaluronan interaction motif. These studies are presented in Chapter 15 of this volume.

Another molecule that contain the PTR-repeat is brain-enriched hyaluronan-binding protein [18,19], which has been identified in brain, particularly during glia development. It is only expressed in the nervous system. The sequence of the protein shares extensive homologies with the aggrecan family of hyaluronan-binding proteins. The cDNA encodes a 371 amino acid-long protein, including the signal peptide. The protein does not contain a membrane-spanning region and therefore likely has a role in the extracellular matrix. Interestingly, it is up-regulated with the development of the more mature extracellular matrix in the central nervous system and therefore may have a role in organizing the hyaluronan during this developmental process.

GHAP (glial hyaluronate binding protein) has been isolated from brain. Currently available data indicate that the protein represents the N-terminal globular domain of versican. It is possible that it has been liberated from the rest of the molecule by metalloproteinase cleavage [20].

Hyaluronectin was originally isolated from brain by Delpech and collaborators [21]. Its relationship to other binding proteins is not known, although it appears to be similar to GHAP.

SHAP (serum derived hyaluronate associated protein) has been found in several locations, e.g. synovial fluid. It represents the heavy chains of the IαI and can bind to hyaluronan. It has actually been shown that one linkage is covalent in nature [22]. This linkage appears to be via an esterification of the carboxyl group of the C-terminal aspartic acid with the C-6 hydroxy group of the N-acetylhexosamine of hyaluronan.

Although no functional role of the formation of the complex has been established, its formation occurs concomitantly with the release of bikunin, which contains the trypsin inhibitor activity of IαI. Whether this has implications for proteolytic inhibition to occur is not clear.

P-32 is a hyaluronan-binding protein that has been isolated from kidney and fibroblasts [23]. Although the specificity of the binding of the protein to hyaluronan has not been demonstrated, its sequence has been determined and found to represent a factor co-purifying with the pre-mRNA splicing factor SF2. The protein contains a sequence suggested to represent a hyaluronan-binding domain, that is two basic amino acids flanking a seven amino acids stretch [24]. The data suggest that the protein is phosphorylated at several casein kinase 2 sites. The binding strength of the protein to hyaluronan, as well as to several matrix proteins, e.g. laminin, fibronectin and collagen IV, is in the range of 10^{-9} M. Immunolocalization identified the protein at the cell surface. A role for the protein in regulating the splicing activity of SF2 has been suggested.

HABP102 has been isolated from rat chondrosarcoma [25]. The protein appears distinct from other hyaluronan-binding proteins presented since it reacted on Western blots neither with antibodies against link protein, aggrecan, hyaluronan-binding region nor hyaluronectin. It has not been shown with what specificity the protein binds to hyaluronan.

A number of previously well-characterized proteins have also been shown to bind to hyaluronan. These include collagen VI, fibronectin, C1q and hemopexin.

Collagen VI has been shown to bind to hyaluronan [26,27]. It appears that the binding has relevance for the formation of collagen VI microfibres and could facilitate polymerization of the collagen *in vitro*. Binding appeared rather specific for hyaluronan, although keratan sulphate showed some, minor ability to inhibit hyaluronan binding. The binding to hyaluronan appears to reside in the collagen VI N-terminal globular domain, since recombinant domains N2–N9 bind tightly to hyaluronan. Thus, hyaluronan appears to have an important role in organizing the extracellular matrix, not only by binding to the large proteoglycan but also to other major structural components of the matrix.

Studies of hyaluronan binding to fibronectin suggest an interaction with the N-terminal part, although the binding is apparently rather weak. It may be that future studies of fibronectin fibrillar assemblies will be better suited for examining these interactions, since the long hyaluronan filament would extend over many fibronectin units, thereby allowing polyvalent interactions.

C1q appears to have a capacity for interaction with hyaluronan since a complex of the two could be isolated from synovial fluid of patients with rheumatoid arthritis [28].

Electrophoretic technology has been used to indicate that hemopexin may bind to hyaluronan [29]. No details of specificity have been provided.

In conclusion, a number of components, primarily found in the extracellular environment, have been shown to bind to hyaluronan. In relatively few cases specificities have been determined and show either a specificity for a decasaccharide sequence of hyaluronan in the case of extracellular matrix components or specificity for hexasaccharide with cell surface-binding proteins. There are also a number of proteins where specificity has not been determined. Often these proteins show limited similarity to those where interactions have been well characterized.

From the multitude of hyaluronan-interacting molecules with rather different properties, it appears attractive to conclude that the polysaccharide may have very important roles in defining interactions and in helping organizing the macromolecular assemblies of extracellular matrix. In addition, hyaluronan appears to have roles in establishing contacts between cells and surrounding matrix by, for example, binding hyaluronan either to cell surface receptors or having hyaluronan bound to, for example, the synthetase at the cell surface.

References

1. Mörgelin, M., Heinegård, D., Engel, J. and Paulsson, M. (1994) Biophys. Chem. 50, 113–128
2. Heinegård, D. and Oldberg, Å. (1989) FASEB J. 3, 2042–2051
3. Doege, K.J., Sasaki, M., Kimura, T. and Yamada, Y. (1991) J. Biol. Chem. 266, 894–902
4. Kimura, J.H., Hardingham, T.E., Hascall, V.C. and Solursh, M. (1979) J. Biol. Chem. 254, 2600–2609
5. Kimura, J.H., Hardingham, T.E. and Hascall, V.C. (1980) J. Biol. Chem. 255, 7134–7143
6. Ratcliffe, A., Hughes, C., Fryer, P.R., Saed-Nejad, F. and Hardingham, T. (1987) Collagen Relat. Res. 7, 409–421
7. Björnsson, S. and Heinegård, D. (1981) Biochem J. 199, 17–29
8. Oegema, Jr., T.R. (1980) Nature 288, 583–585
9. Sandy, J.D. and Plaas, A.H.K. (1989) Arch. Biochem. Biophys. 271, 300–314
10. Sah, R. L-Y., Grodzinsky, A.J., Plaas, A.H.K. and Sandy, J.D. (1990) Biochem. J. 267, 803–808
11. Bayliss, M.T., Ridgway, G.D. and Ali, S.Y. (1984) Biosci. Rep. 4, 827–833
12. Sommarin, Y. and Heinegård, D. (1983) Biochem. J. 214, 777–784
13. Prehm, P. (1984) Biochem J. 220, 597–600
14. Mörgelin, M., Paulsson, M., Heinegård, D., Aebi, U. and Engel, J. (1995) Biochem. J. 307, 595–601
15. Mörgelin, M., Engel, J., Heinegård, D. and Paulsson, M. (1992) J. Biol. Chem. 267, 14275–14284
16. Mörgelin, M., Paulsson, M., Malmström, A. and Heinegård, D. (1989) J. Biol. Chem. 264, 12080–12090

17. Aspberg, A., Binkert, C. and Ruoslahti, E. (1995) Proc. Natl. Acad. Sci. U.S.A. **92**, 10590–10594
18. Jaworski, D.M., Kelly, G.M. and Hockfield, S. (1995) J. Neurosci. **15**(2), 1352–1362
19. Jaworski, D.M., Kelly, G.M. and Hockfield, S. (1994) J. Cell Biol. **125**, 495–509
20. Perides, G., Biviano, F. and Bignami, A. (1991) Biochim. Biophys. Acta **1075**, 248–258
21. Delpech, B. (1982) J. Neurochem. **38:4**, 978–984
22. Zhao, M., Yoneda, M., Ohashi, Y., Kurono, S., Iwata, H., Ohnuki, Y. and Kimata, K. (1995) J. Biol. Chem. **270**, 26657–26663
23. Gupta, S., Batchud, R.B. and Datta, K. (1991) Eur. J. Cell Biol. **56**, 58–67
24. Deb, T.B. and Datta K. (1996) J. Biol. Chem. **271**, 2206–2212
25. Crossman, M.V. and Mason, R.M. (1990) Biochem. J. **266**, 399–406
26. Kielty, C.M., Whittaker, S.P., Grant, M.E. and Shuttleworth, C.A. (1992) J. Cell Biol. **118**, 979–990
27. Specks, U., Mayer, U., Nischt, R., Spissinger, T., Mann, K., Timpl, R., Engel, J. and Chu, M-L. (1992) EMBO J. **11**, 4281–4290
28. Prehm, P. (1995) Ann. Rheum. Dis. **54**, 408–412
29. Hrkal, Z., Kuzelova, K., Muller-Eberhard, U. and Stern, R. (1996) FEBS Lett. **383**, 72–74

13

Hyaluronan binding function of CD44

Jayne Lesley

Cancer Biology Laboratory, The Salk Institute, P.O. Box 85800, San Diego, CA 92186-5800, U.S.A.

Introduction

CD44 is a major cell surface receptor for hyaluronan (HA). It is expressed on a variety of cell types, including most haematopoietic cells, some neuronal cells, endothelial cells, keratinocytes and chondrocytes. The contribution of CD44 to cell function may differ in different cell types. In haematopoietic cells it is thought to be involved in cell circulation patterns, in neuronal cells in directed migration and in chondrocytes and keratinocytes in assembly of the extracellular matrix and perhaps in degradation of HA (for reviews, see [1–4]).

There is now a large body of literature associating changes in CD44 expression with particular malignancies. These changes include up-regulation or down-regulation of expression of specific CD44 isoforms and changes in the pattern of isoforms expressed. It is not clear how these changes contribute to malignancy of particular tumour cell types. Changes in cell adhesion and migration and/or changes in the assembly and degradation of the extracellular matrix may all be involved.

Our laboratory's interest in CD44 arose from observations that CD44 expression was developmentally regulated on thymocytes. CD44 is expressed on the earliest progenitors that migrate to the thymus from the bone marrow, then lost on early thymic precursors and re-expressed on some mature thymocytes and peripheral T-cells. It is dramatically up-regulated on activated lymphocytes and remains stably elevated on memory T-cells. A similar pattern of changes in expression is found in B-cell and myeloid lineages and this pattern is thought to relate to changes in the circulatory properties of leucocytes during their development (see [3]).

In addition to developmentally regulated changes in CD44 expression, changes in CD44 function play an important role in its contribution to leucocyte circulation. Like other cell adhesion molecule families that are involved in leucocyte circulation, such as some of the integrins and selectins, ligand binding function is regulated in a cell-specific manner. All of these adhesion molecules contribute to cell adhesion and migration, all are expressed on a broad population of haematopoietic cells and all recognize ligands that are broadly distributed. The ability to participate in specific cell–cell and cell–substrate interactions in this receptor- and ligand-rich milieu depends in part on the cell's ability to regulate the ligand-binding function of its adhesion receptors.

The contribution of CD44 to metastasis may well have to do with tumour cells overriding the normal regulatory mechanisms that limit the interaction of CD44 with its ligand HA. Changes in levels of expression, in the isoforms expressed and in post-translational modifications of CD44 observed in tumour cells may result in subversion of the normal regulation.

Description of the molecule

To understand how CD44 binds HA and modulates this binding activity we must appreciate some of the features of the molecule (see [3]). CD44 is a type I transmembrane cell surface glycoprotein, consisting of a single polypeptide chain. The most prevalent isoform, termed the 'haematopoietic' or 'standard' isoform, includes none of the so-called 'variant' exons, which give rise to higher M_r isoforms by alternate splicing. The 341 amino acid polypeptide (for the mature, haematopoietic isoform, referred to as CD44H [5]) has an external domain of 248 amino acids that contains numerous sites for possible post-translational addition of carbohydrate side chains (see Figure 1). Other possible post-translational modifications of CD44 include phosphorylation of serines in the cytoplasmic domain and palmitoylation of a cysteine in the transmembrane domain.

The 21 amino acid transmembrane and 72 amino acid cytoplasmic domains are very highly conserved among mammalian species (~90% homology). The amino terminal ~170 amino acids in the external domain are also highly conserved (~80% homology among mammalian species) and include a sequence of about 100 amino acids (indicated in solid black in Figure 1) with sequence similarity to other hyaluronan binding proteins such as aggrecan, link protein and TSG6. (The structure of this sequence, termed the 'link module', has recently been determined for TSG6 and shown to resemble the C-type lectin domain ([6], and see Chapter 15 by A. Day and A.A. Parkar). Six cysteines, believed to form three disulphide bonds, and five potential N-glycosylation sites (designated N1–N5 in Figure 1) are conserved in the amino-terminal half of the external domain. Two clusters of basic aa (expanded in Figure 1) have been implicated in HA binding [7]. The more N-terminal cluster, located within the 'link module', contains a critical arginine (Arg-41) residue. Mutation of this residue to alanine abolishes HA binding altogether. The second cluster of basic amino acids is outside the link module. Here mutation of multiple basic residues results in a reduction in HA binding [7].

The membrane proximal portion of the extracellular domain (stippled in Figure 1) is relatively unconserved among mammalian species. Within this region (indicated by the arrow in Figure 1) is the site of potential insertion of additional amino acid sequences determined by alternative splicing of various combinations of eleven 'variant exons' located in the middle of the 21 exon CD44 gene [8–10]. Use of multiple variant exons can create isoforms with external domains more than double the size of the external domain of the haematopoietic isoform and can contribute additional sites for potential carbohydrate modification including N-linked and O-linked glycosylation sites and sites for glycosamino glycan addition. Higher M_r variant isoforms nevertheless contain the conserved N-terminal part of

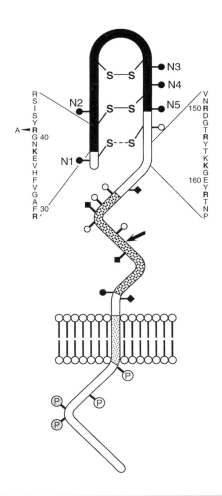

Figure 1

Features of the haematopoietic isoform of CD44

Numbering of the amino acids (shown in single-letter code) and position of the features is based on the sequence for human CD44H (including a 20 amino acid leader sequence) as described in [5]. Potential N-glycosylation sites are indicated by solid circles and numbered from the amino-terminus. Potential O-glycosylation sites are indicated by ○ ; potential sites for chondroitin sulphate attachment by closed diamonds; potential sites for phosphorylation of serines in the cytoplasmic domain by a circled P. The two shaded phosphorylation sites have been shown to be required for constitutive phosphorylation in several cell lines [31]. Solid black area indicates 'link module' sequences. Stippled area indicates the region of minimal sequence similarity among mammalian species. The arrow indicates the site of insertion of additional sequences determined by variant exon splicing.

the external domain, including the link module and basic amino acid clusters implicated in HA binding, and the transmembrane and cytoplasmic domains present in the haematopoietic isoform.

HA binding by CD44 is regulated

The HA binding function of CD44 is illustrated in Figure 2. The murine T-lymphoma AKR1 does not express CD44 or bind fluorescein-conjugated HA (Fl-HA). After transfection with cDNA of the murine haematopoietic isoform of CD44, the cells express CD44 on the cell surface and bind Fl-HA. The HA binding is blocked by CD44-specific monoclonal antibody (Figure 2) and by unlabelled HA, but not by other glycosaminoglycans (including chondroitin sulphate-4 and -6, heparin, heparan sulphate, keratan sulphate and dermatan sulphate) [11,12]. However, certain specific chondroitin-sulphate-containing proteoglycans do bind CD44 and compete with HA binding [11].

The main point I wish to make about the HA binding function of CD44 is that it is regulated by the cells in which CD44 is expressed. Although we can demonstrate that CD44 can mediate HA binding as shown in Figure 2, many cell lines, tumour cells and normal cells that express high levels of CD44 do not bind HA. For example, murine myeloid progenitors in the bone marrow express very high levels of CD44 but do not bind HA in flow cytometry assays using Fl-HA or in adhesion assays to immobilized HA [13]. Resting lymphocytes of human or mouse also do not bind HA.

Some cells can be rapidly induced to bind HA by certain CD44-specific mAbs. For example, some T-cell lymphomas and a subset of murine peripheral

Figure 2

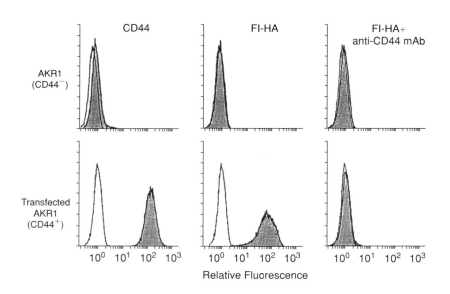

Flow cytometry histograms of CD44 expression and Fl-HA binding of AKR1 cells untransfected and transfected with wild-type murine CD44H

CD44 expression is determined by staining with fluorescein-conjugated mAb IM7. Lines without shading on each panel show the fluorescence of unstained cells, which is normalized to 1.0 mean relative fluorescence. In column 3, binding of Fl-HA is inhibited by CD44-specific mAb KM 201 [14].

T-cells that do not constitutively bind HA are induced to bind by the CD44-specific mAb IRAWB14 [13,14]. To understand how rapid induction of HA binding might occur we have studied the properties of this antibody mediated induction [15]. (i) It affects pre-existing CD44 molecules on the cell surface. (ii) Induction occurs immediately upon binding of the antibody. (iii) Intracellular signalling is not required. Antibody-mediated activation can affect CD44 in metabolically inert cells that have been preserved by fixation [15] and purified CD44-Rg fusion molecules in a cell-free system [16]. (iv) Induction requires multivalent antibody. Monovalent Fab fragments of inducing antibody are not able to induce [15].

This latter observation suggests that induction may require cross-linking or clustering of CD44 molecules into a multivalent configuration. In addition to clustering, a conformational change in the external domain of CD44 may be involved [16].

We have used flow cytometry to evaluate binding of Fl-HA in the presence and absence of inducing mAb IRAWB14 by an assortment of CD44$^+$ cell lines and normal haematopoietic cells. We have defined three activation states of CD44 as shown in Table 1. Some CD44$^+$ cell types and cell lines do not bind HA in the presence or absence of inducing mAb and are referred to as being in an 'inactive' state. Change from this state may require cell activation and/or differentiation. Cells with 'inducible' CD44 bind HA only in the presence of inducing mAb like IRAWB14. This state might allow for rapid and reversible induction of HA binding upon appropriate stimulation *in vivo*. However, we have to date been unable to identify physiological factors that can rapidly activate HA binding in inducible cells similar to the action of inducing mAb. 'Active' CD44 binds HA constitutively without a requirement for inducing mAb. Cell activation by the appropriate *in vitro* or *in vivo* stimulation can result in the development of active HA binding subpopulations in some normal cell populations. For example, strong allogeneic stimuli *in vivo* can induce CD44-mediated HA binding in subsets of murine B and T-cells [17,18]. Unlike mAb-mediated induction, however, this activation requires hours to days and may involve *de novo* biosynthesis of new CD44 molecules. Cells that have CD44 in an active state usually show enhanced HA binding in the presence of IRAWB14.

HA-binding phenotype	Regulation	Example	Table 1
Inactive	Negative?	RAW253 (pre-B line)	
		L-cells (fibroblast line)	
		Normal resting B-cells	
Inducible	mAb and (?) Physiological factors	SAKRTLS 12 and EL4 (T- lymphomas)	
		Some normal T-cells	
Active	Positive?	BW5147 (T-ymphoma)	
		S194 (Myeloma)	
		Activated B- and T-cells	

Three activation states of the CD44/HA receptor

Table 2   • Association with other molecules on the cell surface
• Interaction with the cytoskeleton
• Inside-out signalling mediated through the cytoplasmic domain
• Post-translational modifications, e.g. glycosylation, phosphorylation
• Variation in isoform expression
• Changes in cell surface distribution

Proposed mechanisms of regulating CD44 HA-binding function

Mechanisms of regulating CD44 function

The main focus of our laboratory's efforts in recent years has been to understand what determines the difference in HA-binding function of CD44 in different cellular environments. We know that the same cDNA construct for the haematopoietic isoform of murine CD44, which confers active HA binding when transfected into the T-lymphoma AKR1, is inactive when transfected into the RAW253 pre-B-cell line or L-cell fibroblasts, and is inducible when transfected into the T-lymphomas Jurkat or SAKRTLS 12 ([3,19] and our unpublished work). Therefore the cellular environment in which CD44 is expressed determines the activation state. Table 2 lists some of the possible mechanisms of regulation of CD44 function that we have considered.

These mechanisms are not mutually exclusive and may overlap or interact. For example, changes in phosphorylation of the cytoplasmic domain of CD44 may control inside-out signalling or interaction of CD44 with the cytoskeleton. Cytoskeletal associations may influence cell surface distribution. Differences in glycosylation of the external domain of CD44 may affect its association with other molecules on the cell surface or its ability to adopt a particular cell surface distribution. Changes in CD44 isoform expression may confer new ligand binding properties and new glycosylation patterns and thus new regulatory opportunities. We have repeatedly attempted to isolate CD44-associated cell surface molecules, by co-precipitation under different conditions of detergent extraction and by chemical cross-linking, without success to date.

Are the differences in HA binding due to differences in the HA binding properties of the CD44 molecules themselves, or owing to other factors in the cellular environment? When comparing CD44 molecules that exhibit differences in HA-binding function, it is preferable that the cellular backgrounds in which they are expressed are as similar as possible. Thus, we have derived two sets of cell lines that express CD44 in each of the three activation states on a common genetic background [20]. As shown in Figure 3, parental cell lines RAW253 (a pre-B line) and L-cells (a fibroblast line) are in the inactive state. We used fluorescence-activated cell sorting with Fl-HA in the presence of inducing mAb IRAWB14 to enrich for rare mutant cells that bound HA in the presence of inducing mAb. After several rounds of sorting we were able to clone stable cell lines that were in the inducible state. Reconstruction experiments have shown that this method of selection is able to isolate rare variant cells that are present in the initial population at a frequency of about 10^{-5} to 10^{-6}, or the frequency of a single mutation event [21]. From the inducible mutant cell lines, a further round of repeated fluorescence-activated cell sorting with Fl-HA alone allowed the cloning of cells

Figure 3

Derivation of cell lines in three activation states

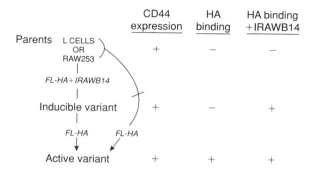

Scheme for selection of mutant cell lines representing different HA-binding functions of CD44 [20]

---

in the 'active' state, which could bind HA in the absence of inducing mAb. We could not derive 'active' mutant cell lines directly from the inactive parents by sorting with Fl-HA only. Since the 'inducible' variant had to be derived as an intermediate to the 'active' state, we believe that two mutation events separate inactive parent cells and the active variant, which would have been present in the starting parental population at too low a frequency (e.g. 10^{-10} to 10^{-12}) to be isolated by cell sorting. It is hoped that these two sets of cell lines will eventually allow isolation of genes involved in the regulation of CD44 function.

Meanwhile, we have looked for differences in the CD44 molecules themselves. Southern blotting of RT-PCR products generated using primers that flank the site of potential insertion of sequences derived from variant exons showed that there was no change in the CD44 isoform expressed as a result of selection for changes in CD44 HA-binding function. Both parental cell lines and their derivatives express the haematopoietic isoform of CD44. However, slight differences in the molecular weight of CD44 between parental and variant cell lines were seen in both the L-cell and RAW253 series when immunoprecipitated CD44 was run on long SDS/PAGE gels. Since there was no change in the CD44 isoform, these differences were the result of post-translational modifications. CD44 from inactive parent cells was of a higher apparent M_r than CD44 from inducible cells. CD44 from active cells was lower in apparent M_r than that of inducible and parent cells. Digestions with N-glycanase and treatment with tunicamycin to inhibit N-glycosylation indicated that N-glycosylation differences accounted for most of the variation in M_r, as deglycosylated forms from parental and variant cells were more similar in M_r. Thus, inactive cells were more heavily N-glycosylated than inducible cells, and both inactive and inducible cells were more heavily N-glycosylated than active cells.

Effects of inhibition of glycosylation on HA binding function

Could N-glycosylation differences account for the differences in HA-binding function? Treatment of parental and inducible cell lines for 2 to 3 days in 5 μg/ml tunicamycin resulted in increased ability of these cell lines to bind HA. Inducible cells of both lineages could bind HA constitutively. Inactive (parental cells) became completely inducible and some of the cells bound HA constitutively. (Table 3 and [20]) This indicated that N-glycosylation of cell surface glycoproteins inhibited HA binding in inactive and inducible cell lines. Inhibition of O-glycosylation with BZαGalNAc (benzyl 2-acetamido-2-deoxy-α-D-galactopyranoside) did not change the HA binding state of inducible or inactive cells.

Tunicamycin treatment alters N-glycosylation of the entire cell complement of glycoproteins. Though there were clearly changes in the glycosylation of CD44 as a result of tunicamycin treatment, it was not certain that these accounted for the changes in HA-binding. To analyse the role of N-glycosylation of CD44 itself, we have set out to ablate specific N-glycosylation sites of CD44 by mutation. So far, we have mutated (by substitution of alanine for asparagine) the two most N-terminal N-glycosylation sites, termed N1 and N2, which flank the first basic amino acid cluster involved in HA recognition (see Figure 1). These mutations do not affect HA binding of CD44 when they are expressed in AKR1, in which CD44 is in an active state. To look for positive effects of ablation of N-glycosylation sites on HA binding, we need to assay these constructs in cells that are in an inactive or inducible state. A CD44-negative mutant of the inducible variant of RAW253 was selected by cytotoxic ablation and cell sorting. When wild-type CD44 was expressed in this cell line, it was in the inducible state. Elimination of N1 alone was sufficient to convert the inducible CD44 into an active function in these cells. Cells expressing CD44 with the N2 mutation were still inducible, like those expressing wild-type CD44. Thus the most amino-terminal N-glycosylation site of CD44 is modified in a specific manner in the inducible cell environment so that constitutive binding of HA is prevented. When CD44 is expressed in AKR1 cells this N1 site is also glycosylated, but in these cells, the manner of glycosylation does not interfere with constitutive HA binding. These results show that cell-specific glycosylation of CD44 itself can regulate HA-binding function. A number of studies by ourselves and others present data on the effects of inhibition of glycosylation and mutation of glycosylation sites on the function of CD44. These are discussed in detail elsewhere [22] and clearly show that cell-specific glycosylation of CD44 has profound effects on the function of CD44. In some cells, glycosylation is necessary for HA binding

Table 3

| | Untreated | | Tunicamycin-treated | |
	Non-induced	Induced	Non-induced	Induced
Parent	1.1	1.2	5.5	10.8
Inducible	1.1	4.0	3.7	11.1
Active	10.7	6.6	14.4	13.7

Tunicamycin treatment of L-cell variant lines

Relative mean fluorescence of FL-HA-treated cells. Negative control without fluorescein = 1.0.

function. In other cells, such as the inactive and inducible cell lines described above, specific glycosylation patterns modulate the HA binding function of CD44 in a negative manner.

Effect of cell surface distribution of CD44 on HA binding

Another factor that might contribute to regulation of CD44/HA binding function was suggested by the observation that mAb-mediated induction of HA binding required multivalent antibody [15]. This suggested that antibody needed to cross-link CD44 molecules into a di- or multi- valent cluster. This idea is consistent with a number of other observations: that HA is a highly multivalent ligand, that most other HA binding proteins have two tandem 'link module' motifs while CD44 has only one, and that a number of other cell surface adhesion molecules are much more efficient at ligand recognition when in a multivalent configuration [23].

Mutant CD44 constructs lacking all of the cytoplasmic domain except the six most membrane proximal amino acids (referred to as 'tailless' CD44) bind HA very poorly when expressed in AKR1 cells where wild-type CD44 is in an active state. Multivalent mAb IRAWB14 rapidly induces HA binding in AKR1 cells expressing 'tailless' CD44 [13]. Since the tailless deficiency was overcome by cross-linking CD44, we postulated the cytoplasmic domain might contribute to HA binding by promoting a multivalent distribution of CD44 on the cell surface. To test whether multimerization of CD44 could overcome the tailless defect, a CD44 construct was designed that would be expressed as a dimer on the cell surface [24]. The transmembrane domain of CD44 was replaced with the transmembrane domain of the zeta chain of the T-cell receptor-associated CD3 complex, which forms a disulphide bonded dimer within the membrane. Both full-length and tailless CD44 dimers were expressed in AKR1 cells and assayed for HA binding. While cells expressing tailless CD44 failed to bind HA, cells expressing the equivalent level of cell surface tailless dimer bound HA. In fact dimeric CD44, whether full length or tailless, bound HA better than wild-type CD44 (i.e. bound HA at lower levels of CD44 expression than were required for equivalent binding by wild-type CD44), as shown in two-colour flow cytometry analysis of CD44 expression and HA binding by these lines (Figure 4). This figure also shows that there is a close relationship between the level of CD44 expression on the cell surface and HA binding activity in these cells. However, in each cell line a minimal or 'threshold' level of CD44 must be reached before HA binding is detected. Even in the case of tailless CD44, HA binding is observed if CD44 levels are high enough. The threshold for dimeric CD44 is lower than for wild-type CD44, which is lower than for tailless CD44. These observations are consistent with the postulate that multimeric aggregates of CD44 favour HA binding. Dimers could occur by chance at very high levels of tailless CD44. Aggregation may be favoured by the presence of the cytoplasmic domain, thus lowering the threshold for wild-type CD44. Preformed dimers clearly have a lower threshold still.

Liu and Sy [25] have found disulphide-bonded dimeric CD44H in inducible cells that have been activated to bind HA by phorbol ester, but not in

Figure 4

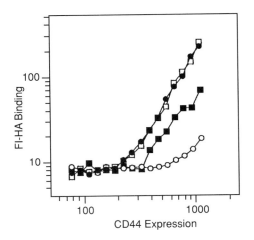

Two-colour flow cytometry analysis of AKR1 cells transfected with CD44 constructs

Cells were labelled with Fl-HA and biotinylated Fab fragment of CD44-specific mAb IM7 plus phycoerythrin-streptavidin [24]. Data were plotted by calculating the mean Fl-HA fluorescence (y-axis) over five-channel intervals of phycoerythrin fluorescence representing CD44 expression (x-axis). Representative cloned cell lines were transfected with wild-type CD44 (■), tailless CD44 (○), dimeric full length CD44 (□) and dimeric tailless CD44 (●).

unactivated cells. This dimerization was dependent on a cysteine in the CD44 transmembrane domain. We did not detect disulphide bonded dimers of wild-type CD44 in AKR1 cells that were active in HA binding [24]. We also found no requirement for the transmembrane cysteine in AKR1 cells, since CD44 with the transmembrane domain substituted with that of CD45 (which has no intramembrane cysteine) bound HA as well as wild-type CD44 in these cells in an active state [26].

Role of the CD44 cytoplasmic domain

Although it is not yet demonstrated that wild-type CD44 is in a multimeric configuration in cells in an active state, we speculated that there were sequences within the cytoplasmic domain that promoted di- or multi-merization of CD44 on the cell surface, perhaps by interacting with soluble cytoplasmic proteins or with cytoskeletal components, or by self-association. Mutant CD44 constructs with partial truncations or with internal deletion mutations in the cytoplasmic domain were assayed by transfection into AKR1 cells [24]. All of these constructs bound HA as well as wild-type CD44, including a truncation expressing only the 16 most membrane proximal amino acids of the cytoplasmic domain and an internal deletion mutant that omitted the membrane proximal amino acids between 3 and 12 amino acids from the membrane. Thus, no specific sequence was found that accounted for the cytoplasmic domain's contribution to HA binding.

Other functions of the CD44 cytoplasmic domain include basolateral localization in polarized epithelial cells [27], which has been shown to depend on a di-hydrophobic leucine-valine motif [28], and promotion of cell migration on HA substrates [29,30], which requires two serines that are involved in phosphory-lation [30]. These phosphorylation sites are not required for basolateral sorting in polarized cells [31] or for HA binding [30,32].

A portion of the CD44 on the surface of fibroblasts is insoluble upon extraction with TritonX-100, which is often interpreted as indicating an association with the Triton-insoluble cytoskeleton through the cytoplasmic domain. However, tailless CD44H expressed in fibroblasts retained its TritonX-100 insoluble component [26,27], while CD44 with a substituted transmembrane domain, either from the zeta chain of the T-cell receptor-CD3 complex or from CD45, became completely soluble in TritonX-100 [26]. This indicated that the TritonX-100 insolubility of CD44 in fibroblasts was a property of the transmembrane domain. Further experiments indicated that the transmembrane domain-dependent Triton insolubility was the result of association with TritonX-100-insoluble lipids [26,33].

We have found no evidence that TritonX-100-insolubility of CD44 in some cell types is correlated with HA binding function. We find no TritonX-100-insoluble CD44 in a variety of lymphoid cell lines representing all three HA binding activation states. Among fibroblast lines, which do have TritonX-100-insoluble CD44, L-cells are inactive and NIH- and Swiss-3T3 cells are in an active state. In the series of L-cells representing each of the three activation states of CD44, there is no difference in the proportion of CD44 that is TritonX-100 insoluble.

Conclusion

The HA binding function of CD44 is strictly regulated in haematopoietic cells where CD44 is thought to contribute to the complex patterns of leucocyte differ-entiation and circulation. Regulation is probably the rule in other cell types that express CD44 as well. Disruption of normal regulatory mechanisms may account for the frequent observation of changes in CD44 expression associated with many different tumours. The ability to isolate variant cell lines in different activation states with respect to the HA binding function of CD44 (see Figure 3) indicates that a limited number of mutation events can result in changes in CD44 function that might contribute to malignancy. These cell lines provide a system for studying mechanisms and isolating genes involved in regulating CD44 function.

We have examined a number of potential mechanisms for modulating CD44 function (see Table 2). These studies show that cell-specific differences in N-glycosylation of CD44, the presence of the cytoplasmic domain, and the cell surface distribution of CD44 are all factors that contribute to HA binding, and thus represent possible mechanisms of regulation.

This work was supported by National Institute of Allergy and Infectious Diseases grant A1-31613 and was carried out in collaboration with Robert Hyman, Nicole English and Astrid Perschl.

References

1.   Haynes, B.F., Telen, M.J., Hale, L.P. and Denning, S.M. (1989) Immunol. Today **10**, 423–428
2.   Haynes, B.F., Hale, L.P., Patton, K L., Martin, M.E. and McCallum, R. M. (1991) Arthritis Rheum. **34**, 1434–1443
3.   Lesley, J., Hyman, R. and Kincade, P. (1993) Adv. Immunol. **54**, 271–335
4.   Underhill, C. (1992) J. Cell Sci. **103**, 293–298
5.   Stamenkovic, I.M., Amiot, M., Pesando, J. M. and Seed B. (1989) Cell **56**, 1057–1062
6.   Kohda, D., Morton, C.J., Parkar, A.A., Hatanak, H., Inagaki, F. M., Campbell, I. D. and Day, A.J. (1996) Cell **86**, 767–775
7.   Peach, R., Hollenbaugh J.D., Stamenkovic, I. and Aruffo A. (1993) J. Cell Biol. **122**, 257–264
8.   Screaton, G.R., Bell, M.V., Jackson, D.G., Cornelis, F.B., Gerth, U. and Bell, J.I. (1992) Proc. Natl. Acad. Sci., U.S.A. **89**, 12160–12164
9.   Tölg, C., Hofmann, M., Herrlich, P. and Ponta, K. (1993) Nucleic Acids Res. **21**, 1225–1229
10. Yu, Q. and Toole, B.P. (1996) J. Biol. Chem. **271**, 20603–20607
11. Toyama-Sorimachi, N., Sorimachi, H., Tobita, Y., Kitamura, F., Yagita, H., Suzuki, K. and Miyaska, M. (1995) J. Biol. Chem. **270**, 7437–7444
12. Lesley, J., Hyman, R. and Kincade, P. (1993) Adv. Immunol. **54**, 271–335
13. Lesley J. and Hyman R. (1992) Eur. J. Immunol. **22**, 2719–2723
14. Lesley, J., He, Q., Miyake, K., Hamann, A., Hyman, R. and Kincade, P. (1992) J. Exp. Med. **175**, 257–266
15. Lesley, J., Kincade, P. and Hyman, R. (1993) Eur. J. Immunol. **23**, 1902–1909
16. Zheng, Z., Katoh, S., He, Q., Oritani, K., Miyake, K., Lesley, J., Hyman, R., Hamik, A., Parkhouse, R., Farr, A. and Kincade, P. (1995) J. Cell Biol. **130**, 485–495
17. Lesley, J., Howes, N., Perschl, A. and Hyman, R. (1994) J. Exp. Med. **180**, 383–387
18. Murakami, S., Miyake, K., Abe, R., Kincade, P.W. and Hodes, R. J. (1991) J. Immunol. **146**, 1422–1427
19. Liao, H-X., Lee, D. M., Levesque, M.C. and Haynes, B.F. (1995) J. Immunol. **155**, 3938–3945
20. Lesley, J., English, N., Perschl, A., Gregoroff, J. and Hyman, R. (1995) J. Exp. Med. **182**, 431–437
21. Hyman, R., Trowbridge, I., Stallings, V. and Trotter J. (1982) Immunogenetics **15**, 413–420
22. Lesley, J., Hyman, R., English, N., Caterall, J.B. and Turner, G. (1997) Glycoconj. J. **14**, 611–622
23. Dustin;, M.L. and Springer, T.A. (1991) Annu. Rev. Immunol. **9**, 27–66
24. Perschl, A., Lesley, J., English, N., Trowbridge, I. and Hyman, R. (1995) Eur. J. Immunol. **25**, 495–501
25. Liu, D. and Sy, M-S. (1996) J. Exp. Med. **183**, 1987–1999
26   Perschl, A., Lesley, J., English, N., Hyman, R. and Trowbridge, I. (1995) J. Cell Sci. **108**, 1033–1041
27. Neame, S.J. and Isacke, C.M. (1993) J. Cell Biol. **121**, 1299–1310
28. Sheikh, H. and Isacke, C.M. (1995) Eur. J. Immunol. **25**, 1883–1887
29. Thomas, L., Byers, R.H., Vink, J. and Stamenkovic, I. (1992) J. Cell Biol. **118**, 971–977
30. Peck, D. and Isacke, C. M. (1996) Curr. Biol. **6**, 884–890
31. Neame, S.J. and Isacke, C. M. (1992) EMBO J. **11**, 4733–4738
32. Uff, C.R., Neame, S.J. and Isacke, C..M. (1995) Eur. J. Immunol. **25**, 1883–1887
33. Neame, S.J., Uff, C.R., Sheikh, H., Wheatley, S.C. and Isacke, C. M. (1995) J. Cell Sci. **108**, 3127–3135

14

Hyaluronan receptors and the role of the B(X₇)B recognition motif in signalling

Z. Lin, G. Hou, R. Harrison and E. A. Turley*

Cardiovascular Research, The Hospital for Sick Children, 555 University Ave., Toronto, Ontario M5G 1X8, Canada

Introduction

Hyaluronan (HA), a large glycosaminoglycan composed of repeating disaccharides of N-acetylglucosamine and β-glucuronic acid, is ubiquitously present in the extracellular matrix (ECM) [1]. The production of HA has been linked to a variety of diseases as well as developmental and physiological processes [1,2]. Besides functioning as an important ECM component of skin, cartilage and brain, HA is also known to regulate cell locomotion during morphogenesis, wound repair, tumour invasion and immune response [3]. Most of the cellular functions of HA seem to be mediated through the interactions of HA and its cell surface receptors, which include CD44 [4–6] and RHAMM (receptor for HA-mediated motility) [7–9]. Both CD44 and RHAMM have been implicated in several cellular processes. CD44 is known to be involved in immune responses [10,11] and tumour progression [12–14], while RHAMM plays roles in cellular events such as locomotion [15], injury response [16,17] and proliferation [8]. Indeed, it has recently been demonstrated that HA:RHAMM signalling controls the cell's entry into mitosis by regulating synthesis of CDC2 and cyclin B1 proteins [18], and, not surprisingly, overexpression of RHAMM is transforming [8].

      Both CD44 and RHAMM have been sequenced and molecularly characterized. CD44 is a broadly distributed cell surface glycoprotein consisting of a transmembrane domain and a cytoplasmic tail [19,20]. Although the CD44 gene can be alternatively spliced, thus producing various isoforms in different cell types, all isoforms contain an identical transmembrane domain and a cytoplasmic tail [19]. Similarly, RHAMM is also a widely present protein consisting of various isoforms [21], including variant 4, which is the only isoform that possesses transforming activity. However, RHAMM has no transmembrane domain in spite of the fact that it is present extracellularly, on the cell surface and intracellularly [9].

      Both CD44 and RHAMM bind to HA on the cell surface and *in vitro*, although CD44 binds to HA with five times greater affinity than RHAMM does [22]. The binding of either of the cell surface receptors to HA has been proven to be crucial for some of the functions mediated by HA, since antibodies generated

*To whom correspondence should be addressed.

from CD44 or RHAMM can abolish the functions of HA in a metastatic process [14] or cell locomotion [22], and since mutation of one of the binding sites for HA in these receptors ablates their ability to signal [8,31].

In the following sections, we will discuss the structures and role of a binding motif common to CD44 and RHAMM, which is believed to be crucial for not only HA binding but also mediating functions of HA.

The B(X$_7$)B HA-binding motif in CD44 and RHAMM

The HA binding motif in HA receptors has been identified by Yang and co-workers [23]. Two binding domains near the carboxyl terminus of RHAMM containing two clusters of basic amino acids, with 10 and 11 amino acids, respectively, were originally identified as the HA binding domains in RHAMM [24]. Each of the binding domains contributes equally to the HA binding ability of RHAMM. When the two binding domains are mutated by site-directed mutation, the capability of RHAMM to bind HA is completely lost. Although there is no apparent homology between the two domains, a common structural feature is present [23]. The finding has led to the identification of a HA binding motif in most HA receptors and other HA binding proteins molecularly characterized to date. The binding capability of the motif, B(X$_7$)B, where B represents arginine or lysine residues while X$_7$ are 7 amino acid residues containing no acidic residues, at least one basic residue and mostly hydrophobic residues, has been carefully examined. Three such motifs have been identified in CD44, two located in the extracellular portion and one in the cytoplasmic domain [23]. One of these is critical for maintaining CD44's HA-binding capability [25,26]. When each of the motifs is linked to the non HA-binding N-terminal region of RHAMM (amino acids 1–125) by homologous recombination, the resultant fusion protein achieves HA-binding activity [23]. These data suggest that the B(X$_7$)B is the minimal requirement for HA binding capability of proteins, including RHAMM, CD44 and linked proteins. However, other HA-binding mechanisms may also be present among proteins that bind HA. One such example is a Link module consisting of 2 α helices and 2 β sheets arranged around a large hydrophobic core [27–29]. In CD44, aggrecan and TSG6 (tumour necrosis factor-stimulated gene-6), the HA-binding motif is located within the Link module [27]. The relationship between the binding motif and the Link module is still not known; however, the B(X$_7$)B motif is absolutely critical for generating signals from RHAMM that regulate cell motility and proliferation.

HA:RHAMM signals cell motility

RHAMM was originally found to be essential for HA-promoted locomotion in ras-transformed cells [22]. Anti-RHAMM antibody but not anti-CD44 antibody blocked the cell's response to HA. Further investigation of the motility of H-ras-transformed 10T1/2 mouse fibroblast cells has been performed using various concentrations of HA and anti-RHAMM antibody [15]. Both exogenous HA and

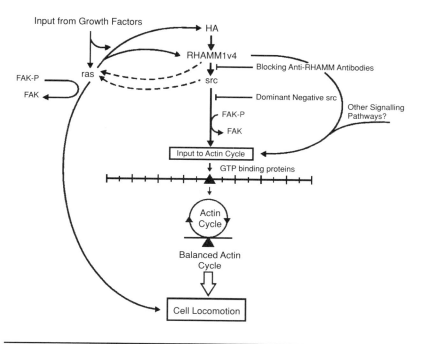

Figure 1

Model: HA/RHAMM/src signalling promotes cell motility [30]

low concentrations of a stimulatory anti-RHAMM antibody can significantly promote cell motility and this is accompanied by the transient tyrosine phosphorylation of several proteins, notably focal adhesion kinase, pp125[FAK]. The time course of the enhanced followed by decreased tyrosine phosphorylation of FAK parallelled a rapid assembly and disassembly of focal adhesions as detected by vinculin immunofluorescence. Upon treatment of the cells with tyrosine kinase inhibitors or with anti-phosphotyrosine antibody through microinjection, HA-stimulated cell motility as well as the increased activity of transient tyrosine phosphorylation was inhibited. These data suggest that HA regulates cell locomotion by signalling through the transient tyrosine kinase activity mediated by RHAMM. Recently, a role of src, a kinase that phosphorylates FAK, in regulating cell locomotion has been directly demonstrated [30]. In fibroblasts derived from mice lacking src [src(−/−)], cells move much slower than the wild-type fibroblasts. The expression of c-src but not the kinase-deficient src in these mutant cells restores cell locomotion. RHAMM is also required for the restoration of src (−/−) cell locomotion, which is indicated by the fact that motility of cells expressing c-src is reduced to the src (−/−) level by anti-RHAMM antibody while the cell locomotion of src (−/− remains unaffected by the same antibody. Furthermore, the dominant negative src significantly inhibits RHAMM-dependent ras and serum-regulated motility. These data are consistent with the hypothesis that src acts downstream of RHAMM but upstream of ras in the signalling process (Figure 1).

The importance of the HA-binding motif in regulating this process has been demonstrated by expression of the RHAMM mutated in its $B(X_7)B$ motif [8]. HA binding capability of the mutant RHAMM is ablated. Interestingly, this mutation also abolishes the cell's ability to reduce FAK phosphorylation in response to RHAMM as well as to generate a signal from ras to enhance cell mobility and proliferation. These results suggest, paradoxically, that RHAMM acts upstream of src but downstream of ras. The relationship between src and ras in generating a motility signal is currently being studied.

Conclusion

It has become clear that HA and its receptors have important roles in regulating various cellular processes. HA and its cell receptors also play important roles in cellular processes other than cell locomotion. As mentioned earlier, HA and RHAMM variant 4 (RHAMMv4) are involved in regulating cell proliferation [8]. Overexpression of RHAMMv4 in fibroblast cells causes the cells to transform. Conversely, interaction between HA and CD44 is crucial for the regulation of tumour development [31]. In this case, a CD44 mutant-transfected human melanoma with arginine 41 altered to alanine by site-directed mutation has no HA-binding capability and thus fails to trigger tumour formation, while the wild-type CD44-transfected cell does. However, the molecular mechanisms involved in these HA signalling events remain unclear. Further studies on the molecular mechanisms of HA signalling are expected to contribute not only to the understanding of these processes but also to the potential medical applications of HA.

References

1.   Laurent, T.C. and Fraser, J.R.E. (1992) FASEB J. 6, 2397–2404
2.   Turley, E.A. (1992) Cancer Metastasis Rev. 11, 21–30
3.   Evered, D. and Whelan, J., eds.(1989) The Biology of Hyaluronan, Ciba Foundation Symposium No. 143, Wiley, Chichester
4.   Culty, M., Miyake, K., Kincade, P.W., Silorski, E., Butcher, E. and Underhill, C. (1990) J. Cell Biol. 111, 2765–2774
5.   Lesley, J., He, Q., Miyake, K., Hammann A., Hyman, R. and Kincade, P.W. (1992) J. Exp. Med. 175, 257–266
6.   Gunthert, U. (1993) Curr. Top. Microbiol. Immunol. 184, 47–63
7.   Hardwick, C.,Hoare, K., Owens, R., Hohn, H.P., Höök, M., Moore, D., Cripps, V., Austen, L., Nance, D.M. and Turley, E.A. (1992) J. Cell Biol., 117, 1343–1350
8.   Hall, C.L., Yang B., Yang, X., Zhang, S., Turley, M., Samuel, S., Lange, L.A., Wang, C., Curpen, G.D., Savani, R.C., Greenberg, A.H. and Turley, E.A. (1995) Cell 82, 19–28
9.   Entwistle, J., Hall, C.L., and Turley, E.A. (1996) J. Cell Biochem. 61, 569–577
10.   Denning, S.M., Le, P.T., Singer, K.H. and Haynes, B.F. (1990) J. Immunol. 144, 7–15
11.   Huet, S., Groux, H., Caillou, B., Valentin, H., Prieur, M. and Bernard, A. (1989) J. Immunol. 143, 798–801
12.   Günthert, U. Hofmann, M., Rudy, W., Reber, S., Zöller, M., Haubmann, I., Matzku, S., Wenzel, A., Ponta, H. and Herrlich, P. (1991) Cell 65, 13–24
13.   Rudy, W., Hofmann, M., Schwartz-Albiez, R., Zöller, M., Heider, K.H., Ponta, H. and Herrlich P. (1993) Cancer Res. 53, 1262–1268
14.   Reber, S., Matzku, S., Günthert, U., Ponta, H., Herrlich, P. and Zöller, M. (1990) Int. J. Cancer 46, 919–927
15.   Hall, C.L., Wang, C., Lange, L.A. and Turley, E.A. (1994) J. Cell Biol. 126, 575–588
16.   Savani, R.C. and Turley, E.A. (1995) Int. J. Tissue React. 17, 141–151
17.   Savani, R.C., Khalil, N. and Turley, E.A. (1995) Proc. West. Pharmacol. Soc. 38, 131–136

18. Mohaptra, S., Yang, X., Wright, J.A., Turley, E.A. and Greenberg, A.H. (1996) J. Exp. Med. **183**, 1663–1668
19. Screaton, G.R., Bell, M.V., Jackson, D.G., Cornelis, F.B., Gerth, U. and Bell J.I. (1992) Proc. Natl. Acad. Sci. U.S.A. **89**, 12160–12164
20. Sherman, L., Sleeman J., Herrlich, P. and Ponta, H. (1994) Curr. Opin. Cell Biol. **6**, 726–733
21. Entwistle, J., Zhang, S., Yang, B., Wong, C., Li, Q., Hall, C.L., A, J., Mowat, M., Greenberg, A.H. and Turley, E.A. (1995) Gene **163**, 233–238
22. Turley, E.A., Moore, L.A.D. and Hoare K. (1993) Exp. Cell Res. **207**, 277–282
23. Yang, B., Yang, B., Savani, R.C. and Turley, E.A. (1994) EMBO J. **13**, 286–296
24. Yang, B., Zhang, L. and Turley, E.A. (1993) J. Biol. Chem. **268**, 8617–8623
25. Goetinck, P.F., Stirpe, N.S., Tsonis, P.A. and Carlone, D. (1987) J. Cell Biol. **105**, 2403–2408
26. Peach, R.J., Hollenbaugh, D., Stamenkovic, I. and Aruffo, A. (1993) J. Cell Biol. **122**, 257–264
27. Kohda, D., Morton, C.J., Parkar, A.A., Hatanaka, H., Inagaki, F.M., Campbell, I.D. and Day, A.J. Cell **86**, 767–775
28. Perkins, S.J., Nealis, A.S., Dudhia, J. and Hardingham, T.E. (1989) J. Mol. Biol. **206**, 737–753
29. Hardingham, T.E. and Fosang, A.J. (1992) FASEB J. **6**, 861–870
30. Hall, C.L., Lange, L.A., Prober, D.A., Zhang, S. and Turley, E.A. (1996) Oncogene **13**, 2213–2224
31. Bartolazzi, A., Peach, R., Aruffo, A. and Stamenkovic I. (1994) J. Exp. Med. **180**, 53–66

15

The structure of the Link module: a hyaluronan-binding domain

Anthony J. Day* and Ashfaq A. Parkar

Department of Biochemistry, University of Oxford, South Parks Road, Oxford OX1 3QU, U.K.

Introduction

The interactions of hyaluronan with hyaluronan-binding matrix proteins and receptors have a crucial role in the formation and stability of extracellular matrix and in regulating many aspects of cell behaviour during development, morphogenesis, tumorigenesis and inflammation [1,2]. These interactions are often mediated by a common protein domain of approximately 100 amino acids, termed a Link module, which is also referred to as a proteoglycan tandem repeat [3,4]. This module has a characteristic consensus sequence consisting of 4 disulphide-bonded cysteines (where the first is connected to the fourth and the second is connected to the third) as well as other invariant and highly conserved residues [5]. This is illustrated in Figure 1, which shows a sequence alignment of Link modules from the ubiquitous hyaluronan receptor CD44, cartilage link protein, aggrecan, and tumour necrosis factor (TNF)-stimulated gene-6 (TSG-6). A brief review of the biological roles of these proteins is given below.

Link protein is an essential component of cartilage and, together with hyaluronan and the proteoglycan aggrecan, forms huge multimolecular aggregates that provide the tissue with its load-bearing properties (see [6,7]). Link protein and the hyaluronan-binding region of aggrecan (termed the G1 domain) are homologous and each consists of an N-terminal immunoglobulin domain followed by two contiguous Link modules [5,8]. In link protein both Link modules have been demonstrated to interact with hyaluronan independently [9]. It is long established that a hyaluronan decasaccharide is the minimum size of glycosaminoglycan that can interact strongly with aggrecan or link protein [10,11]. However, the size of hyaluronan oligomer that interacts with an individual Link module in these proteins is not known.

CD44 has a single Link module, close to its N-terminus, that is involved in hyaluronan binding in association with other regions of the protein [12]. In this case the minimum size of hyaluronan recognized is a hexasaccharide (see [1]). On cartilage chondrocytes, CD44 has a role in both the assembly of the pericellular matrix and the local turnover of hyaluronan (see [1]). In arthritis CD44 is up-regulated on many synovial cell types [13] and is involved in the infiltration of inflammatory leucocytes on which it is also expressed [14]. Recently it has been shown that the rolling of lymphocytes on vascular endothelium is mediated via the interaction of CD44 with hyaluronan [15].

*To whom correspondence should be addressed.

Figure 1

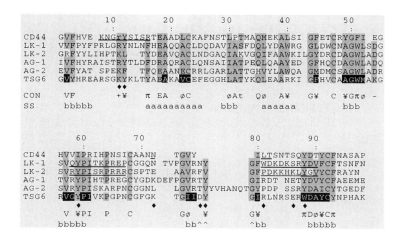

Sequence alignment of the Link module superfamily

The TSG-6 Link module is aligned with those from CD44, cartilage link protein (LK-1 and LK-2), and the GI domain of aggrecan (AG-1 and AG-2), where all the sequences are from human proteins. Amino acid numbering is given for the TSG-6 Link module where secondary structure (SS) elements (a, α helix; b, β strand; ^, bulge) and core residues (black background) are indicated. A Link module consensus sequence (CON) shows the presence of identities (single letter code) or conservative replacements (+ = K or R; − = D or E; t = S or T; π = aromatics F,Y or W; ø = aliphatic hydrophobics A, M, I, L or V; ¥ = π or ø) in at least five of the sequences (grey shading). Regions implicated in hyaluronan binding are underlined (see text), and R41, which has a critical role in CD44, is shown in lower case. Amino acids that form the putative hyaluronan-binding surface in TSG-6 are denoted by ◆.

Expression of TSG-6 mRNA in fibroblasts, chondrocytes and synovial cells is induced by the inflammatory cytokines, interleukin-1 (IL-1) or TNF (see [16]). In addition, the protein (which contains a single Link module [17]) has been detected in synovial fluids from arthritis patients but not from normal individuals [18]. As IL-1 and TNF are known to have a central role in the induction of cartilage breakdown in arthritis (see [19]), TSG-6 may have a role in these processes [17,18]. However, in chondrocytes at least, TSG-6 mRNA levels are also up-regulated by transforming growth factor-β1, which is thought to have a role in cartilage repair [16]. In this regard, full-length recombinant TSG-6 has been reported to potentiate plasmin inhibition by inter-α-inhibitor, with which TSG-6 forms a covalent complex [20]. In addition, it has also been shown to have an anti-inflammatory effect *in vivo* by inhibiting neutrophil infiltration in a murine air pouch model [20]. However, it has not been established whether these phenomena are related. Recombinant expression in *Escherichia coli* has also been used to produce the TSG-6 Link module individually [21], which has allowed the determination of the first tertiary structure for this domain and investigation into its hyaluronan-binding properties [22].

The structure of the Link module

The tertiary structure of the Link module was determined in solution at pH 6.0 using nuclear magnetic resonance spectroscopy (NMR) as described by Kohda and co-workers [22]. Figure 2 shows the structure, which has a compact fold consisting of two α helices and two anti-parallel β sheets (SI and SII), where the N- and C-termini protrude from the same face of the molecule. The positions of these secondary structure elements in the amino acid sequence are indicated on Figure 1. The structure has a large well-defined hydrophobic core composed of 21 amino acids (see Figure 1) with residues in the β3 and β6 strands having a particularly important role in stabilizing the structure and providing much of the central scaffold. The disulphide linked Cys-23 and Cys-92, which connect the α1 helix and β6 strand, are completely buried within the protein. From the sequence alignment in Figure 1 it can be seen that the residues that define the core in TSG-6 are highly conserved in Link modules from CD44, link protein and aggrecan (and also the related proteoglycans versican, neurocan and brevican – not shown). In addition, there are no insertions or deletions in regions of secondary structure. Therefore, the structure determined for the Link module of TSG-6 defines the consensus fold for this module superfamily.

The Link module structure has the same fold as that of the C-type lectin domain (found in the selectin family), with identical topology and a very similar

Figure 2

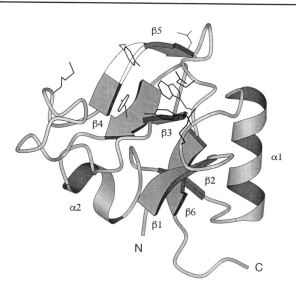

The tertiary structure of the Link module from human TSG-6

This consists of 2 α helices and 2 anti-parallel triple-stranded β sheets (SI and SII) that contribute to a large hydrophobic core. The SI sheet is made up of strands β1, β2 and β6 and the SII sheet of β3, β4 and β5. The β5 strand contains a bulge that is indicated by the absence of shading. The side chains of residues that comprise the putative hyaluronan binding site on TSG-6 (see Figure 1) are shown (note that D77 is obscured by the β5 strand). This figure was made using the programme MOLSCRIPT [32].

organization of secondary structure [22]. The structural similarity between these modules strongly suggests that they have a common evolutionary origin, which provides a possible explanation for the similar roles of CD44 and the selectins in the rolling of leucocytes on the vascular endothelium prior to extravasation at sites of inflammation.

The interaction of hyaluronan with TSG-6

The recombinant Link module from TSG-6 has been demonstrated to bind specifically to hyaluronan by use of a microtitre plate assay [22]. We have also investigated the pH-dependency of the TSG-6 interaction with hyaluronan and compared it with that of aggrecan (G1 domain) and link protein (A.A. Parkar, J.D. Kahmann, S.L.T. Howat, M.T. Bayliss and A.J. Day, unpublished work). In Figure 3 the effect of pH on TSG-6 and aggrecan binding is shown. In the case of aggrecan the pH-dependency is very similar to that determined previously [23,24], where there is minimal binding at pH 3.5 with close to maximal binding between pH 6.5 and 7.5. This is consistent with aggrecan's structural role in normal cartilage, which has been calculated to have a pH of about 6.9 (see [25]). For the TSG-6 Link module there is also minimal binding to hyaluronan at pH 3.5 but with maximal binding seen around pH 6.0 (i.e. the pH at which the tertiary structure was determined). The level of binding over this pH range correlates very

Figure 3

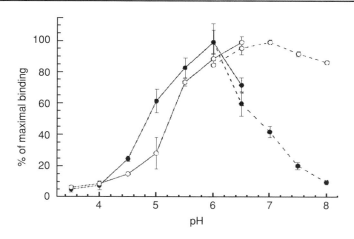

A comparison of the pH-dependencies of hyaluronan binding by TSG-6 and aggrecan

The interaction of biotinylated-hyaluronan with wells coated with the G1 domain of porcine aggrecan (○) was examined at different pHs using the microtitre plate assay described by Kohda and co-workers [22]. A similar assay was used for TSG-6 (●), where mono-biotinylated-Link module binding to hyaluronan-coated wells was measured. The assays were performed in 100 mM NaCl with either 50 mM sodium acetate (continuous lines – over pH range 3.5 to 6.5) or 50 mM Hepes (dotted lines – pH 6.0 to 8.0) as buffer. Values are plotted as mean binding (n = 3) ± standard error of the mean.

well with the percentage of folded Link module present as determined using 1-dimensional NMR (data not shown). However, on further increasing the pH from 6.0 to 8.0 the level of hyaluronan binding reduces to almost zero. A very similar effect is seen in assays with biotinylated-hyaluronan binding to Link module-coated wells (not shown). Using NMR we have demonstrated that this pH-mediated reduction in binding is not due to a gross structural change in the protein (i.e. it is not unfolding) and, therefore, is likely to be due to a local structural change in a functional element (see below). This suggests that pH may be an important factor in the regulation of TSG-6 activity. It is interesting to note that breakdown of cartilage matrix constituents may be initiated by a reduction in pH in the perichondrocytic region [26,27]. In this regard, IL-1 (which induces TSG-6 expression in joint cells) has been shown to increase both nitric oxide and lactic acid production by chondrocytes, which is very likely to lead to such a reduction in pH [27].

The structural basis for hyaluronan binding

Elucidation of the tertiary structure of the Link module has allowed the identification of a putative hyaluronan-binding surface on TSG-6 (see [22]). This is composed of positively and negatively charged residues (Lys-11, Lys-72, Asp-77, Arg-81 and Glu-86) arranged around a solvent-exposed hydrophobic patch made up of Tyr-12, Tyr-59, Tyr-78 and part of Trp-88. Similar surfaces are likely to be formed in other members of the Link module superfamily. The positions of these residues in the Link module sequence and structure are shown in Figures 1 and 2, respectively. Chemical modification studies have implicated basic amino acids in binding of hyaluronan where these are likely to form ionic interactions with carboxylate anions of the glycosaminoglycan (see references in [22]). Hydrophobic interactions are also likely to be involved as they have been found to have a general role in protein–carbohydrate recognition (see [22]), and hyaluronan has large clusters of CH groups forming patches of a highly hydrophobic nature [28]. From Figure 1 it can be seen that the amino acids that form this putative hyaluronan binding surface are brought together from different parts of the sequence. Most of these residues come from regions of the Link module (shown underlined on Figure 1) that have been previously implicated in hyaluronan binding by CD44 or link protein [12,29–31]. In particular, Peach and co-workers [12] have demonstrated that mutation of Arg-41 of CD44 (equivalent to Lys-11 in the TSG-6 Link module – see Figure 1) to alanine completely abolishes hyaluronan binding. This amino acid is located next to one of the surface-exposed tyrosine residues (Tyr-12) on the loop between β1 and α1 (see Figure 2). Interestingly, there is an aspartic acid (Asp-89) buried below this loop that may be involved in restraining this region when in its protonated state. It is possible, therefore, that the unusual pH-dependency of hyaluronan binding by TSG-6 (see above) could be related to the pK_a of this acidic residue.

In full-length recombinant TSG-6, two amino acids within the Link module (Glu-6 and Lys-13) have been identified that may play a role in the inhibition by TSG-6 of neutrophil migration to sites of inflammation [20].

Mutation of Glu-6 to lysine reduced significantly the anti-inflammatory effect of TSG-6, whereas changing Lys-13 to glutamine abolished it completely. Glu-6 is part of the β1 strand (hydrogen bonded to Asp-89 on β6) and its mutation is, therefore, likely to cause a structural perturbation of the module and may not have a direct functional role. Lys-13, however, is found on the loop that contains the putative hyaluronan binding residues Lys-11 and Tyr-12 (see above) and, thus, could be either directly involved in hyaluronan binding or affect the orientation of these residues. This suggests to us that the anti-inflammatory effect of TSG-6 may relate to its hyaluronan-binding function rather than its ability to increase the plasmin inhibitory activity of inter-α-inhibitor as hypothesized by Wisniewski and co-workers [20].

Conclusions

Determination of the tertiary structure of the TSG-6 Link module is providing new general insight into the roles of hyaluronan in matrix assembly and cell migration. Work is in progress to define the exact molecular nature of the protein–hyaluronan interaction and to further investigate the functional activities of this important arthritis-associated protein.

We thank Professor Iain D. Campbell for his support and encouragement, and Drs. Craig J. Morton and Caroline M. Milner for useful discussions and reading of the manuscript. A.J.D. and A.A.P. acknowledge the generous support by the Arthritis and Rheumatism Council (Grants D0067, D0086 and D0540).

References

1.   Knudson, C.B. and Knudson, W. (1993) FASEB J. **7**, 1233–1241
2.   Sherman, L., Sleeman, J., Herrlich, P. and Ponta, H. (1994) Curr. Opin. Cell Biol. **6**, 726–733
3.   Bork, P., Downing, A.K., Kieffer, B. and Campbell, I.D. (1996) Q. Rev. Biophys. Biochem. **29**, 119–167
4.   Perkins, S.J., Nealis, A.S., Dudhia, J. and Hardingham, T.E. (1989) J. Mol. Biol. **206**, 737–753
5.   Neame, P.J., Christner, J.E. and Baker, J.R. (1986) J. Biol. Chem. **281**, 3519–3535
6.   Neame, P.J. and Barry, F.P. (1993) Experientia **49**, 393–402
7.   Mörgelin, M., Paulsson, M., Heinegård, D., Aebi, U. and Engel, J. (1995) Biochem. J. **307**, 595–601
8.   Doege, K., Sasaki, M., Horigan, E., Hassel, J.R. and Yamada, Y. (1987) J. Biol. Chem. **262**, 17757–17767
9.   Grover, J. and Roughly, P.J. (1994) Biochem. J. **300**, 317–324
10. Hardingham, T.E. and Muir, H. (1973) Biochem. J. **135**, 905–908
11. Hascall, V.C. and Heinegård, D. (1974) J. Biol. Chem. **249**, 4242–4249
12. Peach, R.J., Hollenbaugh, D., Stamenkovic, I. and Aruffo, A. (1993) J. Cell Biol. **122**, 257–264
13. Haynes, B.F., Hale, L.P., Patton, K.L., Martin, M.E. and McCallum, R.M. (1991) Arthritis Rheum. **34**, 1434–1443
14. Mikecz, K., Brennan, F.R., Kim, J.H. and Glant, T.T. (1995) Nat. Med. **1**, 558–563
15. DeGrendele, H.C., Estess, P., Picker, L.J. and Siegelman, M. H. (1996) J. Exp. Med. **183**, 1119–1130
16. Maier, R., Wisniewski, H-G., Vilcek, J. and Lotz, M. (1996) Arthritis Rheum. **39**, 552–559
17. Lee, T., Wisniewski, H-G. and Vilcek, J. (1992) J. Cell Biol. **116**, 545–557
18. Wisniewski, H-G., Maier, R., Lotz, M., Lee, S., Lee, T.H. and Vilcek, J. (1993) J. Immunol. **151**, 6593–6601
19. Feldmann, M., Brennan, F.M. and Maini, R.N. (1996) Cell **85**, 307–310
20. Wisniewski, H-G., Hua, J-C., Poppers, D.M., Naime, D., Vilcek, J. and Cronstein, B.N. (1996) J. Immunol. **156**, 1609–1615

21. Day, A.J., Aplin, R.T. and Willis, A.C. (1996) Protein Expr. Purif. **8**, 1–16
22. Kohda, D., Morton, C.J., Parkar, A.A., Hatanaka, H., Inagaki, F. M., Campbell, I.D. and Day, A.J. (1996) Cell **86**, 767–775
23. Hardingham, T.E. and Muir H. (1972) Biochim. Biophys. Acta **279**, 401–405
24. Hardingham, T. E. (1979) Biochem. J. **177**, 237–247
25. Wilkins, R.J. and Hall, A.C. (1995) J. Cell Physiol. 164, 474–481
26. Woessner, F. and Gunja-Smith, Z. (1991) J. Rheumatol. **18** (Suppl. 27), 99–101
27. Stefanovic-Racic, M., Stadler, J., Georgescu, H.I. and Evan, C.H. (1994) J. Cell Physiol. **159**, 274–280
28. Mikelsaar, R-H. and Scott, J.E. (1994) Glycoconj. J. **11**, 65–71
29. Goetinck. P.F., Stirpe, N.S., Tsonis, P.A. and Carlone, D. (1987) J. Cell Biol. **105**, 2403–2408
30. Yang, B., Yang, B.L., Savani, R.C. and Turley, E.A. (1994) EMBO J. **13**, 286–296
31. Zheng, Z., Katoh, S., He, Q., Oritani, K., Miyake, K., Lesley, J., Hyman, R., Hamik, A., Parkhouse, R.M.E., Farr, A.G. and Kincade, P.W. (1995) J. Cell Biol. **130**, 485–495
32. Kraulis, P.J. (1991) J. Appl. Crystallogr. **24**, 946–950

16

The structure and function of inter-α-inhibitor and related proteins

Erik Fries* and Anna M. Blom
Department of Medical and Physiological Chemistry, University of Uppsala, Biomedical
Center, Box 575, S-751 23 Uppsala, Sweden

Inter-α-inhibitor (IαI), or inter-alpha-trypsin inhibitor as the protein was originally named, was isolated from serum more than 30 years ago [1,2]. However, it was only at the beginning of the 1990s that a biological process was found in which IαI played a role: the formation of the hyaluronan-containing coat that surrounds many cell types. This property has made IαI interesting from a cell biological point of view. In addition, the analysis of the structure and biosynthesis of this protein has revealed peculiar characteristics that are of general biochemical relevance.

Structure

The basic structure of IαI remained unknown long after its isolation and it was only through the use of recombinant DNA techniques in the 1980s that it became clear that IαI consists of three polypeptides – coded for by three separate genes [3]. Two of these polypeptides are of similar mass (~80 kDa) [4] and amino acid sequence, and are referred to as heavy chain 1 and 2 [5]. The third polypeptide, which is much smaller (16 kDa) and carries an 8 kDa chondroitin sulphate chain [6], consists of two tandemly arranged domains, each with proteinase inhibitor activity [7]. The structures of these domains are of the Kunitz type and the whole polypeptide was therefore recently given the name bikunin [8]. As will be discussed in detail below, all three polypeptides of IαI are synthesized in hepatocytes and during their transport to the cell surface the heavy chains become covalently linked to the chondroitin sulphate chain of bikunin. Figure 1 shows schematically the structure of IαI. There is also another plasma protein with a structure similar to that of IαI: pre-α-inhibitor (PαI; Figure 1). In addition to bikunin, this protein contains only one heavy chain, which is homologous to those of IαI [4,9]. Whereas IαI is the major bikunin-containing protein in man, PαI is the most abundant form in rat and bovine serum [10, 11].

The structure of the link between bikunin and the heavy chains has been studied by mass spectroscopy. This analysis showed that C-6 of an internal GalNAc residue of the chondroitin sulphate chain forms an ester bond with the α-carbon of the C-terminal amino acid residue of the heavy chains [12,13]. Furthermore, the esterified GalNAc residue was found to be unsulphated, a finding consistent with the observation that the chondroitin sulphate chain has all

---

*To whom correspondence should be addressed.

Figure 1

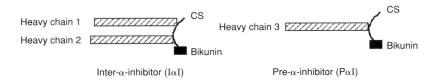

Heavy chain 1

Heavy chain 2

CS

Bikunin

Inter-α-inhibitor (IαI)

Heavy chain 3

CS

Bikunin

Pre-α-inhibitor (PαI)

Schematic representation of IαI and PαI

These proteins consist of one or two homologous polypeptides of ~80 kDa (the heavy chains) and bikunin. Bikunin carries an 8 kDa chondroitin sulphate chain (CS) that is covalently linked to the C-terminal amino acid residues of the heavy chains.

its sulphate groups located close to the bikunin polypeptide [14]. Whether this uneven distribution of the sulphate groups is essential for the coupling reaction remains to be seen.

Functions

As mentioned above, bikunin is a proteinase inhibitor. Its two domains have different specificities and bikunin therefore inhibits a wide range of proteolytic enzymes [15]. However, bikunin binds more weakly to those enzymes that have been studied than do other more abundant serum inhibitors [16] and it has therefore been proposed that bikunin does not function as a general proteinase inhibitor. Bikunin has also been shown to act as a growth factor for endothelial cells [17] and to inhibit the rise of cytoplasmic $[Ca^{2+}]$ upon lipopolysaccharide stimulation [18]. In view of the fact that the plasma level of bikunin is relatively stable [19,20], the physiological significance of these findings is unclear. In plasma, more than 90% of bikunin occurs in the form of IαI and PαI [21] and it is thus possible that local release of bikunin from these proteins by proteolytic cleavage might be of importance [22,23]. Free bikunin in plasma has a short lifetime and approximately half of the clearance occurs by glomerular filtration [24]. In the urine, bikunin is the major proteinase inhibitor and has therefore also been called urinary trypsin inhibitor. What appears to be a fragment of bikunin has been identified as the major inhibitor of kidney stone formation [25].

During an ovulatory burst, the cells surrounding an oocyte – the cumulus cells – start producing a hyaluronan-containing extracellular matrix, leading to an expansion of the cumulus-cell–oocyte complex. This process can also be made to occur *in vitro*; in the presence of serum, the cells will remain as a complex during the expansion, whereas in its absence, the cells will disperse [26]. In 1992 a serum protein was isolated from fetal bovine serum that could substitute for whole serum in stabilizing the cell complex [27]. Shortly afterwards, the protein was identified as PαI [11] and IαI was also shown to have the same effect [28]. Evidence that these observations were physiologically significant was provided by the finding that IαI appears in the ovary during the expansion process, apparently through leakage from surrounding blood vessels [29].

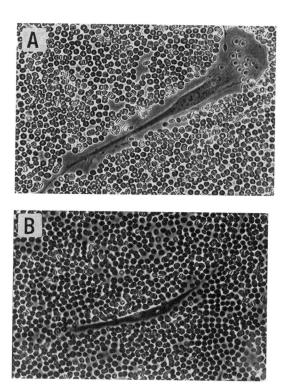

Figure 2

Detection of the hyaluronan-containing coat on fibroblasts

The cells were grown in plastic dishes and a suspension of fixed erythrocytes was added; (A) a normal fibroblast and (B) one that was treated with hyaluronidase just before the addition of the erythrocytes – note the zone of exclusion around the cell in (A). As discussed in the text, hyaluronidase-treated cells will reform their coats if serum or IαI is present in the medium.

Many cells in culture, as for example fibroblasts, have been found to be surrounded by a coat that excludes large particles such as bacteria or erythrocytes [30] (Figure 2A). The coat contains hyaluronan, as shown by the fact that it disappears when the cells are treated with hyaluronidase (Figure 2B). If cells that have been deprived of their coats are incubated in the presence of serum the coats form again [31]. The studies on the cumulus-cell–oocyte complexes prompted us to ascertain whether IαI would have a role in the formation of these coats. First we established that when hyaluronidase-treated fibroblasts were cultured in medium containing 15% human serum, some 35% of the cells formed coats [32]. We then found that approximately the same number formed coats when the serum was replaced by IαI of the corresponding serum concentration. In the same study we also observed that half-maximal formation of the coats occurred at a serum concentration of 30% [32]. This finding suggests that the extracellular matrix of fibroblasts *in vivo* could be affected by changes in the vascular permeability.

The molecular mechanisms by which IαI supports the formation of the hyaluronan-containing coats are still unclear, but a number of relevant observations have recently been made. Hyaluronan extracted from fibroblasts in culture has been shown to contain covalently bound heavy chains [33]. Furthermore, upon incubation of hyaluronan with serum, the heavy chains have been found to be released from bikunin and become covalently attached to the polysaccharide [33]. Thus it appears that hyaluronan can displace the chondroitin sulphate chain in IαI! Similar studies with cumulus-cell–oocyte complexes have shown that the mural granulosa cells produce a factor that promotes the displacement reaction [34].

It was reported many years ago that hyaluronan isolated from the synovial fluid of patients with rheumatoid arthritis has IαI attached to it [35]. More recently, it was shown that this hyaluronan contains the heavy chains only and that these are covalently linked to the polysaccharide in the same way as they are in IαI or PαI [36]. IαI has also been found to interact with tumour necrosis factor-stimulated gene-6 (TSG-6), a protein secreted by fibroblasts and other cells upon activation with the proinflammatory cytokines interleukin-1 or tumour necrosis factor-α [37]. When IαI and TSG-6 are incubated together, TSG-6 apparently displaces one of the heavy chains and becomes covalently linked to the chondroitin sulphate chain [37]. Interestingly, the antiplasmin activity of bikunin in these complexes is higher than in normal IαI [38]. Animal experiments have shown that TSG-6 has an anti-inflammatory activity and it has been suggested that this effect might be caused by the potentiation of bikunin [38]. The antiplasmin activity has also been implicated as the cause for bikunin inhibiting the metastasis of cancer cells [39,40].

Figure 3

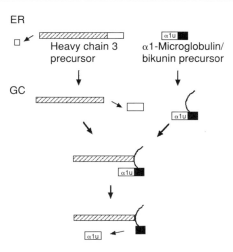

Biosynthesis of PαI

Heavy chain 3 is synthesized with both N-and C-terminal extensions, which are released in the endoplasmic reticulum (ER) and the Golgi complex (GC), respectively. Bikunin is synthesized as a precursor also containing α₁-microglobulin. When the bikunin precursor has acquired the chondroitin sulphate chain, which occurs in the Golgi complex, it becomes covalently linked to the heavy chain 3 and α₁-microglobulin is released. The assembly of IαI occurs in a similar fashion.

Biosynthesis

Just as the studies of the structures of IαI and PαI have yielded some remarkable results, so has the analysis of the biosynthesis of these proteins (Figure 3). Bikunin, to begin with, is synthesized as a precursor also containing α_1-microglobulin [41]; there is no known functional relationship between this protein and bikunin and it is therefore unclear why they are synthesized together. As the precursor reaches the Golgi complex, its chondroitin sulphate chain is completed and shortly afterwards, bikunin and α_1-microglobulin are released by a proteolytic cleavage after a dibasic sequence [42].

The heavy chains are synthesized with both N- and C-terminal extensions, the functions of which are unknown. The N-terminal propeptide seems to be removed in the endoplasmic reticulum [43]; such an early proteolytic processing step is very unusual for secretory proteins. The C-terminal peptide, on the other hand, is released as the protein acquires complex sugar structures [43] – presumably in the *trans*-Golgi. The cleavage occurs between an aspartate and a proline residue and all heavy chains for which the amino acid sequence has been determined have the same sequence around the cleavage site: DPHFII; no proteinase with such a specificity has previously been described [12].

Concluding remarks

IαI and PαI are the only proteins known in which a glycosaminoglycan forms a covalent link between polypeptides, but there may well be similar proteins that have not yet been discovered. In fact, we recently found that bikunin and heavy chain 3 upon co-expression in COS cells became covalently linked [10], which shows that a non-hepatic cell can also have the ability to form such a containing cross-link. The function of bikunin in IαI and PαI is still unclear. It is possible that one purpose of its intracellular coupling is to preserve the capacity of the heavy chains to become linked to hyaluronan until the proteins interact with cells. Except for oocyte maturation, the actual physiological function of the bikunin-containing proteins is still unclear. But the observation that they are required for the formation of the extracellular matrix of certain cells and that their production increases during fetal development [44] suggests that they have a fundamental role in tissue formation.

The work described in this paper was supported by grants from the Swedish Natural Science Research Council and Gustav V:s 80-års fond.

References
1. Heide, K., Heimburger, N. and Haupt, H. (1965) Clin. Chim. Acta **11**, 82–85
2. Steinbuch, M. (1976) in Methods in Enzymology (L. Lorand, ed.), pp. 760–772, Academic Press, Orlando
3. Schreitmuller, T., Hochstrasser, K., Reisinger, P.W.M., Wachter, E. and Gebhard, W. (1987) Biol. Chem. Hoppe-Seyler **368**, 963–970
4. Bourguignon, J., Diarra, M.M., Thiberville, L., Bost, F., Sesboue, R. and Martin, J.P. (1993) Eur. J. Biochem. **212**, 771–716
5. Malki, N., Balduyck, M., Maes, P., Capon, C., Mizon, C., Han, K.K., Tartar, A., Fournet, B. and Mizon, J. (1992) Biol. Chem. Hoppe-Seyler **373**, 1009–1018

6. Sjöberg, E.M. and Fries, E. (1990) Biochem. J. **272**, 113–118
7. Wachter, E. and Hochstrasser, K. (1979) Hoppe-Seyler's Z. Physiol. Chem. **360**, 1305–1311
8. Gebhard, W., Hochstrasser, K., Fritz, H., Enghild, J.J., Pizzo, S.V. and Salvesen, G. (1990) Biol. Chem. Hoppe-Seyler **371**, 13–22
9. Enghild, J.J., Thogersen, I.B., Pizzo, S.V. and Salvesen, G. (1989) J. Biol. Chem. **264**, 15975–15981
10. Blom, A.M., Thuveson, M. and Fries, E. (1997) Biochem. J. **328**, 185–191
11. Castillo, G.M. and Templeton, D.M. (1993) FEBS Lett. **318**, 292–296
12. Enghild, J.J., Salvesen, G., Hefta, S.A., Thogersen, I.B., Rutherfurd, S. and Pizzo, S.V. (1991) J. Biol. Chem. **266**, 747–751
13. Morelle, W., Capon, C., Balduyck, M., Sautiere, P., Kouach, M., Michalski, C., Fournet, B. and Mizon, J. (1994) Eur. J. Biochem. **221**, 881–888
14. Toyoda, H., Kobayashi, S., Sakamoto, S., Toida, T. and Imanari, T. (1993) Biol. Pharm. Bull. **16**, 945–947
15. Swaim, M.W. and Pizzo, S.V. (1988) Biochem. J. **254**, 171–178
16. Potempa, J., Kwon, K., Chawla, R. and Travis, J. (1989) J. Biol. Chem. **264**, 15109–15114
17. McKeehan, W.L., Sakagami, Y., Hoshi, H. and McKeehan, K.A. (1986) J. Biol. Chem. **261**, 5378–5383
18. Kanayama, N., Halim, A., Maehara, K., Kajiwara, Y., Fujie, M. and Terao, T. (1995) Biochem. Biophys. Res. Commun. **207**, 324–330
19. Smith, J.M., Balabanian, M.B. and Freeman, R.M. (1976) J. Lab. Clin. Med. **88**, 904–913
20. Daveau, M., Rouet, P., Scotte, M., Faye, L., Hiron, M., Lebreton, J.P. and Salier, J.P. (1993) Biochem. J. **292**, 485–492
21. Slota, A., Sjöquist, M., Wolgast, M., Alston-Smith, J. and Fries, E. (1994) Biol. Chem. Hoppe-Seyler **375**, 127–133
22. Balduyck, M., Piva, F., Mizon, C., Maes, P., Malki, N., Gressier, B., Michalski, C. and Mizon, J. (1993) Biol. Chem. Hoppe-Seyler **374**, 895–901
23. Kobayashi, H., Gotoh, J., Hirashima, Y. and Terao, T. (1996) J. Biol. Chem. **271**, 11362–11367
24. Sjöberg, E., Blom, A., Sjöquist, M., Wolgast, M., Alston-Smith, J. and Fries, E. (1995) Biochem. J. **308**, 881–887
25. Tang, Y., Grover, P.K., Moritz, R.L., Simpson, R.J. and Ryall, R.L. (1995) Br. J. Urol. **75**, 425–430
26. Salustri, A., Yanagishita, M. and Hascall, V.C. (1989) J. Biol. Chem. **264**, 13840–13847
27. Chen, L., Mao, S.J. and Larsen, W.J. (1992) J. Biol. Chem. **267**, 12380–12386
28. Chen, L., Mao, J.T., McLean, L.R., Powers, R.W. and Larsen, W.J. (1994) J. Biol. Chem. **269**, 28282–28287
29. Powers, R.W., Chen, L., Russell, P.T. and Larsen, W.J. (1995) Am. J. Physiol. **269**, E290–E298
30. Clarris, B.J. and Fraser, J.R.E. (1968) Exp. Cell Res. **49**, 181–193
31. Heldin, P. and Pertoft, H. (1993) Exp. Cell Res. **208**, 422–429
32. Blom, A., Pertoft, H. and Fries, E. (1995) J. Biol. Chem. **270**, 9698–9701
33. Huang, L., Yoneda, M. and Kimata, K. (1993) J. Biol. Chem. **268**, 26725–26730
34. Chen, L., Zhang, H., Powers, R.W., Russel, P.T. and Larsen, W.J. (1996) J. Biol. Chem. **271**, 19409–19414
35. Hutadilok, N., Ghosh, P. and Brooks, P.M. (1988) Ann. Rheum. Dis. **47**, 377–385
36. Zhao, M., Yoneda, M., Okashi, Y., Kurono, S., Iwata, H., Ohnuki, Y. and Kimata, K. (1995) J. Biol. Chem. **270**, 26657–26663
37. Wisniewski, H.G., Burgess, W.H., Oppenheim, J.D. and Vilcek, J. (1994) Biochemistry **33**, 7423–7429
38. Wisniewski, H., Hua, J-C., Poppers, D.M., Naime, D., Vilcek, J. and Cronstein, B.N. (1996) zJ. Immunol. **156**, 1609–1615
39. Kobayashi, H., Shinohara, H., Takeuchi, K., Itoh, M., Fujie, M., Saitoh, M. and Tearo, T. (1994) Cancer Res. **54**, 844–849
40. Kobayashi, H., Fuije, M., Shinohara, H., Ohi, H., Sugimura, M. and Terao, T. (1994) Int. J. Cancer **57**, 378–384
41. Kaumeyer, J.F., Polazzi, J.O. and Kotick, M.P. (1986) Nucleic Acids Res. **14**, 7839–7850
42. Sjöberg, E.M. and Fries, E. (1992) Arch. Biochem. Biophys. **295**, 217–222
43. Thogersen, I.B. and Enghild, J.J. (1995) J. Biol. Chem. **270**, 18700–10709
44. Salier, J.P., Chan, P., Raguenez, G., Zwingman, T. and Erickson, R.P. (1993) Biochem. J. **296**, 85–91

17

Hyaluronan–cell interactions in morphogenesis

Bryan P. Toole

Department of Anatomy and Cellular Biology, Tufts University School of Medicine, 136 Harrison Avenue, Boston, MA 02111, U.S.A.

During embryonic development, regeneration and wound healing the extracellular matrix surrounding migrating and proliferating cells is enriched in hyaluronan. Examples during embryonic morphogenesis include migrating and proliferating mesenchymal cells in the primary corneal stroma, somite cells around the notochord, neural crest cells exiting from the neural tube, and cushion cells invading the cardiac jelly (reviewed in [1]). High concentrations of hyaluronan are also present in regenerating amphibian limb blastemata, regenerating mammalian tendons, fetal mammalian wounds and in the stromal regions surrounding metastatic tumours [1–3]. One way in which hyaluronan facilitates migration is by creating hydrated pathways separating cellular or fibrous barriers to penetration by migrating cells [1,4]. Formation of hydrated pericellular matrices also may facilitate cell rounding during mitosis [5]. In addition to these physicochemical effects, however, hyaluronan directly influences cell behaviour via interaction with cell surface hyaluronan-binding sites or 'receptors'.

Hyaluronan interacts with receptors on the surface of cells

Two major classes of receptor that mediate the effects of hyaluronan on cell behaviour are CD44 [6] and RHAMM (receptor for hylauronan-mediated motility) [7].

CD44 was originally described as a cell surface glycoprotein involved in leucocyte behaviour but it is now clear that CD44 is very widely distributed and most likely involved in numerous physiological processes. Molecular approaches have shown that numerous transmembrane CD44 isoforms arise by alternative splicing [8]. Recently we have characterized several novel isoforms of CD44 that also arise by alternative splicing but lack cytoplasmic and transmembrane domains; consequently these isoforms are secreted as soluble glycoproteins [9]. The most widely distributed isoform, termed CD44H (H = haematopoietic) or CD44s (s = standard), corresponds to the 85 kDa, cell surface hyaluronan receptor described in earlier work [10], but all forms of CD44 contain the hyaluronan-binding domain within their primary sequence. Despite this, however, there are many cases of cells that express CD44 but do not bind hyaluronan [11]. Thus other factors, e.g. clustering of receptors [11–13], glycosylation [14], small alterations in primary sequence [15] or intracellular interactions [11,12,16], are involved in regulating the capacity of CD44-bearing cells to bind hyaluronan.

RHAMM is a hyaluronan-binding protein that is found both intracellularly and on the surface of many cell types. RHAMM is involved in cell migration, cell proliferation and cell transformation [17]. The hyaluronan-binding motif within RHAMM has been identified as [–B(X$_7$)B–], where B is arginine or lysine and X is any non-acidic amino acid. Strong evidence has been presented that this motif also mediates hyaluronan binding, at least in part, to CD44 and link protein [18]. We have identified additional hyaluronan-binding proteins that contain this motif, e.g. Cdc37, a cell cycle regulatory molecule [19].

Hyaluronan–receptor interactions influence cell behaviour

Interaction of endogenous cell surface hyaluronan with CD44 has been shown to mediate divalent cation-independent aggregation of several cell types, especially transformed cell lines. This aggregation occurs since hyaluronan is multivalent and can cross-bridge receptors on adjacent cells [1,4]. Also, in the case of some macrophages and lymphocyte lines, addition of exogenous hyaluronan will induce their aggregation by cross-bridging cell surface CD44 [20,21]. The role of these interactions in immune or transformed cell functions is not yet clear. However, hyaluronan-CD44 cross-bridging may be involved in interactions of lymphocytes with marrow stromal cells [22] and with the venular endothelial cells of lymphoid organs [6], and in mesenchymal cell condensations during embryonic development (see below).

CD44 is frequently associated with removal of hyaluronan during differentiation, e.g. during development of bone [23], lung [24] and hair follicles [25]. One of the morphogenetic events in which this occurs is the condensation of mesenchymal cells accompanying some epithelial–mesenchymal interactions [25,26]. These condensations most likely result from partial removal of peri- and intercellular hyaluronan by CD44-mediated internalization [25]; removal of hyaluronan would lead to elimination of the highly hydrated, intercellular spaces that are characteristic of non-condensed mesenchyme. Cross-bridging of the condensed cells via interaction of residual hyaluronan with receptors on the surface of adjacent cells could then occur, thus stabilizing the condensate. Such events have been demonstrated for condensing mesoderm from the embryonic chick limb [27–29].

Another way in which hyaluronan modulates cell interactions is by assembly of hyaluronan-dependent pericellular matrices or 'coats' [1,4,29]. Three components are required for pericellular matrix assembly by chondrocytes: hyaluronan, a hyaluronan-binding proteoglycan and a cell surface hyaluronan receptor [30,31]. However, it is apparent that CD44 is not required for pericellular matrix assembly by all cell types since, in some cases, the pericellular matrix appears to be stabilized by continued interaction of 'nascent' hyaluronan with hyaluronan synthase [32,33].

During embryonic development, the volume of intercellular space separating cells *in vivo* closely parallels their ability to form hyaluronan-dependent pericellular matrices in culture. For example, during limb development, early mesodermal cells are separated by extensive, hyaluronan-containing matrix *in vivo* and express voluminous, hyaluronan-dependent pericellular matrix in culture. However, subsequent condensation of mesoderm is parallelled by the inability to form these matrices in culture and, as explained above, the onset of hyaluronan-mediated cross-bridging of the condensed cells. Further differentiation of this mesoderm to cartilage is accompanied by extensive matrix formation *in vivo* and recovery of the ability to form extensive hyaluronan-dependent pericellular matrices in culture [29]. Prior to differentiation, myoblasts also are separated from each other by a hyaluronan-rich matrix *in vivo* and express pericellular matrices in culture; however, in this case differentiation and fusion to give myotubes is accompanied by irreversible loss of the ability to form these matrices [34].

Turley and co-workers [17,35] have shown that interaction of RHAMM with hyaluronan promotes cell locomotion. Hyaluronan—RHAMM interaction stimulates tyrosine phosphorylation of several proteins, including a key component of focal adhesions, p125^{FAK}, resulting in increased focal adhesion turnover and promotion of cell motility [36]. However, interaction of hyaluronan with CD44 also stimulates cell migration [37]; thus it seems likely that interaction of hyaluronan with either receptor can stimulate cell movement, but their relative importance would depend on the cell type or other physiological factors.

Several past studies have shown that hyaluronan synthesis and hyaluronan synthase activity vary with the proliferative state of cells, usually increasing during active division and declining at confluence, and that they fluctuate during the cell cycle, peaking at mitosis [5,38–40]. Inhibition of hyaluronan synthesis leads to arrest at mitosis, just before cell rounding and detachment [5]. Vertebrate hyaluronan synthase is a transmembrane protein with its active site at the inner surface of the plasma membrane [33,41–43]; 'nascent' hyaluronan is extruded across the membrane while the polymer is still attached to the synthase complex [44,45]. Thus hyaluronan extrusion on to the cell surface just prior to mitosis most likely creates a hydrated micro-environment that promotes partial detachment and rounding of the dividing cells [5,36,46]. Recent evidence suggests that intracellular hyaluronan-binding proteins, e.g. the cell cycle regulatory factor, Cdc37 [19], and an intracellular form of RHAMM [17,47], may be involved in regulation of these events, possibly via interaction with 'nascent' hyaluronan still attached to hyaluronan synthase.

Disruption of hyaluronan—receptor interactions inhibits tumorigenesis

Since the process of tumorigenesis mimics many of the cellular events of embryogenesis, we and others have also examined the relationship of hyaluronan-receptor interactions to tumour cell behaviour. Many malignant tumours contain high concentrations of hyaluronan [3], and recent studies have demonstrated that hyaluronan—RHAMM [48] and hyaluronan—CD44 [49] interactions are most likely involved in tumour progression.

One mechanism whereby hyaluronan–receptor interactions may be involved in tumorigenesis would be to mediate attachment of tumour cells to surrounding stromal tissues or to other sites of invasion. We have used the mouse ovarian ascites tumour, MOT, and breast ascites tumour, TA3/St, as syngeneic models to examine hyaluronan and CD44 distribution during tumour growth, attachment and invasion *in vivo* [50]. MOT is an ovarian tumour in which the cells attach and grow on the peritoneal surface in a similar fashion to human ovarian carcinomas. TA3/St is a breast carcinoma that also attaches to the peritoneal wall but then rapidly invades the entire mesentery and peritoneal wall. Using biotinylated proteoglycan as a specific probe for visualizing hyaluronan, we observed local hyaluronan accumulation at the initial sites of attachment of tumour cells to the surface of the mesentery and peritoneal wall, followed in time by high levels of accumulation throughout. However, the tumour cell lines synthesize very low amounts of hyaluronan in culture. Therefore, we have concluded that the observed hyaluronan accumulation *in vivo* results from increased synthesis and secretion by mesothelial cells and/or fibroblasts, most likely in response to stimulation by direct interaction with the tumour cells [51] or by their products [52,53]. We also found that tumour cells that initially attached to the mesentery and peritoneal wall were all CD44-positive even though, in the case of the MOT cells, the injected cells were mostly CD44-negative [50].

On the basis of the above results we postulated that initial attachment of tumour cells is mediated, at least in part, via interaction of hyaluronan produced by the mesentery with CD44 on the tumour cell surface. To obtain further evidence for this postulate we transfected the TA3/St cells with cDNAs for soluble CD44 [9], selected stable transfectants and injected these cells into the peritoneum of mice. The transfected cells did not form tumours in the peritoneal wall, whereas TA3/St cells mock-transfected with vector and non-transfected TA3/St cells formed tumours consistently (R. Moore, Q. Yu, I. Stamenkovic and B.P. Toole, unpublished work).

Hyaluronan oligomers may also act as antagonists of tumorigenesis since they would compete for interaction of hyaluronan polymer with CD44 [54]. To test this possibility, we introduced hyaluronan oligomers subcutaneously, at the site of B16 melanoma implantation in nude mice. To reduce losses of oligomer due to rapid diffusion, we used micro-osmotic pumps to bathe the tumours continuously with oligomer, then examined the size of tumours obtained in comparison with control tumours treated with buffer or with chondroitin sulphate. The B16 melanomas grew rapidly in both sets of controls. However, the size of tumours obtained in the presence of hyaluronan oligomer was only ~10% of that of controls (C. Zeng, J. Kuo, S. Kinney, B.P. Toole and I. Stamenkovic, umpublished work).

We conclude from the above experiments that hyaluronan–receptor interactions are essential to the progression of at least some tumour types.

References

1.   Toole, B.P. (1981) in Cell Biology of Extracellular Matrix (Hay, E.D., ed.), pp. 259–294, Plenum Press, New York
2.   Mast, B.A., Diegelmann, R.F., Krummel, T.M. and Cohen, I.K. (1992) Surg. Gynecol. Obstet. **174**, 441–451

3. Knudson, W., Biswas, C., Li, X.Q., Nemec, R.E. and Toole, B.P. (1989) in The Biology of Hyaluronan (Evered, D. and Whelan, J., eds.), Ciba Foundation Symposium No. 143, pp. 150–169, Wiley, Chichester
4. Toole, B.P. (1991) in Cell Biology of Extracellular Matrix, 2nd edn. (Hay, E.D., ed.), pp. 305–341, Plenum Press, New York
5. Brecht, M., Mayer, U., Schlosser, E. and Prehm, P. (1986) Biochem. J. **239**, 445–450
6. Aruffo, A., Stamenkovic, I., Melnick, M., Underhill, C.B. and Seed, B. (1990) Cell **61**, 1303–1313
7. Hardwick, C., Hoare, K., Owens, R., Hohn, H.P., Hook, M., Moore, D., Cripps, V., Austen, L., Nance, D.M. and Turley, E.A. (1992) J. Cell Biol. **117**, 1343–1350
8. Screaton, G.R., Bell, M.V., Jackson, D.G., Cornelis, F.B., Gerth, U. and Bell, J.I. (1992) Proc. Natl. Acad. Sci. U.S.A. **89**, 12160–12164
9. Yu, Q. and Toole, B.P. (1996) J. Biol. Chem. **271**, 20603–20607
10. Underhill, C.B., Green, S.J., Comoglio, P.M. and Tarone, G. (1987) J. Biol. Chem. **262**, 13142–13146
11. Lesley, J., Hyman, R. and Kincade, P.W. (1993) Adv. Immunol. **54**, 271–335
12. Bourguignon, L.Y., Lokeshwar, V.B., Chen, X. and Kerrick, W.G. (1993) J. Immunol. **151**, 6634–6644
13. Sleeman, J., Rudy, W., Hofman, M., Moll, J., Herrlich, P. and Ponta, H. (1996) J. Cell Biol. **135**, 1139–1150
14. Bartolazzi, A., Nocks, A., Aruffo, A., Spring, F. and Stamenkovic, I. (1996) J. Cell Biol. **132**, 1199–1208
15. Dougherty, G.J., Cooper, D.L., Memory, J.F. and Chiu, R.K. (1994) J. Biol. Chem. **269**, 9074–9078
16. Isacke, C.M. (1994) J. Cell Sci. **107**, 2353–2359
17. Entwistle, J., Hall, C.L. and Turley, E.A. (1996) J. Cell Biochem. **61**, 569–577
18. Yang, B., Yang, B.L., Savani, R.C. and Turley, E.A. (1994) EMBO J. **13**, 286–296
19. Grammatikakis, N., Grammatikakis, A., Yoneda, M., Yu, Q., Banerjee, S.D. and Toole, B.P. (1995) J. Biol. Chem. **270**, 16198–16205
20. Green, S.J., Tarone, G. and Underhill, C.B. (1988) Exp. Cell Res. **178**, 224–232
21. Lesley, J., Schulte, R. and Hyman, R. (1990) Exp. Cell Res. **187**, 224–233
22. Miyake, K., Underhill, C.B., Lesley, J. and Kincade, P.W. (1990) J. Exp. Med. **172**, 69–75
23. Pavasant, P., Shizari, T.M. and Underhill, C.B. (1994) J. Cell Sci. **107**, 2669–2677
24. Underhill, C.B., Nguyen, H.A., Shizari, M. and Culty, M. (1993) Dev. Biol. **155**, 324–336
25. Underhill, C.B. (1993) J. Invest. Dermatol. **101**, 820–826
26. Wheatley, S.C., Isacke, C.M. and Crossley, P.H. (1993) Development **119**, 295–306
27. Maleski, M.P. and Knudson, C.B. (1996) Exp. Cell Res. **225**, 55–66
28. Knudson, C.B. and Toole, B.P. (1987) Dev. Biol. 124, 82–90
29. Knudson, C.B. and Toole, B.P. (1985) Dev. Biol. **112**, 308–318
30. Knudson, C.B. and Knudson, W. (1993) FASEB J. **7**, 1233–1241
31. Knudson, C.B. (1993) J. Cell Biol. **120**, 825–834
32. Heldin, P. and Pertoft, H. (1993) Exp. Cell Res. **208**, 422–429
33. Itano, N. and Kimata, K. (1996) J. Biol. Chem. **271**, 9875–9878
34. Orkin, R.W., Knudson, W. and Toole, B.P. (1985) Dev. Biol. **107**, 527–530
35. Turley, E.A. (1992) Cancer Metastasis Rev. **11**, 21–30
36. Hall, C.L., Wang, C., Lange, L.A. and Turley, E.A. (1994) J. Cell Biol. **126**, 575–588
37. Thomas, L., Byers, H.R., Vink, J. and Stamenkovic, I. (1992) J. Cell Biol. **118**, 971–977
38. Tomida, M., Koyama, H. and Ono, T. (1974) Biochim. Biophys. Acta **338**, 352–363
39. Yoneda, M., Shimizu, S., Nishi, Y., Yamagata, M., Suzuki, S. and Kimata, K. (1988) J. Cell Sci. **90**, 275–286
40. Matuoka, K., Namba, M. and Mitsui, Y. (1987) J. Cell Biol. **104**, 1105–1115
41. Watanabe, K. and Yamaguchi, Y. (1996) J. Biol. Chem. **271**, 22945–22948
42. Shyjan, A.M., Heldin, P., Butcher, E.C., Yoshino, T. and Briskin, M. (1996) J. Biol. Chem. **271**, 23395–23399
43. Spicer, A.P., Augustine, M.L. and McDonald, J.A. (1996) J. Biol. Chem. **271**, 23400–23406
44. Philipson, L.H. and Schwartz, N.B. (1984) J. Biol. Chem. **259**, 5017–5023
45. Prehm, P. (1984) Biochem. J. **220**, 597–600
46. Koochekpour, S., Pilkington, G.J. and Merzak, A. (1995) Int. J. Cancer **63**, 450–454
47. Mohapatra, S., Yang, X., Wright, J.A., Turley, E.A. and Greenberg, A.H. (1996) J. Exp. Med. **183**, 1663–1669
48. Hall, C.L., Yang, B., Yang, X., Zhang, S., Turley, M., Samuel, S., Lange, L.A., Wang, C., Curpen, G.D., Savani, R.C., Greenberg, A.H. and Turley, E.A. (1995) Cell **82**, 19–28
49. Bartolazzi, A., Peach, R., Vink, J. and Stamenkovic, I. (1994) J. Exp. Med. **180**, 53–66

50. Yeo, T.K., Nagy, J.A., Yeo, K.T., Dvorak, H.F. and Toole, B.P. (1996) Am. J. Pathol. **148**, 1733–1740
51. Knudson, W., Biswas, C. and Toole, B.P. (1984) Proc. Natl. Acad. Sci. U.S.A. **81**, 6767–6771
52. Decker, M., Chiu, E.S., Dollbaum, C., Moiin, A., Hall, J., Spendlove, R., Longaker, M.T. and Stern, R. (1989) Cancer Res. **49**, 3499–3505
53. Asplund, T., Versnel, M.A. Laurent, T.C. and Heldin, P. (1993) Cancer Res. **53**, 388–392
54. Underhill, C.B., Chi-Rosso, G. and Toole, B.P. (1983) J. Biol. Chem. **258**, 8086–8091

18

Hyaluronan in embryonic cell adhesion and matrix assembly

Cheryl B. Knudson

Departments of Biochemistry and Pathology, Rush Medical College, Rush-Presbyterian-St. Luke's Medical Center, 1653 West Congress Parkway, Chicago, IL, 60612-3864, U.S.A.

Hyaluronan in cell adhesion

During embryonic development we observe the exquisite capability of cells to become organized into tissues. The field of cell adhesion is growing rapidly, and three general classes are recognized: homotypic adhesion, heterotypic adhesion and adhesion via an extracellular linker molecule. As first proposed by Grobstein [1], extracellular matrix macromolecules may function as inductive signals or as extracellular linker molecules mediating cell–cell adhesion by binding to the cell surface. It is this third class of cell adhesion wherein the role of hyaluronan as an extracellular linker molecule to cross-bridge cells, thus facilitating their aggregation, has been investigated.

Tissue formation requires cell-to-cell adhesion. The significance of hyaluronan in cell–cell adhesion during embryonic tissue formation and in the homeostasis of differentiated tissue has been shown in several experimental systems. More of this could be sorted out if we understood the regulation of: (a) the temporal and spatial deposition of hyaluronan; (b) the tissue-specific expression of hyaluronan receptors; and (c) the temporal expression of hyaluronan receptors. The association of hyaluronan with the cell surface can influence the behaviour of cells especially in regard to modulation of cell aggregation. At high concentrations hyaluronan blocks aggregation of several types of cells, whereas at low concentrations it mediates aggregation [2–9]. For example, endogenous cell surface hyaluronan on bone marrow stromal cells can mediate the adhesion of a B-cell hybridoma to these cells [8]. Alternatively, the addition of exogenous hyaluronan at low concentration is required for macrophage aggregation, whereas high concentrations block aggregation [4,7].

Multivalent cross-linking of cells by hyaluronan

A significant model was proposed by Toole [10] to explain the dual role of hyaluronan in mediating aggregation as well as inhibiting aggregation. In this model a minimal requirement for relatively high-molecular-weight hyaluronan was proposed to facilitate the cross-bridging of cells via the multivalent interaction of hyaluronan with receptors on adjacent cells. Residual hyaluronan fragments resulting from hyaluronidase treatment would be of such reduced size that these fragments would be unable to cross-bridge cells. Saturation of binding sites on cells by the over-production of endogenous hyaluronan (also mimicked by the addition of increasing amounts of exogenous hyaluronan) would prevent cell aggregation. This model was based on observations with mouse 3T3 fibroblasts and SV40-transformed 3T3 cells. Aggregation was found to depend on the multivalent cross-bridging of endogenous hyaluronan to hyaluronan receptors present on the adjacent cells [11,12]. The aggregation is inhibited with hyaluronidase treatment, supporting the requirement for hyaluronan. However, at high concentrations of exogenous hyaluronan the receptors on the individual cells become saturated, thus inhibiting cross-bridging [3].

Condensation during chondrogenesis

The work in my laboratory has focused on one cell system that apparently exhibits both temporal and spatial regulation of hyaluronan deposition and turnover as well as hyaluronan receptor expression – prechondrogenic limb bud mesenchyme. During limb development, cells make surface contact with one another during the condensation event, triggering the process of chondrogenesis [13,14]. Prior to the condensation process, the early limb mesenchymal cells are separated by an extensive extracellular matrix with hyaluronan being the predominant glycosaminoglycan. With the initiation of condensation, the hyaluronan distribution becomes patterned. The uniform distribution seen at earlier stages persists in the limb periphery, but there is a progressive decline in hyaluronan within the prechondrogenic and premyogenic condensations [15–17]. Coincident with the onset of condensation is the appearance of specific binding sites for hyaluronan on limb mesenchyme cells [18].

Condensation of embryonic limb bud mesenchyme can be studied with a variety of *in vitro* assays, some short term and some longer term, which include culture conditions that promote chondrogenesis. Initially we focused on cell adhesion assays and a cell suspension aggregation assay to study cell–hyaluronan–cell cross-bridging in this embryonic cell system. An intercellular adhesion assay [19] was adapted to avian limb mesenchyme; radio-labelled cells in suspension were added to an unlabelled, confluent monolayer of homologous cells for specific time intervals. After rinsing, the radioactivity in the monolayer was determined. Thus this assay tests the ability of cells in suspension to adhere to a homotypic monolayer. In this assay we detected stage specificity of adhesion; earlier stage cells did not adhere, whereas cells derived from early to late condensation stages show adhesion [20]. This adhesion is apparently mediated by endogenous hyaluronan. When we studied stage 24 limb bud cells pretreated with *Streptomyces* hyaluronidase, adhesion was reduced to background. This adhesion could be restored with exogenous hyaluronan (Sigma, Grade I; 0.1–0.5 mg/ml) yet adhesion was blocked with apparently saturation levels (1 mg/ml) of exogenous hyaluronan [20].

Intercellular aggregation of chondrogenic limb mesenchyme is stage-dependent, trypsin-sensitive and dependent on the presence of hyaluronan. Pre-condensation mesenchyme, which does not express hyaluronan binding activity, did not exhibit adhesion in our monolayer assay. Decreasing mesenchymal adhesion in the presence of increasing amounts of exogenous hyaluronan suggests that saturating the binding sites blocks intercellular adhesion by preventing two cells from binding to (and being cross-bridged by) the same hyaluronan molecule. Treatment of both the monolayer and cell suspension with *Streptomyces* hyaluronidase eliminated adhesion; exogenous hyaluronan was required to obtain ideal adhesion, whereas excess hyaluronan decreased adhesion. Results also suggest that there may be differences in the hyaluronan pericellular matrix of cells in monolayer culture and of cells in suspension that may influence cell-to-cell adhesion. In the absence of hyaluronan, adhesion was not restored with the addition of the glycosaminoglycan chondroitin sulphate or with hyaluronan hexasaccharides.

As a tool we use small hyaluronan oligosaccharides to disrupt the interaction of native hyaluronan with cells. Hyaluronan hexasaccharides have been demonstrated to be the smallest effective oligosaccharide to displace hyaluronan from cells. Incubation of cells with hexasaccharides displaces pre-bound [³H]hyaluronan from cell surfaces [21,22], and competes with binding when added together with [³H]hyaluronan to intact cells [23] or isolated cell membranes [24]. Hyaluronan hexasaccharides were used as a tool for probing cell surface hyaluronan receptor function [25] as the minimum size of hyaluronan oligosaccharide required to effectively compete for the binding of native hyaluronan to its cell surface receptor [26].

The possible involvement of hyaluronan as an extracellular linker molecule in the cell–cell adhesion event of pre-chondrogenic mesenchyme was investigated using two other model systems. Using a cell aggregation assay developed by Bee and von der Mark [27], most cells within a single-cell suspension of stage 22 mesenchyme incubated in serum-free medium become part of cell aggregates within 3 h. In the presence of hyaluronan hexasaccharides, the aggregates formed initially disperse into single cells (Figure 1). Therefore, larger hyaluronan molecules must be present for cross-bridging receptor sites on adjacent cells since, in fact, single cells were displaced from the cell aggregates formed in the suspension aggregation assay in the presence of hyaluronan hexasaccharides. The presence of hyaluronan hexasaccharides apparently can prevent the multiple receptor interactions that occur with native hyaluronan macromolecules.

Figure 2 depicts a model for cellular aggregate formation based on our work and that reported by others. Hyaluronan of high molecular weight is able to cross-bridge cells by multivalent binding to multiple receptors on adjacent cells. Addition of hyaluronan hexasaccharides to these cellular aggregates can displace the interactions of the native hyaluronan macromolecules with these cell receptors. Thus, as receptors become occupied with these small hyaluronan oligosaccharides, cells within the aggregate disperse. Competition between hyaluronan of different sizes could modulate tissue formation during organo-genesis.

Precartilage condensations similar to those formed *in situ* during chondrogenesis occur in micromass cultures of limb mesenchyme [28]. The cellular aggregates that form in early micromass cultures of limb mesenchyme subsequently differentiate into cartilage nodules [29,30]. This model system, developed by Ahrens et al. [29], has been used by many investigators to probe mechanisms of chondrogenesis. Since hyaluronan hexasaccharides did not

Figure 1

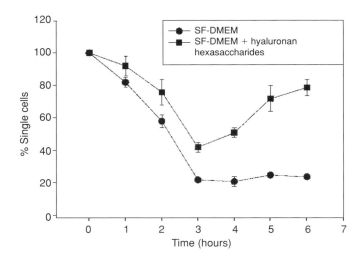

Stage 22 limb bud mesenchymal cell suspension aggregation assay

Four hours after isolation, a single-cell suspension formed aggregates in serum-free Dulbecco's modified Eagle's medium (SF-DMEM); (●). When aggregation was monitored in the presence of hyaluronan hexasaccharides (■) initial aggregates were observed within 3 h. Subsequently, in the presence of hexasaccharides single cells were displaced from the aggregates. Bars indicate the range from the mean percentage of triplicate determination. Redrawn from [20].

support aggregation of condensation stage prechondrogenic mesenchyme in our adhesion assays and, since hyaluronan hexasaccharides can displace receptor-bound endogenous hyaluronan from cells [21,22], the influence of including hyaluronan hexasaccharides in the micromass chondrogenesis system was investigated. A delay in the formation of condensations, shown by Hoffman Modulation optics and PNA-staining, occurred when hyaluronan receptor

Figure 2

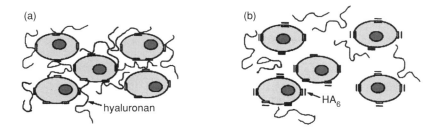

Cell–hyaluronan–cell cross-bridging

In this model of cell aggregate formation, cells are cross-bridged by the multivalent interaction of hyaluronan with receptors on adjacent cells (a). In the presence of hyaluronan hexasaccharides (HA$_6$), the binding of native hyaluronan to cell receptors is displaced. This competition results in disaggregation of cells (b).

function was disturbed by the addition of hexasaccharides [20]. Ultimately, the addition of hyaluronan hexasaccharides resulted in a temporal delay of chondrogenic differentiation of these mesenchymal cells in the micromass cultures. We found decreased matrix deposition (evaluated by Alcian Blue staining, keratan sulphate deposition and immunolocalization of type I versus type II collagen) during micromass chondrogenesis of limb bud mesenchyme cultured in the presence of hyaluronan hexasaccharides [20].

Using a lymphoblastic cell line, Pessac and Defendi [31] found that 50 kDa hyaluronan had a strong aggregation effect whereas oligosaccharides from a testicular hyaluronidase digest were not effective in aggregating these cells. In addition, Raja and co-workers [32] determined that a hyaluronan molecule of 150 monosaccharides in length (30 kDa) was the minimum size to cross-bridge two liver endothelial cell hyaluronan receptors. Thus in other systems in which hyaluronan-dependent cell adhesion occurs, size dependency is critical.

The use of HA_6 does not completely differentiate between potential hyaluronan receptors (RHAMM, IVd4, CD44) but rather competes for the binding of hyaluronan to any potential hyaluronan receptor. Nonetheless, this property is helpful in differentiating the specific binding of hyaluronan to cell surface receptors from its binding to extracellular matrix macromolecules.

Hyaluronan-anchored pericellular matrices

Figure 3 depicts our model for pericellular matrix assembly in which hyaluronan–receptor interactions retain hyaluronan decorated with matrix hyaladherins [26]. The addition of hyaluronan hexasaccharides has the capacity for matrix displacement. Although several cell types have hyaluronan-dependent pericellular matrices, our recent studies have focused on chondrocytes. After their isolation we cultured chondrocytes within alginate beads [33]. Depolymerization with sodium citrate releases chondrocytes with an intact pericellular matrix [34],

Figure 3

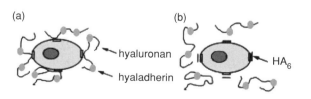

(a)   (b)   hyaluronan   hyaladherin HA_6

Retention of pericellular matrix via hyaluronan

In this model, two major components of the pericellular matrix, hyaluronan and hyaladherins, are indicated. Hyaladherins, which form aggregates with hyaluronan, are anchored to the cell surface via the interaction of the hyaluronan with cell surface receptors (a). In the presence of hyaluronan hexasaccharides (HA_6), the binding of native hyaluronan to cell receptors is displaced, resulting in the displacement of hyaladherins as well (b).

as revealed by Fraser's particle exclusion assay [2]. This pericellular matrix is displaced by incubation with hyaluronan hexasaccharides [34].

Embryonic tibial chondrocytes (day 12 chicken embryos) were labelled from 0 to 24 h with [³H]acetate. A 60 min incubation of these chondrocytes with hyaluronan hexasaccharides resulted in displacement of both newly synthesized chondroitin sulphate proteoglycans and hyaluronan from the pericellular matrix (Figure 4). We looked at earlier time points with labelling from 0 to 6 h after removal of cell surface hyaluronan with *Streptomyces* hyaluronidase. At the earliest time points, newly synthesized hyaluronan could not be displaced by hyaluronan hexasaccharides, whilst after a 6 h pulse, more than 70% of newly synthesized hyaluronan could be so displaced from the chondrocyte pericellular matrix (Figure 5).

Chondrocyte CD44 – hyaluronan binding

We have identified CD44 as a receptor on articular chondrocytes [34,35]. Does binding of hyaluronan to chondrocytes alter CD44? For these studies we used adult bovine articular chondrocytes released following 5 days of culture in alginate beads as 'matrix-intact' chondrocytes in comparison with those same cells treated for 1 h with *Streptomyces* hyaluronidase resulting in 'matrix-depleted' cells. From both cell populations, immunoprecipitation of total CD44 with the monoclonal antibody IM7.8.1 (Pharmagen) resulted in a doublet band at 80–90 kDa on a 6% SDS/PAGE. When the Western blot from the immunoprecipitated CD44 was probed for phosphothreonine, only the faster migrating band of the doublet was reactive. In addition, a 71 kDa phosphoprotein was detected with the anti-phosphothreonine antibody that co-immunoprecipitated with CD44 but

Figure 4

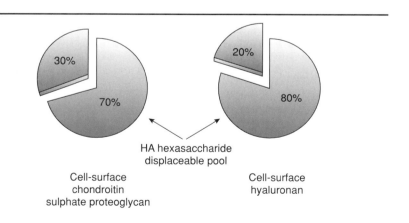

Embryonic chondrocyte pericellular matrix

Matrix components were pulse-labelled for 24 h, and the pericellular matrix pools were isolated. Of the total cell surface hyaluronan, 80% is in the hyaluronan hexasaccharide displaceable pool, and 70% of the chondroitin sulphate proteoglycan is also in the hyaluronan hexasaccharide displaceable pool.

Figure 5

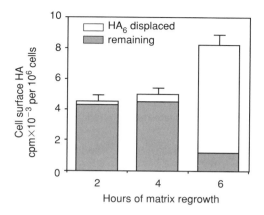

Movement of newly synthesized hyaluronan to the displaceable pool

Cell surface hyaluronan synthesized in the presence of radiolabel was analysed at three timepoints following matrix removal. Of the total cell surface hyaluronan present at 2 or 4 h of culture, very little could be displaced by hyaluronan hexasaccharides (HA₆), and remained anchored (grey bar) to the chondrocytes. However, by 6 h, 85% of the hyaluronan was in the hyaluronan hexasaccharide displaceable pool (open bar).

was not immunoreactive to IM7.8.1 on the Western blot. No bands were detected at either 71 kDa or 80–90 kDa from immunoprecipitation with an isotype control antibody. The phosphothreonine signal on CD44 was decreased on matrix-depleted chondrocytes. However, when matrix-depleted chondrocytes were incubated for 5 to 10 min with exogenous hyaluronan prior to immunoprecipitation, the phosphothreonine signal appeared up-regulated.

Could CD44 function as a mechanoreceptor for chondrocytes in addition to integrins [36] and annexin V [37]? Through differential detergent extraction of chondrocytes we obtained two CD44 pools; the first more soluble pool was released with NP-40 with the residual cytoskeleton pellet solubilized with Emphigen and DNase I treatment. CD44 was predominant in the cytoskeletal-associated pool of the matrix-intact chondrocytes whereas a decrease in this pool was observed from matrix-depleted chondrocytes.

Summary

Hyaluronan facilitates the formation of cellular aggregates by cross-linking cells through the multivalent binding to receptors on adjacent cells. Chondrogenic mesenchyme exhibits this hyaluronan-dependent cell aggregation. In addition, matrix retention during chondrogenesis and by differentiated chondrocytes is mediated via hyaluronan binding to cell surface receptors. CD44 is one hyaluronan receptor present on chondrocytes. CD44 exhibits increased actin association on matrix intact chondrocytes. In addition, the binding of hyaluronan to chondrocytes alters the phosphorylation of CD44.

This paper is dedicated to the memory of my teacher, Stephen Meier. Special thanks to Warren Knudson for many helpful discussions and to Torvard Laurent for his support of my participation in this conference. The diligent work of former and current members of my laboratory, Michael P. Maleski, Larrisa C. Harwick, Dean J. Aguiar, Carita T. Constable and Ghada A. Nofal has added to our understanding of hyaluronan. This work was supported by grants AR39507 and AR39239 from the National Institutes of Health, U.S.A.

References

1. Grobstein, C. (1967) Natl. Cancer Int. Monogr. **26**, 279–299
2. Fraser, J.R.E. and Clarris, B.J. (1970) Clin. Exp. Immunol. **6**, 211–225
3. Underhill, C.B. and Dorfman, A. (1978) Exp. Cell Res. **117**, 155–164
4. Love, S.W., Shannon, B.T., Myrvik, Q.N. and Lynn, W.S. (1979) J. Reticuloendothel. Soc. **25**, 269–282
5. Forrester, J.V. and Lackie, J.M. (1981) J. Cell Sci. **50**, 329–344
6. Wright, T.C., Underhill, C.B., Toole, B.P. and Karnovsky, M.J. (1981) Cancer Res. **41**, 5107–5113
7. Green, S.J., Tarone, G. and Underhill, C.B. (1988) Exp. Cell Res. **178**, 224–232
8. Miyake, K., Underhill, C.B., Lesley, J. and Kincade, P.W. (1990) J. Exp. Med. **172**, 69–75
9. Milstone, L.M., Hough-Monroe, L., Kugelman, L.C., Bender, J.R. and Haggerty, J.G. (1994) J. Cell Sci. **107**, 3183–3190
10. Toole, B.P. (1981) in Cell Biology of the Extracellular Matrix (Hay, E.D., ed.), pp. 259–294, Plenum Press, New York
11. Underhill, C.B. and Toole, B.P. (1981) Exp. Cell Res. **131**, 419–423
12. Underhill, C.B. (1982) J. Cell Sci. **56**, 177–189
13. Thorogood, P.V. and Hinchliffe, J.R. (1975) J. Embryol. Exp. Morphol. **33**, 581–606
14. Solursh, M. (1984) in The Role of Extracellular Matrix in Development (Trelstad, R.L., ed.), pp. 277–303, A.R. Liss, New York
15. Kosher, R.A., Savage, M.P. and Walker, K.H. (1981) J Embryol. Exp. Morphol. **63**, 85–98
16. Singley, C.T. and Solursh, M. (1981) Dev. Biol. **84**, 102–120
17. Knudson, C.B. and Toole, B.P. (1985) Dev. Biol. **112**, 308–318
18. Knudson, C.B. and Toole, B.P. (1987) Dev. Biol. **124**, 82–90
19. Walther, B.T., Ohman, R. and Roseman, S. (1973) Proc. Natl. Acad. Sci. U.S.A. **70**, 1569–1573
20. Maleski, M.P. and Knudson, C.B. (1996) Exp. Cell Res. **225**, 55–66
21. Knudson, C.B. (1995) Trans. Ortho. Res. Soc. **20**, 74
22. Nemec, R.E., Toole, B.P. and Knudson, W. (1987) Biochem. Biophys. Res. Commun. **149**, 249–257
23. Underhill, C.B. and Toole, B.P. (1979) J. Cell Biol. **82**, 475–484
24. Underhill, C.B., Chi-Rosso, G. and Toole, B.P. (1983) J. Biol. Chem. **258**, 8086–8091
25. Knudson, C.B. (1993) J. Cell Biol. **120**, 825–834
26. Knudson, C.B. and Knudson, W. (1993) FASEB J. **7**, 1233–1241
27. Bee, J.A. and von der Mark, K. (1990) J. Cell Sci. **96**, 527–536
28. Solursh, M., Ahrens, P.B. and Reiter, R.S. (1978) In Vitro **14**, 51–61
29. Ahrens, P.B., Solursh, M. and Reiter, R.S. (1977) Dev. Biol. **60**, 69–82
30. Oberlender, S.A. and Tuan, R.S. (1994) Development **120**, 177–187
31. Pessac, B. and Defendi, V. (1972) Science **175**, 898–900
32. Raja, R.H., McGary, C.T. and Weigel, P.H. (1988) J. Biol. Chem. **263**, 16661–16668
33. Hauselmann, H.J., Aydelotte, M.B., Schumacher, B.L., Kuettner, K.E., Gitelis, S.H. and Thonar, E.J.A. (1992) Matrix **12**, 130–136
34. Knudson, W., Aguiar, D.J., Hua, Q. and Knudson, C.B. (1996) Exp. Cell Res. **228**, 216–228
35. Chow, G., Knudson, C.B., Homandberg, G. and Knudson, W. (1995) J. Biol. Chem. **270**, 27734–27741
36. Durr, J., Goodman, S., Potocnik, A., von der Mark, H. and von der Mark, K. (1993) Exp. Cell Res. **207**, 235–244
37. King, K.B., Chubinskaya, S., Reid, D.L., Madsen, L.H. and Mollenhauer, J. (1997) J. Cell. Biochem. **65**, 131–144

19

Hyaluronan in malignancies

Warren Knudson

Departments of Biochemistry and Pathology, Rush Medical College, Rush-Presbyterian-St. Luke's Medical Center, 1653 West Congress Parkway, Chicago, IL, 60612-3864, U.S.A.

Historical perspective.

The association of hyaluronan with malignant tumours has been observed for nearly as long as the chemical identification of hyaluronan itself. Kabat [1] in 1939 described a mucinous, viscous fluid associated with virally induced chicken sarcomas. The viscous fluid was composed predominantly of a polysaccharide having an optical rotation, sensitivity to crude pneumococcus hyaluronidase and content of glucuronic acid similar to the polysaccharide just recently described by Karl Meyer, namely 'hyaluronic acid' [2]. In reality, the presence of hyaluronan in tumours was probably well known even at the turn of the century, well before techniques for definitive detection of hyaluronan were established. For example, Wells [3] in his 1907 text of *Chemical Pathology* discusses a 'mucin' associated with malignant breast carcinoma, "... the stroma, probably contains a connective tissue mucin, analogous to that of umbilical cord. This mucin is acid in reaction, is precipitated by acetic acid, and has an affinity for basic dyes."

Table 1 documents examples of tumours reported to be enriched in hyaluronan. As can be seen, hyaluronan is elevated in a wide variety of malignancies, both of epithelial as well as connective tissue origin –in humans as well as animals. The most striking prevalence of tumour-associated hyaluronan is observed in cases of mesothelioma and nephroblastoma (Wilms' tumour). The presence of hyaluronan in breast carcinoma has also been extensively investigated. The publication years are shown in Table 1 to illustrate how analysis of hyaluronan in many of these tumours has been revisited time and again, as the techniques for detection of hyaluronan have improved. The years encompass detection/quantification of hyaluronan by elemental and viscosity analyses, cellulose acetate electrophoresis methods, quantification of reducing sugars following hyaluronidase digestion, HPLC of hyaluronan disaccharides and, in more recent years, specific hyaluronan-binding-protein-based ELISAs, RIAs and morphological probes. The latter ELISA methods have also been applied to the analysis of small needle biopsies of benign/cancerous breast lesions [41]. Regardless of the methodology used (each 'state-of-the-art' for their time), hyaluronan continues to be a prominent extracellular matrix component associated with many tumours.

Interest in tumour-associated hyaluronan intensified after a role for hyaluronan in embryonic development was established. Toole [42,43] and others [44–47] observed that high concentrations of hyaluronan were deposited within

Table 1

Tumour type	Species	Year reported	References (respectively)
Mesothelioma	Human	1979,1985,1988,1989,1992	[4–9]
Mammary carcinoma	Human	1958,1976,1992,1993	[10–13]
Nephroblastoma (Wilms' tumour)	Human	1970,1978,1989,1990,1995	[14–18]
Lung carcinoma	Human	1981,1989	[19,20]
Basal cell carcinoma	Human	1990	[21]
Pancreatic carcinoma	Human	1956	[22]
Parotid gland carcinoma	Human	1981	[23]
Prostatic carcinoma	Human	1984	[24]
Hepatic carcinoma	Human	1975	[25]
Brain tumours	Human	1982,1993	[26–28]
Oesophageal, gastric, colonic carcinoma	Human	1978,1996	[29,30]
Chicken sarcoma (e.g. Rous)	Chicken	1939,1942,1954	[1,31,32]
Fibrosarcoma	Rat	1966	[33]
Melanoma	Mouse	1985	[34]
Lymphosarcoma	Mouse	1992	[35]
Mammary carcinoma	Mouse	1984,1996	[36,37]
Ovarian carcinoma	Mouse	1996	[37]
Mammary carcinoma	Dog	1979	[38]
V2 carcinoma	Rabbit	1979,1985	[39,40]

Examples of tumours reported to be highly enriched in hyaluronan

embryonic tissues coincident with the onset of cell migrations and/or periods of extensive cellular proliferation (see also Toole, Chapter 17). Further, these elevated hyaluronan levels were found to drop precipitously during subsequent stages of cytodifferentiation, coincident with cessation of cell migration and the establishment of differentiated cell phenotypes. This paradigm also occurred in adult tissues during wound healing [48–50] and early stages of limb regeneration in the amphibian [51,52]. The model suggested that the presence of hyaluronan, so closely timed with cellular migration and proliferation events, somehow was functioning to *facilitate* the process. The original theories included expansion, hydration and a general 'loosening' of the connective tissue, most likely owing to the unique hydrodynamic properties of hyaluronan [53,54]. More recent studies documenting hyaluronan-specific receptors (e.g. CD44 and RHAMM), present on many of the cells embedded within these hyaluronan-enriched matrices, suggest additional ways hyaluronan may facilitate migration, via the involvement of cell–matrix interactions [55]. A logical extension of this hyaluronan-in-morphogenesis model was that the elevated levels of hyaluronan associated with tumours, already well documented, were serving a similar function as they do in embryonic matrix – providing a highly hydrated environment conducive to cell migration and proliferation. An early attempt to test for such a role was made by

Toole and co-workers [39] using the rabbit V2 carcinoma. When implanted in the nude mouse, the rabbit V2 carcinoma grows as a well-encapsulated, benign tumour with little cellular invasion into the adjacent connective tissue. However, in the syngeneic rabbit host, the same V2 carcinoma cells are highly invasive and metastatic. Hyaluronan levels were substantially elevated (3–4-fold) in the invasive rabbit tumour as compared with the tumour established in the nude mouse. Additionally, within the invasive form of the tumour (rabbit host) the adjacent connective tissue contained three times more hyaluronan than the central tumour mass (tumour parenchyma). No such difference was observed in the non-invasive form of the V2 carcinoma (nude mouse host) where the surrounding connective tissue exhibited nearly the same low concentration of hyaluronan as the central tumour mass. This was one of the first demonstrations that the level of hyaluronan does, in fact, correlate with cell migration or invasiveness of a tumour. But, does this correlation reflect 'matrix-engineering' directed by the infiltrating cells or does aggressive cellular invasion trigger a host response (elevation in hyaluronan) in an effort to impede further migration?

It is often asked whether the increased deposition of hyaluronan is part of a mechanism by the host to 'wall-off' an invasive tumour. Unfortunately, only indirect evidence is available to address the question of whether hyaluronan plays a 'facilitation' or 'host defensive response' role during tumour invasion. First, by analogy, if tumour invasion fits with the embryonic/regeneration paradigm, hyaluronan is elevated during or coincident with cell migration (particularly the migration of less well-differentiated cell types), and is removed coincident with cessation of cell movements. Second, in the example of the V2 carcinoma a seemingly successful defence of tumour invasion was mounted and maintained by the nude mouse host, with little need for an elevation in hyaluronan, even within the dense capsular connective tissue surrounding the tumour. A corollary question is whether high concentrations of hyaluronan, in general, inhibit or promote cellular events such as migration or proliferation. For certain cell types such as neutrophils, the presence of high concentrations of hyaluronan significantly inhibit their migration [56]. Yet for many other cell types, such as fibroblasts, hyaluronan-rich matrices appear to promote migration [57]. Part of this dichotomy may depend on the level of expression of hyaluronan receptors by a particular cell type. As will be discussed below, migratory ability may also be modulated by the level and type of hyaluronan-binding proteoglycans (hyaladherins) associated with hyaluronan.

More recent studies regarding hyaluronan in malignancies

More recent studies on the association of hyaluronan and malignancies have been concerned primarily with the morphological localization of hyaluronan within sections of tumour tissues. This focus has followed on the heels of technical development of morphological probes that specifically detect hyaluronan within fixed or frozen tissue sections. Because hyaluronan is essentially non-immunogenic, the use of anti-hyaluronan antibodies has not been previously available. However, many indirect methods have evolved using particular connective tissue proteins that naturally bind hyaluronan with high affinity and specificity, e.g. aggrecan-like proteoglycans and cartilage link protein. Our laboratory reported previously on the localization of hyaluronan within frozen sections of human lung carcinoma using chondroitinase-digested aggrecan proteoglycan [58]. Enzymic treatment of this high-affinity hyaluronan binding protein generates an unsaturated disaccharide 'neo-epitope' that is recognized by a commercially available antibody, 2-B-6. Far better staining of tissue sections has been achieved in recent years using a biotinylated hyaluronan binding region (HABR)/link protein complex. HABR is derived from partial proteolytic digestion of aggrecan isolated from bovine articular cartilage. De la Torre and co-workers [13] used such a probe to demonstrate intense hyaluronan staining within sections of human breast carcinoma. Most of the hyaluronan was found deposited within the stromal elements of the tumour. The nests of malignant infiltrating cells were essentially negative. This study was especially helpful to our understanding of tumour-associated hyaluronan because: (1) it presented side-by-side serial sections of tissue stained with either biotinylated probe or Alcian Blue dye; (2) it showed an example section of normal breast tissue (mostly faint reactivity) and; (3) it highlighted a section of tumour that included the tumour *margin*. The latter clearly showed that the intense staining for hyaluronan was associated with the tumour-involved stroma, with a marked diminution of staining within the adjacent normal breast tissue (containing both epithelial and connective tissue elements). Using a similar probe, Yeo and co-workers [37] found intense hyaluronan staining of lining mesentery following attachment and growth of murine ovarian carcinoma cells. The mesenteric cells were suggested as the source of the hyaluronan as the ovarian cells have little capacity to synthesize hyaluronan, at least when assayed *in vitro*. The authors found similar findings following inoculation of mice with metastatic breast carcinoma cells. A study by Wang and co-workers [30] (published in the same journal issue as Yeo and co-workers), examined hyaluronan distribution within normal and cancerous human gastrointestinal tissues. Stratified squamous epithelia of normal oesophagus exhibited prominent hyaluronan staining, whereas the simple epithelia of the stomach and large intestine were essentially negative. The epithelial localization of hyaluronan was maintained only in well-differentiated oesophageal cancers, disappearing completely in more poorly differentiated tumours. Like the breast and ovarian cancers described above, the stroma of all of the gastrointestinal tumours displayed intense positive staining for hyaluronan stained as compared with normal connective tissue. Again, the oesophageal (poorly-differentiated), stomach and colon carcinoma cells were negative for hyaluronan. Nonetheless, one should not generalize from the above examples that the source of hyaluronan in all tumours is the stromal connective tissue. For example, in Wilms' tumour investigators using a similar biotinylated HABR morphological probe clearly demonstrated hyaluronan localized within the epithelial blastemal cells, with little reaction in the associated stromal compartment [17].

Tumour-derived hyaluronan stimulatory factors

One outcome of the morphological studies is that, for many tumours (particularly solid tumours of lung, breast, stomach, colon and even mesothelium), hyaluronan accumulation is predominantly localized within the tumour-adjacent connective tissue. In reality, these studies validate older work on carefully dissected tumour tissues [39,59]. Several explanations have been proposed to explain the spatial localization of hyaluronan. Many of the tumour cells, particularly those of epithelial-derived carcinomas, have little capacity to synthesize hyaluronan [58]. Thus, the most likely explanation is that 'signals' derived from the invading malignant cells communicate and direct the synthesis of hyaluronan by normal cells within the surrounding connective tissue (e.g. fibroblasts). Similar signalling mechanisms have been shown to regulate the synthesis of matrix-degrading enzymes that participate in tumour invasion [60,61]. *In vitro* co-cultures of various tumour cells together with normal fibroblasts have demonstrated the potential for tumour-directed control of hyaluronan synthesis by fibroblasts [58]. In some of these systems the tumour-derived stimulatory activity is expressed as a soluble, diffusible factor. In other systems direct cell–cell interaction is required for stimulation to occur. For example, conditioned medium from human mesothelioma cells contains soluble factors that stimulate hyaluronan synthesis in normal fibroblasts or mesothelial cells [62]. Antibodies to basic fibroblast growth factor (bFGF) as well as the BB-isoform of PDGF significantly inhibit this stimulation. For several years our laboratory has been characterizing a non-secreted, membrane-associated hyaluronan stimulatory factor present on a human lung carcinoma cell line, LX-1. These LX-1 cells stimulate hyaluronan synthesis in normal human fibroblasts when co-cultured together or following the addition of LX-1 membrane extracts to the fibroblasts [58]. We prepared monoclonal antibodies directed against a semi-purified LX-1 membrane protein preparation and were able to isolate two hybridoma clones producing antibodies that completely blocked stimulation of hyaluronan synthesis in co-cultures [63]. The exact identity of the stimulatory factor recognized by these antibodies is currently being investigated. However, we have obtained preliminary data concerning the distribution of this factor in various human tumour tissues. The antigen recognized by these two blocking antibodies is present on a variety of cultured tumour cells (e.g. cells derived from human mammary, colon and lung carcinoma) but no reactivity was observed on cultured human connective tissue cells such as fibroblasts. The antibodies were also found to stain formaldehyde-fixed, paraffin-embedded sections of human breast carcinoma. Figure 1(a), depicts an example of human infiltrating ductal carcinoma stained with a biotinylated HABR/link protein probe for hyaluronan. Like the work described above (De la Torre and co-workers [13]), the connective tissue stroma stains intensely for hyaluronan, whereas the nests of infiltrating tumour cells are essentially negative. Figure 1(b) represents a typical control for these types of proteoglycan-based probes, by the application of HABR/link protein complex in the presence of excess hyaluronan oligosaccharide (24-mer) as a control. The oligosaccharide specifically blocks the hyaluronan binding active site of the HABR/link protein probe. Figure 1(c) depicts immunostaining of a serial section with one of the blocking antibodies to

Figure 1

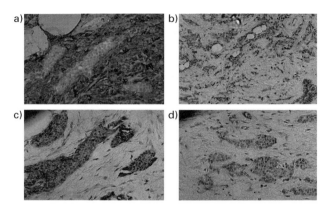

Localization of hyaluronan and hyaluronan stimulatory factor within sections of human breast carcinoma

Serial sections were made from one sample of formaldehyde-fixed, paraffin-embedded human breast carcinoma (infiltrative ductal carcinoma). Sections were deparaffinized and incubated with biotinylated HABR/link protein complex (a); HABR/link protein complex in the presence of hyaluronan oligosaccharide (b); Anti-LX-1 hyaluronan stimulator factor antibody (c) or, an equivalent IgG concentration (as compared with c) of IgG2b murine isotype control antibody (d). Following incubations washed sections were incubated with either Streptavidin peroxidase or peroxidase-conjugated goat-antimouse IgG, followed by reaction with diamino-benzidine substrate. Sections were then counterstained with Mayer's haematoxylin. All panels represent black and white reproductions of colour slides. Nests of infiltrative carcinoma cells shown in (a) have been digitally lightened to remove blue haematoxylin staining present in the colour slides. The blue counterstain of cells in control sections (c and d) has not been adjusted. Cells in (c) and (d) appear dark grey due to haematoxylin blue (present in the colour slides), not due to peroxidase-positive brown.

the membrane-bound stimulatory factor (Figure 1d depicts the IgG isotype control). As can be seen, the staining pattern essentially reflects an inverse image of the pattern observed in Figure 1(a). That is, the hyaluronan stimulatory factor is expressed on the surface of the infiltrating carcinoma cells (Figure 1c), and hyaluronan becomes deposited within the adjacent stroma (Figure 1a). More work is needed to clone, characterize and identify this stimulatory factor. However, the factor's localization serves to illustrate how hyaluronan accumulation in malignancies may be regulated. That is, hyaluronan in malignancies may reflect 'matrix engineering' that is directed locally by invading malignant cells.

As an additional point, it should be remembered that signalling between adjacent heterologous cell types is not a unique phenomenon of tumour tissues. Similar epithelial–mesenchymal interactions appear to regulate hyaluronan synthesis during embryonic limb development (where ectoderm-derived bFGF and TGFβ (transforming growth factor-β) activities are involved [64]). A factor related to TGFβ probably also participates in oocyte-induced stimulation of cumulus cell hyaluronan synthesis during cumulus expansion (see Hascall,

Chapter 8 of this volume and [65]). Thus, invasive tumour cells may revisit or re-utilize signalling mechanisms previously active during morphogenesis.

Does hyaluronan content in malignancies have diagnostic or prognostic value?

In some cancers such as mesothelioma and nephroblastoma, hyaluronan levels are elevated to such an extent that the elevation can be detected in the serum [8] or urine [18]. In a study of 37 mesothelioma patients, Dahl and co-workers [8] found elevated serum hyaluronan in all patients at time of presentation. However, those that presented with serum hyaluronan at, or above, 250 µg/l showed significant likelihood to fall into the progressive disease group, whereas those with lower values typically fell into a group that responded to therapy. They concluded that serum hyaluronan in this malignancy was predictive of progressive disease. In a study of 107 Wilms' tumour patients, urine hyaluronan was significantly elevated in 74% of the patients preoperatively as compared with normal control volunteers [18]. Hyaluronan values were significantly reduced postoperatively, returning to near normal in disease-free patients – more elevated in relapse/persistence patients. The authors also described a significant correlation between initial hyaluronan levels and clinical tumour staging. Initial studies on malignant breast carcinoma by Delpech and co-workers [66] suggested a significant elevation in serum hyaluronan in patients with disseminated metastatic disease as compared with malignant disease without metastasis or benign disease of the breast. However, subsequent studies [67] with a larger population of patients showed no prognostic significance of serum hyaluronan in breast cancer.

In a small number of studies, attempts have been made to quantify hyaluronan content within the primary tumour and compare these values with tumour grade or stage. For example, Delpech and co-workers [28] demonstrated in a sampling of 35 cases of malignant brain tumours (astrocytomas, gliomas and meningiomas) that hyaluronan levels were substantially elevated as compared with normal brain tissues, but showed little statistical change with tumour grade. The hyaluronan levels in brain tumours grades II through IV were all elevated and resembled concentrations found in fetal brain. However, the hyaluronan-binding proteoglycan found in association with these tumours, termed hyaluronectin, changes more dramatically with tumour grade – being low in fetal, higher in adult and progressively lower in tumours of increasing tumour grade. It has been suggested that hyaluronectin represents a degradation fragment of PG-M/versican, a hyaluronan-binding proteoglycan found in many connective/neural tissues [68]. The ratio of hyaluronectin to hyaluronan, on average, ranges from approximately 0.3 to 0.4 in fetal brain, rising to 1.3–6.5 in normal adult tissues. The ratios begin to decline to 0.8–4.3 in grade II and III tumours, then down to 0.01–0.4 in grade IV tumours [28]. A somewhat similar trend was found by the same group in a sampling of 71 breast cancer patients [12]. Hyaluronan levels within the primary tumour tissue were substantially elevated as compared with control normal tissues, but showed little change within breast cancer malignancies, grades I–III. Again, however, the ratio of hyaluronectin to

Figure 2

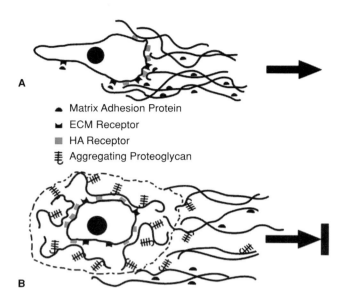

A

- ▲ Matrix Adhesion Protein
- ◣ ECM Receptor
- ■ HA Receptor
- 丰 Aggregating Proteoglycan

B

The effect of matrix composition on hyaluronan-mediated cell migration

A, Multiple cell–matrix interactions mediated via a variety of receptors are likely involved in cell migration. Hyaluronan may provide a hydrated milieu conducive to cell migration, and/or facilitate cell migration directly via interaction with cell surface hyaluronan receptors (e.g. CD44 and RHAMM). B, When migrating cells encounter new matrix environments enriched in both hyaluronan and hyaluronan-binding proteoglycans, cell migration is halted, perhaps owing to the assembly of pericellular matrices. Such matrices may sterically prevent cell–matrix interactions required for cell migration to proceed. Reproduced and modified from [55] with permission.

hyaluronan displayed a trend towards a decrease in ratio with increasing tumour grade. Certainly, far more patients will have to be examined before a general statement concerning changes in hyaluronan concentration with tumour grade or stage can be made. The above examples merely illustrate one possible outcome: that although hyaluronan will often be found elevated, for many tumours this level may not change with tumour progression. These data also suggest that more relevance should be made of the total glycosaminoglycan matrix composition of tumour tissues, i.e. content of hyaluronan and proteoglycan. As with the description of hyaluronan in tumour tissues, alterations in sulphated proteoglycan content within tumours is also not a new observation. In 1949, Sylvén described changes in "ester sulphuric acids in the stroma connective tissue (of carcinomas)" [69]. In his summary of 1100 cases of human carcinomas (all types) he noted that human carcinomas of 'low growth rate' exhibited large amounts of ester-sulphate protein complex; carcinomas of 'moderate growth rate', moderate amounts; and carcinomas of 'high growth rate' had small amounts. In other words, again a trend towards decreased proteoglycan with increased malignancy.

As a final comment, it is interesting to speculate on the biological signif-icance of proteoglycan to hyaluronan ratios within the extracellular matrix. Perris and co-workers [70] have demonstrated that directed migration of neural crest cells on laminin- or fibronectin-coated surfaces is completely inhibited by the exogenous addition of either PG-M/versican or aggrecan proteoglycan. Leech hyaluronidase and hyaluronan oligosaccharides counteract the proteoglycan inhibition, indicating that hyaluronan also participates (likely being synthesized by the neural crest cells). From these data a model could be postulated whereby embryonic cells migrate efficiently in an extracellular matrix rich in hyaluronan until they reach an area concomitantly enriched in proteoglycan, as illustrated in Figure 2. Adult tissues, although generally lower in hyaluronan, maintain a high proteoglycan:hyaluronan ratio. Within tumours of increasing malignant nature this ratio is again reversed as proteoglycan content is reduced, mimicking the permissive embryonic extracellular milieu, illustrated in Figure 2A. In our studies, addition of exogenous hyaluronan and proteoglycan to cells expressing CD44 hyaluronan receptors results in assembly of a pericellular matrix. Perhaps the assembly of such a pericellular matrix by migrating cells *in vivo* serves to sterically inhibit cell–matrix interactions necessary for continued migration. The presence or absence of such natural extracellular 'stop signals' in malignancies may be critical to understanding how hyaluronan contributes to the progression of this disease.

Many thanks are given to Drs. Raija and Markku Tammi for their assistance and gift of biotinylated HABR/link protein probe used to stain tissue shown in Figures 1(a) and 1(b). Thanks and credit is also given to James C. Haupt, my graduate student in the Department of Biochemistry, for his staining of sections shown in Figures 1(c) and 1(d). The work reported from our laboratories was supported in part by research grants AR39239 and AR43384 from the National Institutes of Health.

References

1. Kabat, E.A. (1939) J. Biol. Chem. **130**, 143–147
2. Meyer, K. and Palmer, J.W. (1934) J. Biol. Chem. **107**, 629–634
3. Wells, H.G. (1907) in Chemical Pathology, pp. 411–430, W.B. Sanders, Philadelphia
4. Arai, H., Kang, K., Sato, H., Satoh, K., Nagai, H., Motomiya, M. and Konno, K. (1979) Am. Rev. Respir. Dis. **120**, 529–532
5. Kawai, T., Suzuki, M., Shinmei, M., Maenaka, Y. and Kageyama, K. (1985) Cancer **56**, 567–574
6. Dahl, I.M.S. and Laurent, T.C. (1988) Cancer **62**, 326–330
7. Roboz, J., Chahinian, A.P., Holland, J.F., Silides, D. and Szrajer, L. (1989) J. Natl. Cancer Inst. **81**, 924–928
8. Dahl, I.M.S., Solheim, Ø.P., Erikstein, B. and Müller, E. (1989) Cancer **64**, 68–73
9. Heldin, P., Asplund, T., Ytterberg, D., Thelin, S. and Laurent, T.C. (1992) Biochem. J. **283**, 165–170
10. Ozzello, L. and Speer, F. (1958) Am. J. Pathol. **34**, 993–1003
11. Takeuchi, J., Sobue, M., Sato, E., Shamoto, M., Miura, K. and Nakagaki, S. (1976) Cancer Res. **36**, 2133–2139
12. Bertrand, P., Girard, N., Delpech, B., Duval, C., D'Anjou, J. and Dauce, J.P. (1992) Int. J. Cancer **52**, 1–6
13. De la Torre, M., Wells, A.F., Bergh, J. and Lindgren, A. (1993) Hum. Pathol. **24**, 1294–1297
14. Allerton, S.E., Beierle, J.W., Powars, D.R. and Bavetta, L.A. (1970) Cancer Res. **30**, 679–683
15. Hopwood, J.J. and Dorfman, A. (1978) Pediatr. Res. **12**, 52–56
16. Kumar, S., West, D.C., Ponting, J.M. and Gattamaneni, H.R. (1989) Int. J. Cancer **44**, 445–448
17. Longaker, M.T., Adzick, N.S., Sadigh, D., Hendin, B., Stair, S.E., Duncan, B.W., Harrison, M.R., Spendlove, R. and Stern, R. (1990) J. Natl. Cancer Inst. **82**, 135–138
18. Lin, R.Y., Argenta, P.A., Sullivan, K.M., Stern, R. and Adzick, N.S. (1995) J. Pediatr. Surg. **30**, 304–308

19. Horai, T., Nakamura, N., Tateishi, R. and Hattori, S. (1981) Cancer **48**, 2016–2021
20. Li, X.Q., Thonar, E.J.A. and Knudson, W. (1989) Conn. Tissue Res. **19**, 243–253
21. Wells, A.F., Lundin, Å., Michaëlsson, G. and Pontén, F. (1990) Acta Dermatol. Venereol. **71**, 274–275
22. Cudkowicz, G. (1956) Br. J. Cancer **10**, 758–762
23. Takeuchi, J., Sobue, M., Sato, E., Yoshida, M., Uchibori, N. and Miura, K. (1981) Cancer **47**, 2030–2035
24. De Klerk, D.P., Lee, D.V. and Human, H.J. (1984) J. Urol. **131**, 1008–1012
25. Kojima, J., Nakamura, N., Kanatani, M. and Ohmori, K. (1975) Cancer Res. **35**, 542–547
26. Bertolotto, A., Giordana, M.T., Magrassi, M.L., Mauro, A. and Schiffer, D. (1982) Acta Neuropathol. **58**, 115–119
27. Giordana, M.T., Bertolotto, A., Mauro, A., Migheli, A., Pezzotta, S., Racagni, G. and Schiffer, D. (1982) Acta Neuropathol. **57**, 299–305
28. Delpech, B., Maingonnat, C., Girand, N., Chauzy, C., Maunoury, R., Olivier, A., Tayot, J. and Creissard, P. (1993) Eur. J. Cancer **29A**, 1012–1017
29. Symonds, D.A. (1978) Arch. Pathol. Lab. Med. **102**, 146–149
30. Wang, C., Tammi, M., Guo, H. and Tammi, R. (1996) Am. J. Pathol. **148**, 1861–1869
31. Pirie, A. (1942) Br. J. Exp. Pathol. **23**, 277–284
32. Harris, R.J.C., Malmgren, H. and Sylvén, B. (1954) Br. J. Cancer **8**, 141–146
33. Danishefsky, I., Oppenheimer, E.T., Heritier-Watkins, O. and Willhite, M. (1966) Cancer Res. **26**, 229–232
34. Turley, E.A. and Tretiak, M. (1985) Cancer Res. **45**, 5098–5105
35. Liverman, S. and Sawyer, R.H. (1992) Biochem. Biophys. Res. Commun. **114**, 976–984
36. Knudson, W., Biswas, C. and Toole, B.P. (1984) J. Cell Biochem. **25**, 183–196
37. Yeo, T., Nagy, J.A., Yeo, K., Dvorak, H.F. and Toole, B.P. (1996) Am. J. Pathol. **148**, 1733–1740
38. Palmer, T.E. and Monlux, A.W. (1979) Vet. Pathol. **16**, 493–509
39. Toole, B.P., Biswas, C. and Gross, J. (1979) Proc. Natl. Acad. Sci. U.S.A. **76**, 6299–6303
40. Iozzo, R.V. and Müller-Glauser, W. (1985) Cancer Res. **45**, 5677–5687
41. Haupt, J.C., Cobleigh, M., Wolter, J. and Knudson, W. (1995) Proc. Am. Soc. Clin. Oncol. **14**, 115
42. Toole, B.P. and Trelstad, R.L. (1971) Dev. Biol. **26**, 28–35
43. Toole, B.P. (1972) Dev. Biol. **29**, 321–329
44. Markwald, R.R., Funderberg, F.M. and Bernanke, D.H. (1979) Texas Rep. Biol. Med. **39**, 253–270
45. Derby, M.A. (1978) Dev. Biol. **66**, 321–336
46. Pratt, R.M., Larsen, M.A. and Johnston, M.C. (1975) Dev. Biol. **44**, 298–305
47. Anderson, C.B. and Meier, S. (1982) J. Exp. Zool. **221**, 329–335
48. Reid, T. and Flint, M.H. (1974) J. Embryol. Exp. Morphol. **31**, 489–495
49. Hasty, K.A., Smith, G.N.J. and Kang, A.H. (1981) Dev. Biol. **86**, 198–205
50. Mast, B.A., Flood, L.C., Haynes, J.H., Depalma, R.L., Cohen, K., Diegelmann, R.F. and Krummel, T.M. (1991) Matrix **11**, 63–68
51. Toole, B.P. and Gross, J. (1971) Dev. Biol. **25**, 57–77
52. Smith, G.N., Toole, B.P. and Gross, J. (1975) Dev. Biol. **43**, 221–232
53. Toole, B.P. (1991) in Cell Biology of Extracellular Matrix (Hay, E.D., ed.), pp. 305–339, Plenum Press, New York
54. Fisher, M. and Solursh, M. (1977) J. Embryol. Exp. Morphol. **42**, 195–207
55. Knudson, C.B. and Knudson, W. (1993) FASEB J. **7**, 1233–1241
56. Forrester, J.V. and Wilkinson, P.C. (1981) J. Cell Sci. **48**, 315–331
57. Docherty, R., Forrester, J.V., Lackie, J.M. and Gregory, D.W. (1989) J. Cell Sci. **92**, 263–270
58. Knudson, W., Biswas, C., Li, X.Q., Nemec, R.E. and Toole, B.P. (1989) in The Biology of Hyaluronan (Evered, D. and Whelan, J., eds.), Ciba Foundation Symposium No. 143, pp. 150–169, Wiley, Chichester
59. Bouziges, F., Simon-Assman, P., Leberquier, C., Marescaux, J., Bellocq, J.P., Haffen, K. and Kedinger, M. (1990) Int. J. Cancer **46**, 189–197
60. Biswas, C. (1982) in Tumor Invasion and Metastasis (Liotta, L.A. and Hart, I.R., eds.), pp. 405–425, Martinus Nijhoff Publishers, The Hague/Boston/London
61. Graf, M., Baici, A. and Strauli, P. (1981) Lab. Invest. **45**, 587–596
62. Asplund, T., Versnel, M.A., Laurent, T.C. and Heldin, P. (1993) Cancer Res. **53**, 388–392
63. Knudson, W., Haupt, J.C. and Glant, T.T. (1991) J. Cell Biol. **115**, 445a
64. Knudson, C.B., Munaim, S.I. and Toole, B.P. (1995) Dev. Dyn. **204**, 186–191
65. Salustri, A., Yanagishita, M. and Hascall, V.C. (1989) J. Biol. Chem. **264**, 13840–13847
66. Delpech, B., Chevalier, B., Reinhardt, N., Julien, J.P., Duval, C., Maingonnat, C., Bastit, P. and Asselain, B. (1990) Int. J. Cancer **46**, 388–390

67. Ponting, J., Howell, A., Pye, D. and Kumar, S. (1992) Int. J. Cancer **52**, 873–876
68. Knudson, W. and Kuettner, K.E. (1997) in Primer on the Rheumatic Diseases (Wortmann, R.L., ed.), Arthritis Foundation, Atlanta
69. Sylvén, B. (1949) Acta Radiol. **59**, 11–16
70. Perris, R., Perissinotto, D., Pettway, Z., Bronner-Fraser, M., Mörgelin, M. and Kimata, K. (1996) FASEB J. **10**, 293–301

20

Hyaluronan and receptors: retrospective and perspective

Bernard Pessac

CNRS UPR 9035, Développement & Immunité du Système Nerveux Central, 15 rue de l'Ecole de Médecine, 75270 Paris cedex 06, France

Cherchons comme cherchent ceux qui doivent trouver et trouvons comme ceux qui doivent chercher encore (Saint Augustin)

Twenty five years ago

For my MD thesis, I was given the 'simple' goal of studying cell adhesion between normal cells as well as between cancerous cells. I was able to demonstrate that malignant cells form aggregates significantly larger than normal cells. Further experiments showed the presence of cell aggregation factors in the culture media of these cells. The goal of my fellowship at the Wistar Institute was to investigate the properties and mechanisms of action of these 'aggregation factors' (AFs). It appeared that they were polydisperse, essentially high-molecular-weight molecules that were not destroyed by proteases, thus suggesting that they were not proteins. I discussed these results with Leonard Warren and Mary Glick whose laboratory was located at the University of Pennsylvania, right across the street from Wistar. Not unexpectedly, we agreed that the factors under study were sugars or sugar-containing molecules. One of the suggestions was to determine the effect of various enzymes, including hyaluronidase, on the biological effect of these aggregation factors. The very clear-cut result of these experiments was that all types of hyaluronidases, including testicular hyaluronidase, abrogated the aggregating effect of culture medium. Further experiments showed that indeed hyaluronan (HA) was the main aggregation factor released by these culture cells [1].

The next question was: how does HA aggregate cells? Various models could be put forth, one of which being that AFs produced and released by cells bind to molecularly distinct receptors at the cell surfaces.

A prediction of this hypothesis was that the release of the AFs could be dissociated from the presence of receptors. Therefore, aggregating cells should release AFs (i.e. HA) and possess cell surface 'receptors'. In contrast, the absence of cell aggregation could be accounted for by the absence of AFs and cell receptors or of only one of them.

To verify this hypothesis, it was necessary to investigate different cell lines to determine if they could fit these criteria. The Wistar Institute was the ideal laboratory for this type of investigation since it was a main centre of studies on cells in culture. Therefore, all available cells were screened for their aggregating capacity and we selected those that did not aggregate. These non-aggregating cells were then checked for their capacity to release AFs, i.e. essentially HA, and to respond to AFs or HA. As predicted some cell lines did not release and did not bind AF and/or HA while other cell lines did not release any AFs but possessed receptors.

Taken together these experiments indicated that AF (and HA) aggregated cells by binding to cell surface 'receptors' and suggested that AF and the cell surface receptors were not identical but distinct molecules [1, 2].

To investigate the molecular nature of these receptors, cells that only displayed receptors but did not release AF were submitted to various enzymic treatments and then assayed for their capacity to respond to AFs (HA) by aggregation under appropriate conditions. Only proteases had a clear-cut effect. Indeed, cells treated with pure trypsin completely lost their capacity to aggregate. Conversely, as expected, this treatment had no effect on the AFs.

This led us to the conclusion that AF and receptors are indeed distinct molecular entities. Thus while AF are essentially, if not only, glycosaminoglycans and for the most part hyaluronan, the molecules located at the cell surfaces have at least a protein moiety [1,2].

The presence of receptors required for cell adhesion at the surface of cells derived from the haemopoietic and immune systems was at that time a great puzzle. Indeed, it was difficult to explain how cells known to be single could synthesize molecules required for cell adhesion. Of course things look much clearer now since it is most probable that a main category of HA receptors is the CD44 molecule family, which is known to be widely distributed in particular in the haemopoietic and immune systems. Furthermore, it is now known that these cells can adhere to various substrates such as endothelia [3].

The discovery that AF, i.e. HA and their receptors, are totally distinct molecular entities was a turning point in the biology of HA. Indeed up to that time, HA was considered as a molecule more or less floating in the extracellular medium or matrix, in other words HA might have been considered as tangential to cell surfaces. The discovery of receptors for HA suggested a new biological role requiring contact with cells. In this new geometrical distribution HA was then perpendicular to cell surfaces.

In addition, the finding that proteins located at cell surfaces might be involved in the cell adhesion process was the starting point of an explosion of investigations, which rapidly led to the discovery of fibronectin and of N-CAM.

Furthermore, in this seminal work it was shown for the first time that cells from the nervous tissue synthesize and release glycosaminoglycans. In addition, the relationship between glycosaminoglycans and cancer, in particular metastasis, was discussed for the first time.

HA in 2021

If a scientist makes a retrospective on HA research in 25 years from now what will he report? It may be speculated that, then, one shall know what HA looks like *in vivo*. Indeed, HA being present at the surface of most – if not all – cells, it is crucial to determine its structure, shape and interaction with protein under distinct biological conditions [4,5].

Recent experimental evidence indicates that *in vivo* HA is rigidified and perpendicular to cells, owing to its interaction with proteoglycans [6]. These data fit well the colloidal model of de Gennes [7].

In this context, some basic principles governing ligand–receptor interactions may be revisited. Indeed, the constant of association (and/or dissociation) for antigen and antibodies, substrates and enzymes, any type of ligand and its receptor at the cell surface should take into account the extraordinary heterogeneity of the environment in which two molecules have the probability to meet.

Since HA and its receptors are also present in the nervous tissue [8,9], one may speculate that the opening of neurotransmitters and voltage-gated channels might be dependent in some way on the shape and structure of HA.

Last but not least, a major area in which one may expect an explosion of data is the transduction of messages induced by the binding of HA to its receptor. Indeed, very recent data indicate that in lymphocytes the CD44 molecule is linked to a phosphotyrosine kinase, the p56[lck], known to transduce signal from the CD44 molecule [10].

There are exciting years to come!

References

1. Pessac, B. and Defendi, V. (1972) Science **175**, 898–900
2. Pessac, B. and Defendi, V. (1972) Nature **238**, 13–15
3. Sherman L., Sleeman J., Herrlich, P. and Ponta H. (1994) Curr. Opin. Cell Biol. **6**, 726–733
4. Scott, J.E. (1995) J. Anat. **187**, 259–269
5. Kohda, D., Morton, C.J., Parkar, A.A., Hatanaka, H., Inagaki, F.M., Campbell, I.D. and Day, A.J. (1996) Cell **86**, 767–775
6. Lee, G.M., Johnstone, B., Jacobson, K. and Caterson, B. (1993) J. Cell. Biol. **123**, 1899–1907
7. de Gennes, P.G. (1987) Adv. Colloid Interface Sci. **27**, 189–209
8. Yasuhara, O., Akiyama, H., McGeer, E.G. and McGeer, P.L. (1994) Brain Res. **635**, 269–282
9. Eggli, P.S. and Graber, W. (1996) J. Neurocytol. **25**, 79–87
10. Taher, T.E.I., Smit, L., Griffioen, A.W., Schilder-Tol, E.J.M., Borst, J. and Pals, S.T. (1996) J. Biol. Chem. **271**, 2863–2867

21

The viscoelastic intercellular matrix and control of cell function by hyaluronan

Endre A. Balazs

Matrix Biology Institute, 65 Railroad Avenue, Ridgefield, NJ 07657, U.S.A.

To begin with, I would like to take you back half a century to July 17, 1947, when I gave a lecture in Stockholm at the VI International Congress of Cytology. This was the first international congress after the war at which biochemists, cytologists and physiologists met. This was also the first time that I had the opportunity to present the results of my eight years of research on the extracellular matrix and hyaluronan to an international scientific audience. Interestingly, my lecture was the only one at the Congress on the subject of hyaluronan and hyaluronidase. The experiments I described were carried out on embryonic chicken fibroblasts cultured in homologous plasma coagulum. I discussed the effect of hyaluronan and hyaluronidase on the migration and mitosis of first explant fibroblasts in fibrin coagulum. I prepared the hyaluronidase myself from bull testicles, and the hyaluronan from bovine synovial fluid. Hyaluronan alone had no effect on the fibroblasts, but when the enzyme and the hyaluronan were added together, or when hyaluronidase-degraded hyaluronan alone was added, a stimulation of cell migration and mitosis occurred. Heparin strongly inhibited this effect. As a matter of fact, the appropriate heparin concentration completely inhibits the growth of fibroblasts in these cultures in the presence, as well as in the absence, of hyaluronidase. I must emphasize that these cells were not growing on plastic surfaces, but inside of a fibrin clot, which is an *in vivo* model of wound healing and a completely different cell growth model than that which is currently used in tissue culture experiments. Nevertheless, this was the first demonstration that low-molecular-weight (enzyme-degraded) hyaluronan stimulates fibroblast growth and migration in fibrin coagulum. This was also the first experimental report on the biological action of hyaluronan on cells [1].

Migration of cells

It was not until 20 years later that I repeated these experiments with embryonic fibroblasts cultured in plastic dishes. Unlike the cells inside the fibrin coagulum, the cells attached to the plastic surface could not be stimulated to grow by the same hyaluronan–hyaluronidase system. But this time, we discovered a new effect of hyaluronan. If the cells were seeded on the surface of a high molecular weight, elastoviscous solution of hyaluronan, they rounded up and did not migrate or divide. On the other hand, when the same type of fibroblast was seeded on the plastic and started to divide or move, the same elastoviscous hyaluronan solution poured over them had no effect on cell migration or mitosis [2]. This was recently confirmed using an elastoviscous solution of hylan, a hyaluronan derivative with enhanced molecular weight (see Chapter 28 by Larsen).

By the late 1960s, we had succeeded in isolating the non-inflammatory fraction of hyaluronan (NIF-NaHA) from human umbilical cord and from rooster combs [3]. This preparation was sterile, pyrogen-free and had a weight-average molecular weight, M_w, of 3–4 × 10^6. We used this highly purified NIF-NaHA to study its effect on the random migration of lymphoid cells from blood, lymph nodes, thymus, spleen or ascites lymphosarcoma (YAS). This hyaluronan preparation inhibited the *in vitro* random migration of all these cells, provided it had the appropriate elastoviscosity [2]. Two natural hyaluronan-containing fluids of the body, namely the synovial fluid [4,5] and the liquid vitreus of humans and monkeys, completely inhibited migration, and with increasing dilution, the effect decreased and finally disappeared. Viscous gelatin or DNA solutions showed the same inhibitory effect [2]. This effect of elastoviscous media on macrophage migration was confirmed later by using various viscosity agarose solutions [6]. Forrester and Wilkinson also confirmed these findings with a directed (chemotactic) locomotion model using human neutrophils [7]. The inhibitory effect was dependent on the concentration and average molecular weight of hyaluronan. These studies also demonstrated that the hyaluronan effect was not eliminated by adding excess Ca^{2+} and was not related to the polyanionic nature of the molecule, because sulphated glycosaminoglycans had no effect. Recently, the inhibitory effect was also demonstrated on human glioma cell migration [8].

Migration of embryonic smooth muscle cells (second passage, trypsinized cells) into hydrated collagen gels showed stimulation of cell migration by low-molecular-weight hyaluronan (concentration 1 μg/ml), which was cell specific. Cells harvested from embryonic ductus aorticus were stimulated in a concentration-dependent manner, but cells from the aorta were not [9]. Hyaluronan binding sites on the cell surface play an important role in control of locomotion of cells exposed to hyaluronan *in vitro* [10].

Lymphocytes

Elastoviscous solutions of hyaluronan inhibit the transformation of lymphocytes to lymphoblasts [11]. After peripheral blood (human) lymphocytes were stimulated with mitogens (phytohaemagglutinin, pokeweed mitogen, streptomycin O and others), they were placed into solutions of hyaluronan of various average M_w (0.7–2 × 10^6) and concentrations (0.5–2.0 mg/ml). These solutions prevented the transformation to lymphoblasts, as long as the cells were in the 'inductive phase' (G_0–G_1 phase of the first generation cycle). Cells advanced to the stage of DNA synthesis were not inhibited. The effect was not based on a toxic effect on the cells, because after the hyaluronan was removed, the cells completed their transformation. The inhibitory effect was dependent on the elastoviscosity of the hyaluronan solution used, i.e. on the product of concentration and average molecular weight. The inhibitory effect was also dependent on the distance between the cells adhering to the glass surface. Increased crowding of the lymphocytes counteracted the inhibitory effect. Since cell–cell interaction is essential in the inductive phase of transformation, we assumed that the inhibitory effect was the result of the separation of the cells from each other by a continuous molecular network of hyaluronan, which acted as a molecular barrier or produced an exclusion effect. Most importantly, the inhibition of migration and transformation of lymphocytes was not observed when the hyaluronan used in the same or even higher concentration was first degraded by enzymic, oxidative or free radical processes. Low-molecular-weight hyaluronan (<100 000) tetra- and hexasaccharides over a broad concentration range (1–10^4 µg/ml) were also ineffective. The inhibitory effect could not be abolished by adding low-molecular-weight hyaluronan or the monosaccharide moieties of hyaluronan (glucuronic acid, N-acetylglucosamine) to the media. All these observations indicate that the inhibitory effect is due to the molecular network structure or to the elastoviscosity of the solution and not to a specific receptor-mediated reaction. These results were later confirmed by others [12].

Since lymphocyte migration and proliferation were regarded to be essential in the graft-versus-host reaction, we tested this hypothesis using spleen cells from A/Jax adult mice injected into (C_{57} G_1/6J × A/Jax) F male hybrids [2]. Before injection, the donor cells (2.5×10^7 living cells) were suspended in culture media containing hyaluronans of the same concentration (7 mg/ml) but with various M_w (0.06, 0.5; 1.4; 2.2×10^6). Ten days after the intraperitoneal injection of the cell suspension, the mice were killed and the spleen:body weight ratio was determined. The increase of the spleen weight due to infiltration of mononuclear cells was inhibited when the donor cells were suspended in elastoviscous media containing high-molecular-weight hyaluronan (M_w 1.4×10^6 and 2.2×10^6). Histological examination showed that the spleens of these animals were normal. In contrast, the control animals in which the donor cells were suspended in media with no hyaluronan present showed an increased spleen:body weight ratio and extensive perivascular infiltration of mononuclear cells in the spleen. The use of non-elastoviscous solutions made of low-molecular-weight hyaluronan (M_w 0.06 and 0.5×10^6) did not show any effect.

Lymphocytes from individuals immunized with allogenic cells destroy such 'target' cells *in vitro*. The active movement and modulation of lymphocytes are believed to be essential for the cytotoxic effect of these immunologically activated cells. Since both the movement and modulation of lymphocytes are suppressed by elastoviscous hyaluronan solutions, we tested the effect of such hyaluronan solutions on the cytotoxic reaction of cells. Under various experimental conditions, when stimulated lymphocytes were separated from the target cells by elastoviscous hyaluronan solutions, the cytotoxic destruction of these cells was completely prevented. The protection of the target cells from the effector cells depends both on the concentration and average molecular weight of the hyaluronan present in the medium and consequently on the elastoviscosity of the hyaluronan solutions [2,13].

We carried out the study on the effects of elastoviscous hyaluronan solutions on lymphocyte function one step further, to the survival of skin allographs in *in vivo* systems [2]. As donors, C_{57} BL/6J adult male mice were used, with 2–3-month-old male mice (Balb/c) as skin graft recipients. Hyaluronan

(M_w 2.2×10^6) in 2% solution was administered once topically in the graft bed (0.1 ml) and injected subcutaneously under the graft area (0.2 ml). As control, the same volume of physiological saline solution was used, or alternatively 5 mg of hyaluronan was injected intraperitoneally. Only the topically administered hyaluronan extended significantly the survival time of grafts by approximately 30%. We believe that the 'molecular barrier effect' of the elastoviscous hyaluronan slowed down the migration of lymphocytes and thereby delayed the allograph rejection process.

Protection of cells

As mentioned above, the cytotoxic destruction of target cells by lymphocytes is prevented by elastoviscous hyaluronan solutions in the media. Chondrocytes can be also protected *in vitro* from various agents that may cause release of proteoglycans or have toxic effects on these cells by elastoviscous hylan A solution and hylan B gels, which are derivatives of hyaluronan.

The release of ^{35}S-labelled proteoglycan from cartilage explants is stimulated by various agents [interleukin-1, mononuclear cell conditioned media, oxygen-derived free radicals (ODFR) and polymorphonuclear leucocyte-lysate]. This stimulation was eliminated by elastoviscous solutions of hyaluronan and hylan A (M_w $4–7 \times 10^6$, concentration 2 mg/ml) and the effect was related to the elastoviscosity, not to the concentration of the polysaccharide [14]. Chondrocytes (labelled with ^{51}Cr) in a confluent monolayer are sensitive to the damage caused by ODFR. This damaging effect can be measured by the release of ^{51}Cr by the damaged cells. Elastoviscous hylan A solutions protect the cells from oxidative damage. In the same cell system, the toxic effect of polymorphonuclear leucocyte-lysate on chondrocytes was also eliminated when the cells were surrounded by elastoviscous hylan A solutions. In all these cases, solutions with lower elastoviscosity, but containing the same concentration of hylan A, did not provide protection. The effect can be explained by the accumulation of the large hylan A molecules in or on the surface of the pericellular matrix of chondrocytes, forming a molecular barrier (chemical or physical) to the damaging effect of various agents [14].

The generation of ODFR from polymorphonuclear leucocytes (stimulated with phorbol monistate acetone) is also inhibited by hyaluronan, hylan A and hylan B. This effect is dependent on viscoelasticity; dilution of hyaluronan or hylan preparations results in a decreased inhibitory effect. This protective effect may also be explained by the ability of hyaluronan or hylan to form a barrier to the stimulating agent (PMA). Addition of hyaluronan after exposure to PMA (stimulant) has little or no inhibitory effect on ODFR generation [14].

The protective effect of hyaluronan solutions on synovial and cartilage cells was later confirmed by various *in vitro* and *in vivo* experiments. Monolayer cultures of human synovial cells release fibronectin in the presence of some growth factors. This release can be inhibited by hyaluronan solutions (1 µg/ml, M_w 1.9×10^6). Monolayers of rabbit articular cartilage cells release proteoglycans in the presence of interleukin-1. This release can be inhibited by solutions of hyaluronan, depending on molecular weight and concentration [15]. Degenerative changes of the cartilage (erosion, fibrillation, loss of proteoglycans and chondrocytes) after partial lateral meniscectomy and section of the fibular collateral and sesamoid ligaments in rabbits could be 'lessened' by elastoviscous solutions (1%, M_w 1.9×10^6) of hyaluronan administered intra-articularly twice weekly for 4 weeks. The protective effect was less effective when solutions of the same concentration but lower M_w (0.8×10^6) hyaluronan were used [16].

Elastoviscous hyaluronan and hylan solutions and gels have high elasticity, especially at high frequencies (0.2–10 Hz). This elasticity acts as a shock absorber and mechanical protector for cells against physical impacts. It has been demonstrated that monolayers of chondrocytes can be protected from the mechanical impact of a falling weight [17] by a layer of elastoviscous hyaluronan or hylan solution.

Cell differentiation

A similar mechanism most likely operates in the experimental model using chicken leg muscle myoblasts cultured on hyaluronan substrate. Hyaluronan prevented the cells from fusion and subsequent differentiation, while their replication was uninhibited. This could be reversed when the cells were removed and placed in a non-hyaluronan substrate. Like mitogen-stimulated lymphoblast transformation, the elastoviscous hyaluronan solutions acted as a molecular barrier preventing the lymphoblastic or myogenic expression progress [18].

The *in vitro* stimulatory effect of non-elastoviscous hyaluronan solutions was demonstrated in other differentiating cell models [19]. Hyaluronan-coated substrates caused an increase in the number of limb mesenchymal cells differentiating to cartilage-producing cells. This stimulation of chondrogenic differentiation is inversely dependent on the molecular weight of hyaluronan, because the greatest activity was observed using hyaluronan of M_w 2–4×10^5, while hyaluronan of average M_w $>10^6$ inhibited the effect completely.

This finding parallels the effect of hyaluronan on lymphoblast transformation. In these studies elastoviscous solutions of hyaluronan were not used; therefore, the inhibitory effect was not explored.

Bryan Toole's pioneering work on the role of hyaluronan during embryonic development contributed importantly to our understanding of the regulatory role of this molecule in the intercellular matrix. Hyaluronan, by virtue of its large hydrated molecular volume and its capacity to form molecular matrices, can expand the interfibrillar collagen space to accommodate movement of cells, but also can form barriers to cell migration [20]. Signal molecules, interacting with specific cell membrane binding sites, are involved in this intriguing regulatory system that controls cell traffic and perhaps also cell differentiation during embryogenesis and regeneration.

Cell proliferation

The finding that hyaluronan molecules can influence proliferation of fibroblasts *in vitro* was confirmed in experiments using other cell types. Goldberg and Toole used Simian virus-transformed 3T3 cells (SV-3T3) and 3T3 cells as well as rabbit synovial tissue cells cultured for 2 years that maintained their ability to produce hyaluronan *in vitro* [21]. The proliferation, as measured by ^{3}H-thymidine incorporation, was inhibited only by solutions of hyaluronan-containing molecules with average $M_w = 2–3 \times 10^6$ and in concentrations greater than 0.1 µg/ml. Lower-molecular-weight hyaluronan showed stimulation of cell proliferation, especially at lower concentrations (0.0001–0.01µg/ml). These studies confirmed the same pattern of hyaluronan effect on cell activity, namely elastoviscous solutions are inhibitory, but low concentration solutions containing low-molecular-weight hyaluronan are stimulatory.

Hyaluronan synthesis

The intimate connections between mononuclear macrophages and hyaluronan synthesis were first recognized in the vitreus. It was shown that the phagocytic cells of the cortical gel vitreus synthesize, accumulate, excrete and, by pinocytosis, re-utilize hyaluronan. We called these hyaluronan-producing phagocytic cells hyalocytes [22]. The hyaluronan-producing activity of hyalocytes and peritoneal mononuclear phagocytes, as well as chicken embryo fibroblasts (first explants) has been shown to be sensitive to the hyaluronan present in the pericellular environment. The *in vitro* hyaluronan synthesis, measured by incorporation of 6-^{3}H glucosamine by these cells, was stimulated by increasing the concentration (0.2 to 2 µg/ml) of hyaluronan (average M_w 2×10^6) in the culture medium (B. Jacobson and E.A. Balazs, unpublished work; [23]).

Synovial cells from osteoarthritic patients cultured *in vitro* also showed 20 to 100% stimulation when the hyaluronan (M_r 0.75×10^6) concentration in the culture media was greater than 0.5 µg/ml. These observations illustrate that the increased extracellular hyaluronan concentration can stimulate hyaluronan synthesis by a positive feedback mechanism.

The stimulation of hyaluronan synthesis *in vivo* was demonstrated by using various physical injuries. In rabbit joints, the hyaluronan content could be increased 2- to 5-fold by such diverse stimuli as immobilization of the joint, insertion of a sterile needle, destabilization of the joint by cutting intra-articular ligaments and contusions of the cartilage [24]. This suggests that the hyalocytes of the synovial tissue respond with increased synthesis of hyaluronan after wounding of any of the intra-articular tissues (cartilage, synovium, ligaments).

Phagocytic cells

One of the most interesting effects of hyaluronan is that on the phagocytic activities of macrophages [4,25,26]. Phagocytosis can be divided into two phases:

first is the attachment of the particles to the cell surface, and second is their ingestion. The first is not an active cellular process and occurs at low temperatures (2–3 °C), while the second is an active process of living cells and occurs only at 37 °C. Using latex beads and human peripheral blood neutrophils, we found that solutions of hyaluronan of M_w 2.4×10^6 stimulate the attachment, but only at low concentration (<200 μg/ml). The peak of the stimulatory effect is at 10 μg/ml. Hyaluronan with a M_w of 300000 also stimulates the attachment, but with a peak effect at 100 to 200 μg/ml. Hyaluronan solutions made of molecules with M_w of $0.1–2.4 \times 10^6$ had no effect on the attachment at 300 μg/ml or greater concentrations (see Figure 1). At this concentration, the molecules form a continuous molecular network and the solutions become elastoviscous. In effect, the cell surfaces are saturated with hyaluronan.

The ingestion phase of the latex beads is also profoundly influenced by hyaluronan solutions made of molecules with M_w of $0.1–2.4 \times 10^6$ (Figure 2). The solution containing the low-molecular-weight hyaluronan stimulates the ingestion with maximum effect at 100–200 μg/ml, the same range at which the stimulation of attachment is observed. In contrast, the solution made from ten times larger molecules ($M_w = 2.4 \times 10^6$) stimulates only at very low concentrations (<50 μg/ml). At concentrations greater than 150 μg/ml, the effect is inhibitory. The phagocytosis is completely inhibited at a concentration of 400 μg/ml. Here

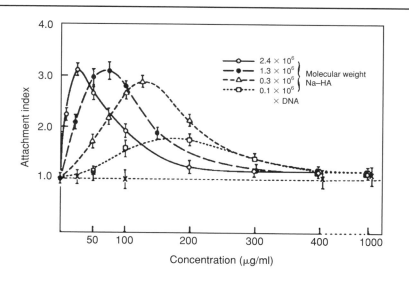

Figure 1

The stimulation of the attachment of latex beads to human neutrophils

Attachment index is defined as the ratio of the number of beads attached per cell in the presence of hyaluronan or DNA to the number of beads attached per cell in the absence of any test material. The 1.1 μm beads were complement-coated. Each data point represents the mean and the standard error of the mean (vertical lines) of 5 experiments, each containing control and test material. Thirty cells were counted in each experiment. Medical grade (sterile, pyrogen-free, non-inflammatory) NIF-NaHA was used. (E.A. Balazs and M. Paul, unpublished work.)

Figure 2

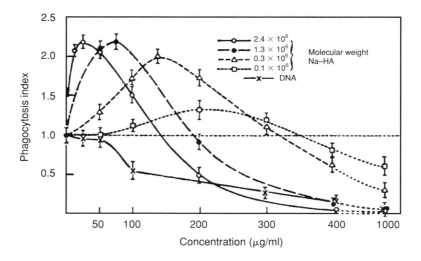

The effect of elastoviscous solutions of hyaluronan and DNA on the ingestion of latex beads by human neutrophils

The phagocytosis index is defined as the ratio of the number of beads ingested by the cells in the presence of hyaluronan or DNA to the number of beads ingested per cell in the absence of any test material. For further details, see legend of Figure 1. (E.A. Balazs and M. Paul, unpublished work.)

again, the inhibitory effect is directly related to the elastoviscosity of the solution. What is interesting is that the attachment is not inhibited by the viscoelastic solutions, but the ingestion is. This means that the elastoviscous solution does not prevent the particles from being attached to the surface of the cell, but prevents them from being ingested.

There is a narrow range of concentration where hyaluronan of high molecular weight neither stimulates nor inhibits phagocytosis. This 'neutral' concentration range for a 2.4×10^6 M_w hyaluronan is around 120 to 150 µg/ml, but for another molecular weight species ($M_w = 1.3 \times 10^6$) this 'neutral point' is around 200 µg/ml. For even smaller M_w species (3×10^5) it is in the range of 300 µg/ml. This is consistent with the overall finding that as the solution of hyaluronan reaches a certain elastoviscosity, it becomes inhibitory. This inhibitory effect overrides the stimulatory effect using hyaluronan of any molecular weight, provided the elastoviscous phase is reached (E.A. Balazs and F. Slahetka, unpublished work).

Mononuclear phagocytes (peritoneal or alveolar) are also sensitive to hyaluronan. Their phagocytic activity is stimulated or inhibited by hyaluronan solutions in the same way as neutrophils (E.A. Balazs and F. Slahetka, unpublished work). The concentration-dependent stimulatory effect and the inhibitory effect of hyaluronan solutions were confirmed using human blood granulocytes [27–29] and human blood monocytes [30,31]. Elastoviscous solutions of hyaluronan (4 µg/ml) also inhibited the uptake of aggregated IgG and the simultaneous release of lysozyme [32] by human peripheral blood polymorphonuclear leucocytes.

Hyaluronan-containing body fluids, like liquid vitreus and synovial fluid, show the same effects as pure hyaluronan solutions. This is demonstrated in Figure 3, where the effect of synovial fluid obtained from healthy horse joints on phagocytosis of mouse peritoneal macrophages is shown. When the fluid is diluted with physiological salt solution the inhibition of phagocytosis is decreased. The only slightly diluted and the undiluted synovial fluid inhibit phagocytosis completely. Identical effects were observed using human synovial fluid or primate liquid vitreus. Plasma fibronectin in low concentrations (10 μg/ml) stimulates phagocytosis, and when added in this concentration to hyaluronan solutions, it abolishes the inhibition effect at up to a concentration of 100 μg/ml (Figure 3) (E.A. Balazs and F. Slahetka, unpublished work).

To elucidate the mechanism of inhibition of phagocytosis by elasto-viscous hyaluronan solutions, we tested the effect of elastoviscous solutions made of purified gelatin and high-molecular weight DNA. Both inhibited the ingestion phase of phagocytosis, but had no effect on the adhesion phase. Most importantly, neither one of them showed any stimulatory effect (Figures 1 and 2) at any concentration. This viscosity (specific viscosity) dependence of the inhibition of macrophage phagocytosis is shown in Figure 4. The specific viscosity of the solutions of DNA and gelatin was varied by concentration, and in the case of

Figure 3

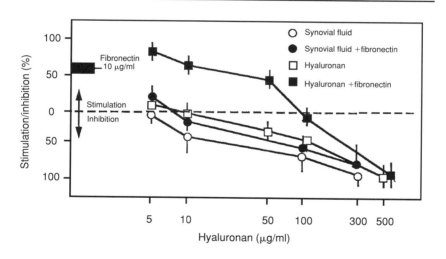

The effect of healthy equine synovial fluid on the phagocytosis of latex beads by mouse peritoneal macrophages

Percentage stimulation and inhibition on the ordinate represents the effect of test materials compared with control (zero line), the absence of test materials. Each data point represents the mean and standard error of the mean (vertical lines) of five experiments. Solutions of various concentrations of hyaluronan were obtained by diluting the synovial fluid with physiological-buffered salt solution. Medical grade (sterile, pyrogen-free, non-inflammatory, M_w: 2.4 × 10^6) NIF-NaHA was used. Plasma fibronectin, when used, was always added to a fluid to obtain a final concentration of 10 μg/ml. The box on the left side represents the fibronectin (10 μg/ml) effect alone.

hyaluronan, the concentration and average molecular weight were varied. One hundred percent inhibition was achieved by hyaluronan solutions with specific viscosities of 6 to 7. The specific viscosity of healthy human blood plasma is around 0.75. Inhibition of phagocytosis was dependent on both molecular weight and concentration. At equivalent viscosities, hyaluronan solutions inhibited phagocytosis to the same degree, independent of the average molecular weight, provided the concentration was adjusted to the appropriate level. This clearly shows that the inhibitory effect is not limited to the chemical nature of the hyaluronan molecule, because elastoviscous solutions made of protein or of nucleic acid have the same effect (Figure 4). Therefore, the inhibitory effect of elastoviscous hyaluronan solutions cannot be mediated by hyaluronan-specific receptors. However, these findings show that the stimulatory effect is specific to hyaluronan and, therefore, it must be mediated through specific hyaluronan receptors (E.A. Balazs and F. Slahetka, unpublished work).

Elastoviscous solutions of hyaluronan, DNA or gelatin also inhibit prostaglandin synthesis and release by peritoneal macrophages during phagocytosis of latex particles. This inhibitory effect was also observed when elastoviscous human or primate liquid vitreus or horse synovial fluid was used [33]. However, hyaluronan solutions of low concentration (50–80 μg/ml) and low

Figure 4

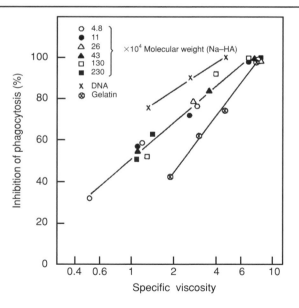

The viscosity-dependent inhibition of the phagocytosis of latex beads by mouse peritoneal macrophages

Hyaluronan solutions were made in phosphate-buffered physiological salt solutions of medical grade hyaluronan (pyrogen-free, sterile, NIF-NaHA) of various average molecular weights. High-molecular-weight DNA and purified bovine gelatin were used. By varying the concentration of the polymers, solutions with different viscosities were obtained.

average M_w (~8 × 10^5) can also be stimulatory *in vitro* by causing the release of interleukin-1 by rabbit peritoneal macrophages, human blood monocytes and polymorphonucleic leucocytes. Lower M_w (6–60 × 10^3) hyaluronan had little or no effect [34].

Cell and substrate adhesion

Early studies showed that hyaluronan promotes [35], inhibits [36], or has no effect [37] on cell adhesion. Later studies provided some clarification for these apparently contradictory results. Forrester and Lackie [38] used hyaluronan with well-defined average molecular weights and neutrophils collected from rabbit peritoneum after glycogen stimulation. They found that the aggregation of these cells was inhibited by hyaluronan solutions, depending on the viscosity. This inhibition was not restricted to hyaluronan because viscous dextran solutions had the same effect. Adhesion of neutrophils to glass surfaces showed the same viscosity-dependent inhibitory effect of hyaluronan solution. The threshold of inhibition was demonstrated only with hyaluronan molecules with average M_w >1 × 10^6 and at a concentration above 0.1 mg/ml for cell aggregation and 0.5 mg/ml for substrate adhesion. During the following decade, the role of hyaluronan and its involvement with other molecules in cell–cell and cell–substrate interaction was well established.

Mechanoreceptors

The cell inhibitory effect of elastoviscous hyaluronan solutions cannot be explained by hyaluronan-specific receptor-mediated mechanisms: first, because this effect is not hyaluronan-specific; second, because the maximum effect is well above the saturation concentration for the hyaluronan solutions; third, because non-elastoviscous hyaluronan solutions of the same concentration do not have any effect; and fourth, because the presence of hyaluronan oligosaccharides and monosaccharides or low-molecular-weight hyaluronan does not abolish the effect. Since the mechanical (rheological) properties of hyaluronan are responsible for this effect, our attention turned to the mechanoreceptors on cell surfaces. The first indication that mechanoreceptors were involved came in the late 1960s, when we injected NIF-NaHA into painful (lame) arthritic joints of race horses [39, 41].

Previously we studied the hyaluronan concentration, molecular distribution and rheology of healthy and arthritic equine, bovine and human synovial fluids (see [40] for review). These studies clearly showed that in traumatic arthritis, rheumatoid arthritis and osteoarthritis, the rheological properties (viscosity, elasticity, pseudoplasticity) of the synovial fluid significantly decrease. This decrease is due to the appearance of low-molecular-mass hyaluronan in the fluid, which causes a shift of the average molecular mass to lower values, resulting in a substantial decrease of the rheological properties of the fluid. Inflammation, when present, further decreases the rheological properties by the flow of water into the joint (effusion), which causes dilution of the hyaluronan. Our hypothesis was that, by replacing the synovial fluid or effusion having pathologically low elastoviscous properties with a hyaluronan solution having rheological properties comparable to those of healthy synovial fluid, we could influence the symptoms of the disease. We found that one or two intra-articular injections of NIF-NaHA eliminated lameness in most cases, and that the race horses could return to training and racing without any other treatment. The effect was on pain, which was reduced shortly after the injection. Most importantly, the lameness did not return even after the injected NIF-NaHA was no longer present in the joint. This was the first time the effect of elastoviscous hyaluronan solutions on pain was observed [41].

The joint has nociceptors, both in the capsule and in the synovial tissue, which are mechanoreceptors, that is they are stimulated by mechanical dislocation of the cell membrane that opens the Ca^{2+} channels [42,43]. These nociceptors are embedded in the intercellular matrix of a collagen fibre network filled with the same hyaluronan that is present in the entire synovial and capsular tissue system, including the matrix of the cartilage surface (*lamina splendens*). This elastoviscous hyaluronan solution can act as a shock absorber or mechanical stabilizer for the nociceptors in the joint. With Carlos Belmonte, we carried out studies on both healthy and inflamed cat joints to test the hypothesis that intra-articular administration of elastoviscous hyaluronan reduces nociceptive activity [44]. The experimental results showed that elastoviscous hylan solutions injected into the joint significantly reduced both the spontaneous ongoing and movement-evoked neural activity in the inflamed joints (pain at rest), and the movement-evoked responses over the natural range of motion in the healthy joints (pain with motion). Non-elastoviscous solutions containing the same concentration of low-molecular-weight hylan did not have any effect. These experimental results supported the theory that the reduction of pain by elastoviscous solutions in the joint is mediated by the mechanosensitive nociceptor (see Chapter 22 in this volume).

The nociceptors are not the only biological system sensitive to mechanical forces and dislocation. Cellular responses to compressive and tensile forces are well known (see [45] for review). Mechanoreceptor systems with very low sensitivity thresholds have been described in several marine animals for depth detection (with sensitivity threshold of 10 g/cm²). The hair cells in the human ear, which are sensitive to mechanical deformation, have a sensitivity threshold of 0.01 g/cm². The sensitivity threshold of chondrocytes is 60 g/cm², which is within physiological conditions and suggests that this mechanoreceptor-transmitted regulation may play an important role in maintaining healthy cartilage structure.

There is experimental evidence that chondrocytes, both in cell cultures and in cartilage explants, detect and transduce small continuous pressures, or cyclical tensile or compression forces, into biosynthetic activity. In general, continuous high mechanical pressure inhibits the synthesis and release of proteoglycan–hyaluronan complexes into the cartilage intercellular matrix. In contrast, low pressure or tensile forces, applied continuously or cyclically, stimulate synthesis. This mechanotransduction may mediate through cell membrane-associated monocilia approximately 1 μm long and cell membrane folds of 1–2 μm in length. These mechanoreceptors protrude into the pericellular matrix.

On the cell surface, they are associated with stretch-activated ion channels and intracellularly with the cytoskeleton [46,47]. Therefore, deformation of these cell-membrane-attached mechanoreceptors caused by tensile or compressive forces can trigger mechanotransduction. This is the same kind of mechanotransduction that operates on nociceptive nerve terminals that are sensitive to mechanical stress produced by motions of the joint.

Pericellular molecular cage (PMC)

To attempt to explain the effect of elastoviscous solutions (hyaluronan, gelatin, DNA, etc.) on a wide variety of cellular functions, one must recall the latest concept of pericellular matrix. Here again, the chondrocytes represent the most widely studied cell model. These cells in primary cultures exhibit well-defined pericellular coats that are approximately equivalent in size to the territorial matrix surrounding these cells *in vivo*. It was shown that the integrity of this PMC depends primarily on the presence of hyaluronan. It appears that hyaluronan is bound to the cell surface via specific receptors and the proteoglycan aggregates are held in the PMC by the hyaluronan [48]. This PMC is estimated to be several μm thick and can be removed and reassembled on the cell surface with hyaluronan [49,50]. It must be conceived as a dynamic structure, because it shows rapid, three-dimensional, random-feathered movements. Greta Lee demonstrated the movements of the pericellular coat of chondrocytes in elegant dynamic experiments and suggested that the hyaluronan–proteoglycan complex is 'grafted' to the surface of the cell in a brush-like configuration similar to surface-grafted polymers [51,52]. Because of the three-dimensional nature of the pericellular matrix and because the molecules form a three-dimensional network, I believe molecular cage is a better descriptive name for it.

The structure of PMC has two fundamental physical properties. First, it acts as a filter, excluding molecules that are greater than the intermolecular space of the cage from its structure, thereby protecting the cell surface receptors from certain signal molecules; and second, it provides structural stability to the mechanoreceptors of the cell surface. If PMC exists, all receptors, including mechanoreceptors, must be embedded in this viscoelastic structure, schematically represented in Figure 5. The PMC, by the exclusion volume effect, must also influence the chemical activity (concentration) of molecules (nutrient and signal molecules) that can penetrate its domain. If the PMC is exposed to an elasto-viscous solution made up of molecules in a random coil configuration, the two surfaces will interact by friction and then by entanglement. This entanglement will result in mechanical stabilization of the PMC and thus provide viscoelastic protection to the cell surface and prevent the dislocation of the mechanoreceptors (Figure 6). This means that as long as the movements of the joint are within the normal range of motion, the mechanoreceptors are protected by the elastoviscous, shock-absorbing milieu and they are not dislocated and no pain is perceived. In an arthritic joint, the mechanoreceptors are sensitized because the normal, protective hyaluronan envelope around the nociceptors is depleted as a result of the inflam-

Figure 5

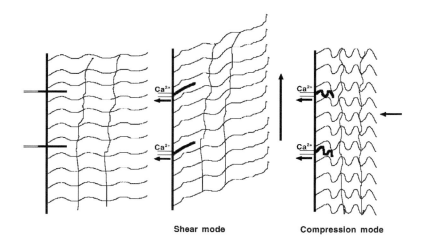

Shear mode   Compression mode

The schematic representation of the pericellular molecular cage under the influence of shear and compression forces

The solid vertical lines represent the cell membranes with Ca^{2+} channels and mechanoreceptors; the thinner lines represent the PMC.

matory process. The sensitized mechanoreceptors register movements within the normal range of motion as pain.

The concept is that mechanoreceptors embedded into the PMC control such cell functions as ingestion of particles by phagocytic cells, cell migration, specific phases of cell division, mechanotransduction of nociceptors and synthesis of extracellular matrix molecules. These mechanoreceptors can be stabilized (desensitized) by the elastoviscous media of the intercellular matrix. This stabilization is mediated by the frictional interaction between the two elastoviscous systems – the PMC and the elastoviscous component of the intercellular matrix.

The two types of intercellular compartments

Whether the intercellular compartment of tissues is filled with liquid or with the semi-solid matrices, they can be divided into two groups in regard to their hyaluronan content. In compartments that are surrounded entirely or partially by endothelial cells, the hyaluronan concentration is always very low. These compartments are filled with fluids (blood, lymph, aqueous or cerebrospinal fluid) or they are solid intercellular matrices between endothelial cells or neural cells. Compartments inside connective tissues, whether they are filled with fluids (synovial fluid, liquid vitreus) or intercellular matrices of various density (gel vitreus, cartilage), always contain a high concentration of hyaluronan. The high and low concentrations of hyaluronan can be defined as saturated (concentrated) or not saturated (diluted) hyaluronan solutions. It is generally accepted that in an adult with a healthy intercellular matrix of connective tissue, the average M_w of

Figure 6

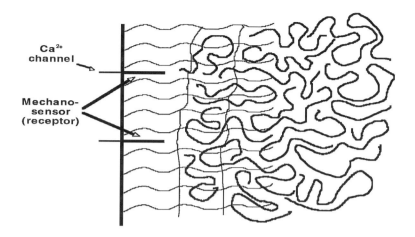

Ca²⁺ channel

Mechano-sensor (receptor)

Schematic representation of the pericellular molecular matrix (thin lines) and its entanglement with the structure of random coil molecules of an elastoviscous polymer solution (thicker lines)

The frictional interaction and entanglement between the two structures has a stabilizing (protective) effect on the mechanoreceptors.

hyaluronan is $4–5 \times 10^6$. Hyaluronan molecules of this size, when hydrated, occupy a sphere with a diameter around 0.5 μm and, therefore, they 'saturate' a solution at a concentration of ~0.2 mg/ml. This means that above this concentration, the molecules become crowded, overlap and become entangled. Consequently, their molecular movements are restricted and they behave as a continuous molecular network. Such 'saturated' solutions exhibit pseudoplastic-flow characteristics, and thus they are elastoviscous, pseudoplastic and their rheological behaviour is highly shear-dependent. However, in blood, lymph, aqueous and cerebrospinal fluid, the hyaluronan concentration is very low (0.01–50 μg/ml) [53–55]. These fluids, therefore, have no elastoviscosity, and rheologically they behave like water.

The concept of the two-compartment system of the intercellular space gives a framework to extend the regulatory role of hyaluronan on cell activities described in *in vitro* studies to *in vivo* systems. Cells in the intercellular compartments that contain low concentration, low-molecular-weight hyaluronan (blood, lymph, aqueous, cerebrospinal fluid, lymphoid tissues, neural tissues, etc.) can be regulated through specific receptors (stimulated or inhibited) by relatively small changes in the concentration and molecular weight of hyaluronan. Cells in these compartments may react to hyaluronan as observed *in vitro* in tissue culture systems. I believe that control of such a regulatory mechanism by hyaluronan is most likely to occur in the lymphomyeloid system.

Conversely, in compartments where the hyaluronan fills the intercellular space as an elastoviscous solution, the regulation is through mechanoreceptors. The elastoviscous solution of hyaluronan stabilizes the PMC by frictional interaction and, in turn, the PMC prevents mechanoperception. This system can be down-regulated by specific molecules (fibronectin, lysozyme, molecules with hyaluronan-binding motif, etc.) that interfere with the mechanoreception or with the structure of PMC, and by a decrease of molecular weight and/or concentration of hyaluronan. The system can be up-regulated by decreasing the concentration of specific molecules that interfere with mechanoreception or by perturbing the normal structure of the PMC, or by increasing the molecular weight and/or concentration of hyaluronan.

The fluid and solid compartment concept of the intercellular matrix can be extended to connective tissues matrices. Figure 7 schematically represents the concept of the three compartments in a soft connective tissue such as skin, intermuscular connective tissue, synovium of the joints and tendons. One solid matrix compartment surrounds the stationary cells (fibrocytes, histocytes, hyalocytes, nociceptors, etc.) and interacts with their PMC. The other solid matrix compartment is the acellular collagen (elastic) fibrous network filled with hyaluronan and proteoglycan molecules. Between these two solid matrices are the liquid channels, where fluid flows are driven by muscle movement, pulsation of blood vessels or negative fluid pressures. These channels represent the principal catabolic pathway of the rapidly metabolizing hyaluronan and proteoglycan molecules and other metabolic products. These channels also serve as the migration path for lymphomyeloid cells from the blood and to the lymph vessels. In other words, this channel drains into the open end of the peripheral lymphatic vessels. These channels will be flooded and expanded in tissue oedema. The three-compartment system of the connective tissue matrix can explain the rapid cellular

Figure 7

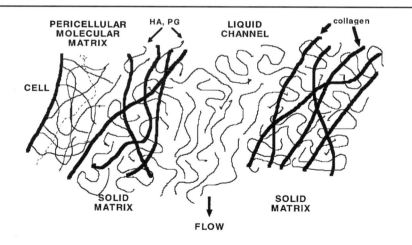

Schematic representation of the three compartments of the extracellular matrix and their relationship

The compartments are the pericellular molecular cage, the solid matrix and the liquid channels. Thin lines represent hyaluronan and proteoglycans; solid thicker lines represent collagen fibres. HA, hyaluronan, PG, proteoglycan.

invasion of the connective tissue in inflammation and in healing of internal wounds, as well as the rapid outflow of hyaluronan in healthy tissues.

Medical aspects

The biological role of hyaluronan as a protective, stabilizing and shock-absorbing elastoviscous structure filling the space between collagen fibres and engulfing cells in the intercellular matrix of connective tissues was discovered nearly four decades ago [56]. The recognition that this structure-stabilizing and shock-absorbing effect could diminish in certain pathological conditions in the vitreus and in the joint led to the first medical therapeutic application of elastoviscous hyaluronan solutions. Several chapters of this book deal with this development and its present status. The use of viscoelastics made of hyaluronan and its derivatives (hylans) as new therapeutic paradigms has clearly had an impact on medicine.

Interestingly, the discovery of the biological effects of hyaluronidase-degraded, low-molecular-weight hyaluronan predates the recognition of the cell-regulatory role of elastoviscous solutions. Yet the cell-stimulating effect of low concentration hyaluronan solutions as a therapeutic agent is not yet used broadly in medicine. There are several reasons for this. Most of the cell-stimulating effects were described in *in vitro* cell culture models. It is difficult to translate *in vitro* biological effects to therapeutics. Large doses of hyaluronan of any molecular mass injected subcutaneously, intramuscularly or intraperitoneally do not adversely affect a healthy animal. Neither could any pathological process in animals be stopped or reversed in a dose-dependent manner by such treatment. In recent clinical studies, parenteral treatment with hyaluronan was tested in order to enhance the host defence mechanism in patients with chronic bronchitis. Weekly subcutaneous injections (for 24 weeks, 7.5 mg/dose) of high M_w ($3-4 \times 10^6$) hyaluronan as compared with placebo, significantly reduced acute exacerbation of the disease [57,58]. It seems that the therapeutic benefits of hyaluronan as a non-elastoviscous solution have a different dimension. The therapeutic effect of elasto-viscous hyaluronan solutions, viscoelastic gels and membranes made of its derivatives is achieved by one to several local applications, and it is often used with surgical procedures (viscosurgery, viscoaugmentation, viscoseparation). In contrast, any generalized therapeutic effect of non-elastoviscous hyaluronan could be achieved only by elevating the hyaluronan concentration for an extended period of time in a given tissue compartment of the lymphomyeloid system (lymph nodes, spleen, bone marrow, etc.).

Conclusion

During the past half century, we have travelled far in the hyaluronan world. During the 1940s, hyaluronic acid was a recently renamed, obscure polysac-charide with a primary structure finally defined, but invisible in the new images of the electron microscope. In the 1950s, it was a voluminous hydrated molecule that filled the space between collagen fibres, stabilizing their structure and regulating the passage and activity of molecules through its domain. In the 1960s, it became an elastoviscous lubricant that protected the joints and stopped cells from moving, proliferating and functioning *in vitro*. In the 1970s, in its highly purified form, hyaluronan became the first viscoelastic substance used as surgical tool as well as a fluid supplement to decrease joint pain. In the 1980s, it became an *in vitro* cell function regulator, a determinant in developing and regenerating tissues and a highly biocompatible polymer that could be derivatized for a broad spectrum of therapeutic applications. And in this decade, hyaluronan and its derivatives are rapidly developing into universal cell regulators of cardinal importance, holding the intercellular matrix together, with certain memory functions and profoundly influencing the function of cells that live in it and pass through it.

The accelerated development and the rapidly growing interest in this molecule raises the question: will future discoveries fulfil our expectations for the universal biological importance and broad therapeutic utility of this ancient molecule produced in the same form by bacteria and humans? Working through five decades with this molecule, from the time when I was the only speaker talking about it at an international congress, to today when all lectures at this conference are addressed to it, I believe that all aspects of the biological importance of hyaluronan are not yet discovered. We have many and often apparently conflicting data, contradicting theories and poorly defined hypotheses. A more complete understanding may come from the integration of the facts already known to us. The special relationship of this molecule with water and with free radicals, its affinity for highly active signal molecules and its rapid turnover are all part of a puzzle yet to be solved. We must also remember that hyaluronan is the only pure polyanion in the extracellular space, just as nucleic acids are the only polyanions in the intracellular space. Does this parallel tell us something? I believe it does.

My warmest thanks to Torvard Laurent for organizing this symposium, and my deepest appreciation to him for dedicating it to my life-long work on hyaluronan. In the late 1940s and early 1950s we worked together on several aspects of hyaluronan. This work laid the foundation for experiments in the fields of radiation sensitivity, physical properties and biological activities of hyaluronan. I was lucky that, as decades went by, Torvard directed most of his and his colleagues' work to hyaluronan research, made numerous fundamental discoveries that stimulated rapid progress in the field, and greatly helped the advancement of my own research. For half a century of friendship and research fellowship, I am most grateful.

I want to express my gratitude to Janet Denlinger, who contributed greatly to the birth of this symposium. She has been my closest scientific collaborator and friend for a quarter of a century. Without her contributions and support, my and my collaborator's work would have never happened.

My warmest thanks to the researchers in many fields of science and medicine who dedicated time, creativity and hard work to advance our understanding of hyaluronan and its function in health and diseases.

Acknowledgments

Algvere, P., Armand, G., Arnott, S., Band, .P., Belmonte, C., Bettelheim, F.A., Bloom, G.D., Bodis-Wollner, I., Boström, H., Bother-By, A.A., Bother-By, C.T., Briller, S.O., Chakrabarti, B., Cleland, R.L., Cowman, M.K., Cremer-Bartels, G., Darzynkiewicz, Z., Dea, I.C.M., Denlinger, J.L., Dohlman, C.H., Duff, I.F., Eckl, E.A., Eisner, G., El-Mofty, A.A.M., Fischbarg, J., Forrester, J.V., Freeman, M.I., Gergely, J., Gibbs, D.A., Goldman, A.I., Graue, E.L., Hanninen, L., Hilal, S.K., Högberg, B., Holmgren, H.J., Howe, A.F, Hultsch, E., Jacobson, B., Jeanloz, R.W., Klöti, R., Larsen, N.E., Laurent, T.C., Laurent, U.B.G., Leshchiner, A.K., Leshchiner, E., Meyer, K., Meyer-Schwickerath, G., Miller, D., Morales, D., Morris, B., Österlin, S., Ozanics, V., Paul, M., Peyron, J.G., Philipson, B.T., Phillips, G.O., Piacquadio, D., Rees, D.A., Regnault, F., Roseman, S., Rydell, N., Scheufele, D.S., Schubert, H.D., Sebag, J., Seppälä, P., Slahetka, M.F., St. Onge, R., Stegmann, R., Sundblad, L., Swann, D.A., Sweeney, D.B., Szirmai, J.A., Toth, L.Z.J., Varga, L., von Euler, J., Wedlock, D.J., Weiss, C., Young, M.D.

References

1. Balazs, E.A. (1947). The influence of extracellular macromolecular polysaccharides on the development and growth of tissues. Congressus VI Internationalis Cytologicus, Stockholm.10–17 July 1947
2. Balazs, E.A. and Darzynkiewicz, Z. (1973) in Biology of a Fibroblast (Kulonen, E. and Pikkarainen, J., eds.), pp. 237–252, Academic Press, London
3. Balazs, E.A., Freeman, M.I., Klöti, R., Meyer-Schwickerath, G., Regnault, F. and Sweeney, D.B. (1972) in Modern Problems in Ophthalmology (Secondary Detachment of the Retina, Lausanne, 1970) (Streiff, E.B., ed.), pp. 3–21, S. Karger, Basel
4. Brandt, K.D. (1974) Clin. Chim. Acta **55**, 307–315
5. Brandt, K. (1970) Arthritis Rheum. **13**, 308
6   Folger, R., Weiss, L., Glover, D., Subjeck, J.R. and Horlas, J.P. (1978) J. Cell Sci. **31**, 245–257
7. Forrester, J.V. and Wilkinson, P.C. (1981) J. Cell Sci. **48**, 315–331
8. Okada, H., Yoshida, J., Sokabe, M., Wakabayashi, T. and Hagiwara, M. (1996) Int. J. Cancer **66**, 255–260
9. Boudreau, N., Turley, E. and Rabinovitch, M. (1991) Dev. Biol. **143**, 235–247
10. Thomas, L., Byers, H.R., Vink, J. and Stamenkovic, I. (1992) J. Cell Biol. **118**, 971–977
11. Darzynkiewicz, Z. and Balazs, E.A. (1971) Exp. Cell Res. **66**, 113–123
12. Anastassiades, T. and Robertson, W. (1984) J. Rheumatol. **11**, 729–734
13. McBride, W.H. and Bard, J.L. (1979) J. Exp. Med. **149**, 507–516
14. Larsen, N.E., Lombard, K., Parent, E.G. and Balazs, E.A. (1992) J. Orthoped. Res. **10**, 23–32
15. Shimazu, A., Jikko, A., Iwamoto, M., Koike, T., Yan, W., Okada, Y., Shinmei, M., Nakamura, S. and Kato, Y. (1993) Arthritis Rheum. **36**, 247–253
16. Kikuchi, T., Yamada, H., Tateda, C. and Shinmei, M. (1993) Rev. Esp. Reumatol. **20**, 318
17. Larsen, N.E., Lombard, K.M. and Balazs, E.A. (1989) The effect of hyaluronan on cartilage and chondrocyte response to mechanical and biochemical perturbation. Orthopaedic Res. Soc., 35th Annu. Meeting, 6–9 February 1989, Las Vegas, NV
18. Kujawa, M.J., Pechak, D.G., Fiszman, M.Y. and Caplan, A.I. (1986) Dev. Biol. **113**, 10–16
19. Kujawa, M.J., Carrino, D.A. and Caplan, A.I. (1986) Dev. Biol. **114**, 519–528
20. Toole, B.P., Goldberg, R.L., Chi-Rosso, G., Underhill, C.B. and Orkin, R.W. (1984) in The Role of the Intercellular Matrix in Development (Hay, E.B., ed.), pp. 43–66, A.R. Liss, New York
21. Goldberg, R.L. and Toole, B.P. (1987) Arthritis Rheum. **30**, 769–778
22. Balazs, E.A., Toth, L.Z.J., Eckl, E.A. and Mitchell, A.P. (1964) Exp. Eye Res. **3**, 57–71
23. Aruffo, A., Stamenkovic, I., Melnick, M., Underhill, C.B. and Seed, B. (1990) Cell **61**, 1303–1313
24. Denlinger, J. (1982) Ph.D. Thesis, Université des Sciences et Techniques de Lille, Lille, France
25. Forrester, J.V. and Balazs, E.A. (1977) Trans. Ophthalmol. Soc. U.K. **97**, 554–557
26. Forrester, J.V. and Balazs, E.A. (1980) Immunology **40**, 435–446
27. Håkansson, L., Hällgren, R. and Venge, P. (1980) Scand. J. Immunol. **11**, 649–653

28. Håkansson, L., Hällgren, R. and Venge, P. (1980) J. Clin. Invest. **66**, 298–305
29. Håkansson, L. and Venge, P. (1987) J. Immunol. **138**, 4347–4352
30. Ahlgren, T. and Jarstrand, C. (1984) J. Clin. Immunol. **4**, 246–249
31. Mazzone, A., Baiguera, R., Rossini, S., Nastasi, G., Tarabini, L., Casali, G. and Ricevuti, G. (1986) Clin. Ther. **8**, 527–536
32. Pisko, E.J., Turner, R.A., Soderstrom, L.P., Panetti, M., Foster, S.L. and Treadway, W.J. (1983) Clin. Exp. Rheumatol. **1**, 41–44
33. Sebag, J., Balazs, E.A., Kulkarni, P.S., Eakins, K. and Bhattacherjee, P. (1981) Invest. Opthalmol. Vis. Sci. (suppl.) **20**, 32
34. Hiro, D., Ito, A., Matsuta, K. and Mori, Y. (1986) Biochem. Biophys. Res. Commun. **140**, 715–722
35. Pessac, B. and Defendi, V. (1972) Science **175**, 898–900
36. Underhill, C. and Dorfman, A. (1978) Exp. Cell Res. **117**, 155–164
37. Knox, P. and Wells, P. (1979) J. Cell Sci. **40**, 70–89
38. Forrester, J.V. and Lackie, J.M. (1981) J. Cell Sci. **50**, 329–344
39. Balazs, E.A. and Denlinger, J.L. (1985) J. Equine Vet. Sci. **5**, 217–228
40. Balazs, E.A. and Gibbs, D.A. (1970) in Chemistry and Molecular Biology of the Intercellular Matrix (Balazs, E.A., ed.), pp. 1241–1254, Academic Press, New York
41. Rydell, N.W., Butler, J. and Balazs, E.A. (1970) Acta Vet. Scand. **11**, 139–155
42. Belmonte, C. (1996) in Neurobiology of Nociceptors (Belmonte, C. and Cervero, F., eds.), pp. 243–257, Oxford University Press, Oxford
43. Schaible, H-G., Neugebauer, V. and Schmidt, R.F. (1989) Semin. Arthritis Rheum. **18**, 30–34
44. Pozo, M.A., Balazs, E.A. and Belmonte, C. (1997) Exp. Brain Res. **116**, 3–9
45. Stockwell, R.A. (1996) Stress and the Chondrocyte. A review. Department of Anatomy, University of Edinburgh, p.13
46. Hall, A.C., Urban, J.P.C. and Gehl, K.A. (1991) J. Orthoped. Res. **9**, 1–10
47. Urban, J.P.G. (1994) Br. J. Rheumatol. **33**, 901–908
48. Goldberg, R.L. and Toole, B.P. (1984) J. Cell Biol. **99**, 2114–2122
49. Knudson, W. and Knudson, C.B. (1991) J. Cell Sci. **99**, 227–235
50. Knudson, C.B. (1993) J. Cell Biol. **120**, 825–834
51. Lee, G.M., Johnstone, B., Jacobson, K. and Caterson, B. (1993) J. Cell Biol. **123**, 1899–1907
52. de Gennes, P.G. (1987) Adv. Colloid Interface Sci. **27**, 189–209
53. Laurent, U.B.G. and Laurent, T.C. (1981) Biochem. Int. **2**, 195–199
54. Laurent, T.C. and Fraser J.R.E. (1986) In Functions of Proteoglycans, Ciba Foundation Symposium 124 (Evered, D. and Whelan, J., eds.) pp. 9–24, Wiley, Chichester
55. Laurent U.B.G., Laurent, T.C., Hellsing, L.K., Persson, L. Hartman, M. and Lilja, K. (1996) Acta Neurol. Scand. **94**, 194–206
56. Balazs, E.A. (1961) in The Structure of the Eye (Smelser, G.K., ed.), pp. 293–310, Academic Press, New York
57. Venge, P., Pedersen, B., Håkansson, L., Hällgren, R., Lindblad, G. and Dahl, R. (1996) Am. J. Respir. Crit. Care Med. **153**, 312–316
58. Venge, P., Rak, S., Steinholtz, L., Håkansson, L. and Lindblad, G. (1991) Eur. Respir. J. **4**, 536–543

22

Modulation by hyaluronan and its derivatives (hylans) of sensory nerve activity signalling articular pain

Carlos Belmonte*‡, Miguel A. Pozo* and Endre A. Balazs†
*Instituto de Neurociencias, Universidad Miguel Hernandez, Campus de San Juan, 03550 San Juan de Alicante, Spain and †Matrix Biology Institute, 65 Railroad Avenue, Ridgefield, NJ 07657, U.S.A.

Introduction

Joints are richly innervated by nerve fibres that end in the synovium, capsule, fat pads and ligaments of the joint [1–3]. A group of articular nerve fibres of large diameter and high conduction velocity respond to innocuous movements of the joint and are presumably involved in signalling the position and normal displacement of the joint [4]. They are called low-threshold mechanoreceptors. Another significant proportion (about 70%) of thin articular nerve fibres end in articular tissues as unspecialized nerve terminals and respond preferentially or exclusively when mechanical forces applied to the joint are strong [4–7]. They have been categorized as joint nociceptors. Nociceptors have the property of changing their excitability when the chemical environment of the nerve terminal is altered, a phenomenon called sensitization [8,9]. Nociceptor sensitization is produced by inflammatory substances (eicosanoids, kinins, amines, cytokines) or ions (K^+, H^+) released in the surroundings of nociceptive terminals by damaged tissues [10]. Sensitization is evidenced by a spontaneous discharge and/or an enhanced sensitivity of the nociceptive fibres to different stimuli [9]. A proportion (about 20%) of articular nociceptors exhibit chemosensitivity but are unresponsive to noxious mechanical forces in the normal joint. They develop mechanical sensitivity when joint tissues are inflamed and have been called 'sleeping' or 'silent' nociceptors. This category of nociceptors may contribute importantly to enhanced pain in the inflamed joint [11,12].

Transduction of injurious stimuli by sensory nerve endings of the joint

Membrane mechanisms underlying the transduction of injurious forces by nociceptive terminals (mechanical, thermal, chemical) into electrical signals (nerve impulses) are still poorly understood. The transduction steps are possibly similar

‡To whom correspondence should be addressed.

to those found in other sensory receptors [13]: a molecular entity located in the nerve ending membrane (membrane channel, receptor molecule) is sensitive to the stimulating energy. At this membrane detector structure, the energy of the stimulus is transduced to another form of energy, either chemical or electrical. This signal is amplified and finally transformed into a change of ionic permeability, with the subsequent production of a series of electrical signals (receptor potential, nerve impulses) in which some of the characteristics of the stimulus (intensity, duration) are encoded. Perireceptor structures, i.e. structures interposed between the stimulus and the sensor element in the nerve membrane, may act as filters for the stimulating energy, modulating the transmission of mechanical forces or the accessibility of chemicals to the transducing area of the receptor membrane [14].

Nociceptors exhibit bare areas devoid of Schwann cells at their nerve terminals, which appear to be the transducing portion of the nerve membrane [12,15]. Molecular structures responsible for mechanotransduction in nociceptors have not been identified. In other cell types, ion channels of the membrane, which are sensitive to mechanical forces (mechanosensory channels) have been described [16–18]. Mechanical forces acting on stretch-activated channels will open them, allowing the passage mainly of sodium and potassium ions. The resulting ionic flow will create a net inward current that depolarizes the nerve terminal. Coupling of the mechanical force to the channel-gating mechanism may be direct or indirect. In the first case, applied tension causes intramolecular arrangements of the channel molecule and a fast (in less than 1 ms) change in channel permeability. When coupling is indirect, a mechanosensitive enzyme triggers the diffusion of an intracellular second-messenger (cyclic nucleotides, fatty acids, calcium) to the channel. This mechanism is slower (from several milliseconds up to seconds) [19,20].

It has been speculated that viscoelastic elements in the extracellular compartment and in the cytoskeletal domain of the receptor membrane play a role in the transmission of force, the determination of sensitivity and the adaptation properties of mechanosensory channels [20]. According to this view, some molecules like integrins, which connect extracellular matrix with the cytoskeleton, and cytoskeleton proteins, like those of the dystrophin-spectrin family, are part of a chain of molecular elements that transmit external mechanical force to a specific membrane channel. Therefore, the possibility exists that the viscoelastic character-istics of the extracellular matrix influence the functional behaviour of mechanosensory channels in nociceptors.

Another relevant property of nociceptor terminals, i.e. their sensitivity to endogenous chemical agents, is based on cellular mechanisms that also are incompletely known [14,21]. Some stimuli (for instance, acidic solutions) after reaching the nociceptor membrane, appear to act directly on non-selective cationic channels and perhaps also on voltage-dependent calcium channels, increasing their permeability [22,23]. Other chemicals that are known to modify nociceptor excitability (BK, PGs, 5HT) act through a receptor-mediated mechanism and the production of second messenger molecules, which will finally cause the phosphorylation of the channel protein and a change in permeability [24,25]. As was the case for the transmission of mechanical forces, the extracellular matrix located around nociceptive endings may modify the effect of chemicals by interfering with the diffusion and/or the accessibility of molecules to the membrane surface.

Figure 1 represents schematically the various mechanisms that may be involved in the transduction of mechanical and chemical forces in joint nociceptors.

Hyaluronan content of the joint

Hyaluronan (sodium hyaluronate) is an important component of the extracellular matrix of all joint tissues, including synovial fluid, synovial tissue and capsule and the surface layer of the cartilage (*lamina splendends*). This highly elastoviscous

Figure 1

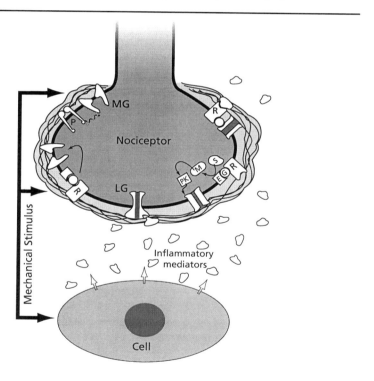

Diagram of a joint nociceptive ending, showing the mechanisms for excitation and/or sensitization by mechanical stimuli

The different types of ion channels that may be gated by noxious mechanical forces are shown schematically: mechanosensory channels (MG) are activated either directly or through membrane and cytoskeleton proteins (P). Mechanical forces also act on mechanosensitive receptor proteins (R). Mechanical injury of surrounding cells releases endogenous chemical mediators. These interact with ligand-gated channels (LG) or with membrane receptor molecules (R) coupled to ionic channels, producing finally a depolarization.

polymer is present in high concentration (3–20 mg/ml) and fills the intercellular space between the collagen fibrillar network, surrounding all of the cells, blood and lymph vessels and neural elements in the tissues of the joint. Hyaluronan, produced by cells in the synovial tissue, flows through the synovial fluid and the intercellular matrix, driven by the movement of the joint, and drains from the matrix into the lymph capillaries [26–30].

A purified fraction of hyaluronan with weight-average molecular weight of $0.5–3.0 \times 10^6$ called the non-inflammatory fraction of hyaluronan (NIF-NaHA), has been used for the treatment of osteoarthritis in human and veterinary medicine [26,28,31,32,34,35]. A derivative of hyaluronan, called hylan, with enhanced average molecular weight ($\sim 7 \times 10^6$), has also been used for the same treatment [35–38]. The elastoviscous solution is injected intra-articularly in order to increase elastoviscosity of the joint fluid and the intercellular matrix of the joint tissues. This treatment was called viscosupplementation, because it enhances the elastoviscosity of the synovial fluid and the tissue matrices which, in arthritic conditions, is abnormally low [33,34]. The therapeutic goal of viscosupplementation is to restore the rheological homoeostasis of the extracellular matrix of the joint. Equine clinical studies showed that the elastoviscosity of the synovial fluid returned to normal levels and simultaneously, the pain associated with the movement of the joint decreased or was completely eliminated. Injection of non-elastoviscous solutions of low-molecular-weight hyaluronan does not cause a decrease of pain [28].

In animal studies it has been shown that intra-articular injection of elastoviscous hyaluronan solution rapidly increases the hyaluronan content of the surrounding tissues, such as the synovial tissue and capsule, where the nociceptors are located [27]. It has been hypothesized [26,29,30,33] that the increase of the elastoviscosity of the intercellular matrix by hyaluronan provides mechanical protection in the form of a rheological mechanical protective envelope around nociceptive endings. This causes a decrease of pain and permits an increase in joint mobility. Thus, according to this interpretation, hyaluronan normally present in the surroundings of nociceptive nerve terminals is an important perireceptor element for the modulation of the tension transmitted to mechanosensory channels of the nociceptor membrane.

Effect of elastoviscous and non-elastoviscous hyaluronan derivatives (hylans) on neural activity of nociceptive fibres of the knee joint

To test the possibility that elastoviscous hyaluronan molecules modulate the sensitivity of mechano-nociceptors of the joint, we studied whether an injection of elastoviscous or non-elastoviscous hylan in the knee joint reduced resting and movement-evoked sensory discharges from normal and arthritic joints in anaesthetized cats.

We recorded single or multi-unit spontaneous and movement-evoked activity from nerve filaments dissected from the medial articular nerve (MAN) [5,6]. This nerve innervates the medial and anteromedial aspects of the knee joint of the cat [39]. In one group of animals, an experimental inflammation was produced by injection of a kaolin/carrageenan solution into the knee joint cavity [7].

In the intact knee joint, sensory nerve fibres were activated with controlled displacements of the joint produced by a pulley and an electrical engine driven by a pulse generator. The knee joint was rotated outwards until a starting position was obtained at which further rotation was opposed by definite resistance of the tissue. In inflamed joints, passive movements were made manually; these consisted of three extensions or rotations within the normal range of the joint, lasting 15–30 s and separated by a 5 min interval. Immediately afterwards, an intra-articular injection of hylan solution was made and ongoing activity was recorded continuously. Passive movements of the joint were also made, at variable times after the elastoviscous and non-elastoviscous hylan injection.

Elastoviscous hylan

To increase the elastoviscosity of the synovial fluid and the intercellular matrix of the synovial tissue and capsule, 0.8% hylan in physiological buffer solution (hylan G-F 20, Sinvisc@ Biomatrix, Inc., Ridgefield, NJ) was used. Hylan is the generic name for cross-linked hyaluronan in which the cross-linking does not affect the carboxylic acid and acetylglucosamine groups of the polysaccharide chain [40]. In the elastoviscous solution of hylan G-F 20, the hylan had an average molecular weight of $\sim 8 \times 10^6$. The rheological properties of the elastoviscous hylan solution used in this study were characterized by a dynamic elastic modulus (G'), which was 89 Pa at 5 Hz, and with a dynamic viscous modulus (G''), which was 13 Pa at 5 Hz. Both moduli are very dependent on the frequency at which they are measured. The two moduli have equal value (14 Pa) at 0.01 Hz.

Intact knee joints

In the normal joint, about one-third of MAN fibres responding only to displacements within the noxious range and with conduction velocities below 20 m/s, exhibited a low frequency, spontaneous activity at rest (below 0.1 s^{-1}). This background discharge increased greatly when the joint was placed in a forced position before starting the stimulation cycles. Connection of the stimulus produced a sudden rise of the discharge rate that attained frequency peaks of up to 30 s^{-1}.

Of 12 single units where the effect of elastoviscous hylan injection was explored, five showed a marked reduction in the number of impulses evoked by the stimulus, as well as a decrease in background activity that developed gradually and lasted for the post-injection recording time (2–4 h). In these fibres, the inhibitory effect appeared between 20 and 40 min after the injection and affected not only movement-evoked discharges but also interstimulus ongoing activity, so that frequency/s values 60 min after injection were significantly reduced in comparison with pre-injection firing rates. On average, mean frequency/s of the impulse discharge evoked by movement decreased to 65% of control values, while firing frequency of the ongoing activity dropped to 45% of pre-injection frequency. Differences with pre-injection firing frequency values were significant

Figure 2

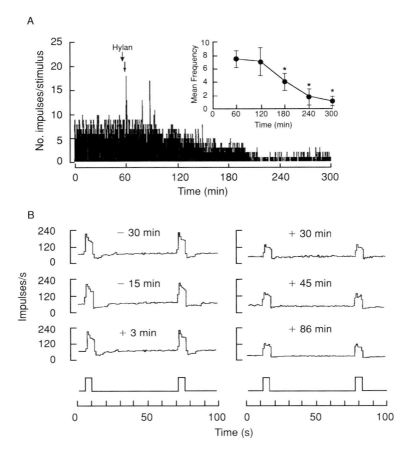

Inhibition of the response of joint afferent fibres of the intact joint by intra-articular injection of elastoviscous hylan

A, Peristimulus time histogram of an A-delta fibre, showing the number of impulses elicited by an outwards rotation of 5 s duration at 2 min⁻¹ applied continuously during 5 h. Hylan was injected (arrow) 1 h after the onset of the mechanical stimulation. Insert: average number of impulses/stimulus during 1 h periods, before and after hylan injection (*significant differences with pre-injection value at $p < 0.05$ or smaller). B, Impulse discharges in a filament of the MAN displaying multi-unit activity, elicited by a forced rotation of the knee joint lasting 10 s and applied at 1 min⁻¹. The upper traces show the frequency of discharge before and at the indicated time, before and after intra-articular injection of elastoviscous hylan, which was made at zero time. The lowest trace represents the output of the function generator that commands the rotation of the knee.

($p < 0.05$ or lower). In seven separate fibres, no consistent changes were observed. Figure 2(A) illustrates an example of the gradual reduction of neural activity produced by a hylan injection in a fibre that fired in response to a 5 s noxious stimulation, applied every 30 s during 4 h. Figure 3(A) depicts the average frequency values of ongoing and movement-evoked nervous activity in another fibre before and at various times after elastoviscous hylan injection.

In nine nerve bundles, multi-unit activity was recorded. In three of them, impulse discharges evoked by noxious movement were reduced to 52% of the control firing rate; ongoing activity during interstimulus periods was also decreased to 32% of control values. In four additional experiments, a reduction of interstimulus ongoing activity to an average value of 42% was also observed, although impulse frequency of the evoked discharges was not significantly altered. Finally, no changes could be detected in another two experiments involving multifibre discharges. Figure 2(B) illustrates the inhibitory effect of a hylan injection on movement-evoked and interstimulus ongoing activity in one of these multi-unit recordings.

Inflamed knee joint

After intra-articular injection of a kaolin–carrageenan solution, single fibres responding initially only to noxious movements developed an irregular discharge,

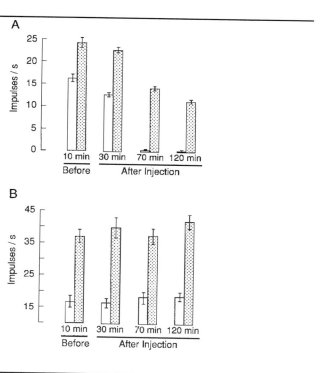

Figure 3

Comparison of the effect of elastoviscous or non-elastoviscous hylan injection on the impulse discharges recorded in two separate joint afferent fibres

Mean frequency/s values were calculated for 5 min periods at the indicated times before and after injection of (A) elastoviscous hylan and (B) non-elastoviscous hylan. Values are means ± SEM. Empty bars: mean discharge rate during interstimulus period. Dotted bars: mean discharge rate during the stimulation cycle. The stimulus consisted of 5 s triangular pulses at 1 min⁻¹. Differences of ongoing and movement-evoked activity values before injection with those at 70 and 120 min after hylan injection are significant (p <0.05 or smaller).

whose firing frequency increased with time, reaching a plateau 2–3 h after carrageenan injection (Figure 4A).

Intra-articular injection of elastoviscous hylan decreased ongoing activity in 60% of nociceptive fibres activated by the carrageenan injection. A typical example is shown in Figure 4(A) (●). On average, ongoing activity of these

Figure 4

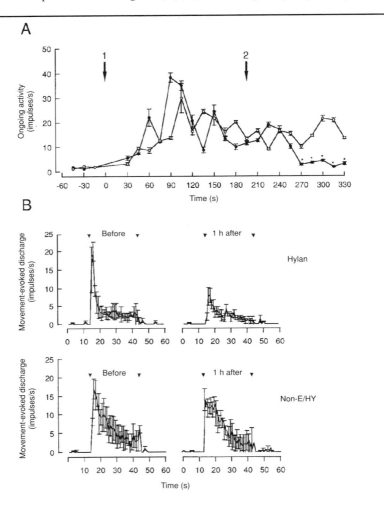

The build-up of ongoing neural activity after injection of kaolin–carrageenan and the effect of intra-articular injection of elastoviscous and non-elastoviscous hylan

A, Mean frequency of the ongoing activity measured during 10 min periods of two A-delta fibres recorded in two separate knee joints (elastoviscous, ●, and non-elastoviscous, ○). Arrow 1, kaolin–carrageenan injection in the knee. Arrow 2, injection of either elastoviscous hylan (●) or non-elastoviscous hylan (○) in the knee joint. Vertical bars are SEM. *p<0.05. B, Mean impulse discharge evoked by three 30 s knee joint extensions made with 5 min intervals, and performed before and 1 h after intra-articular injection of elastoviscous (Hylan) and non-elastoviscous (Non-E/HY) hylan. Vertical bars are SEM. Modified from Pozo et al. [53].

Figure 5

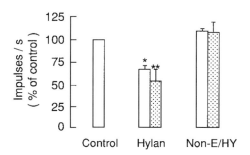

Mean frequency values of ongoing activity and of movement-evoked discharges before and 1 h after injection of elastoviscous and non-elastoviscous hylan

Data were obtained in 8 fibres for elastoviscous hylan and in 6 fibres for non-elastoviscous hylan. Data are expressed as percent of the pre-injection frequency value of ongoing activity (empty colums) and movement-evoked discharge (dotted columns) (** p <0.004). Vertical bars are SEM.

units was reduced to 60% of the control frequency, 1 h after hylan injection (Figure 5). This frequency decrease was significant ($p < 0.01$ or lower). A similar observation was made in multi-unit recordings obtained from inflamed joints. Impulse activity decreased gradually following the injection of elastoviscous hylan to attain a frequency level that was around 55% of the value measured before hylan.

After carrageenan–kaolin injection, joint nerve fibres were activated by innocuous movements. Elastoviscous hylan injection significantly diminished the response to movement in almost 90% of single nociceptive units of the joints suffering the experimental arthritis. One hour after intra-articular injection of elastoviscous hylan they showed a reduction of the mean response to movement to about 50% of the pre-injection values (Figure 4B; Figure 5). Likewise, in multifibre recording experiments, elastoviscous hylan injection also reduced neural responses evoked by passive movements to 52% of pre-injection frequency values.

Non-elastoviscous hylan

A non-elastoviscous hylan G-F 20 solution was obtained by mild acid degradation that did not alter the chemical identity of the molecules, but decreased the average molecular weight and, consequently, the elastoviscosity of solutions made from it. The non-elastoviscous hylan contained the same concentration (0.8%) of hylan polymer dissolved in the same physiological buffer solution as the elastoviscous hylan solution. The chemical composition of the two solutions and the primary structure of the hylan molecules in them were identical. The only difference between the two hylan solutions was in the average molecular weight of the polymer and, consequently, the elastoviscosity of the solution, which in the non-elastoviscous solution was ~30 000. The rheological properties of the non-elastoviscous hylan solution were similar to those of the physiological salt solution.

Non-elastoviscous hylan, injected into the synovial space of intact knee joints in multifibre and single-unit experiments, produced a brief increase in the discharge rate at the moment of injection but failed to modify consistently the evoked responses or the interstimulus ongoing activity during 3 h following the injection (Figure 3B). Similar results were obtained with the injection of physiological salt solution. Likewise, non-elastoviscous hylan injected intra-articularly in the inflamed joint did not influence the enhanced ongoing activity induced by carrageenan–kaolin injection or the impulse responses evoked by passive movements (Figure 4 and Figure 5).

Possible mechanisms for modulation of nociceptor activity by elastoviscous hyaluronan derivatives

The above-described data show that injection of elastoviscous hylan in the cavity of an intact knee joint significantly reduced background and movement-evoked nerve discharges in about 40% of the fine afferent fibres responding to a forced displacement of the intact joint. A similar effect was obtained in a majority of sensory units from knee joints where an acute arthritis was induced experimentally. This effect was not obtained when non-elastoviscous hylan solutions were used. Single units recorded in that study were thin myelinated or unmyelinated fibres that responded distinctly to forced rotation of the joint. Furthermore, they were sensitized after experimental inflammation, becoming spontaneously active and responsive to innocuous movements. Thus, they presumably belong to the types 2 and 3 of joint afferent nerve fibres described by Schaible and Schmidt [5], i.e. fibres exhibiting a high threshold, preferentially activated by noxious movements of the joint and thus serving as nociceptors. Many of such nociceptive afferents, together with fibres that already respond in the normal working range of the joint, were presumably recruited in multi-unit recordings.

In the normal joint, the gradual fall in firing frequency that followed intra-articular injection of hylan took place 1–2 h after the onset of the mechanical stimulation sequence and occurred with elastoviscous hylan but never appeared when physiological saline or non-elastoviscous hylan were injected. Thus, it is unlikely that the reduction of nerve activity observed in a part of the mechanosensory fibres of the intact joint was due to their inactivation by repeated stimulation, and it seems reasonable to associate the inhibitory effect with the presence of hylan in the joint cavity.

This interpretation was confirmed in experiments where an inflammation of the knee joint was induced. It is a well-established experimental observation [7,41] that joint sensory afferent fibres develop an ongoing activity at rest, and an enhanced responsiveness to non-noxious movements, 2–3 h after intra-articular injection of kaolin–carrageenan, i.e. they became sensitized. Injection of elastoviscous hylan in the inflamed joint progressively reduced both the ongoing activity and the movement-evoked discharges. A reduction of multifibre activity by elastoviscous hylan was also observed. In contrast, in both single fibre or in multifibre recordings, ongoing or movement-evoked activity was never affected by intra-articular injection of non-elastoviscous hylan into the inflamed joints.

A majority of, but not all, joint nerve fibres in the inflamed joint were inhibited by intra-articular injection of elastoviscous hylan. Analgesics such as acetylsalicylic acid and indomethacin when injected intravenously in the cat also produced a reduction of nerve discharges in only a fraction of joint nociceptive fibres in the arthritic joint [42]. In spite of such a partial effect on sensory nerve activity, these drugs have a well-documented peripheral analgesic action [43,44]. Therefore, it is reasonable to conclude that the decrease of sensory activity in a significant proportion of joint afferent fibres caused by intra-articular injection of elastoviscous hylan solution will lead to pain reduction in the injected joint.

The flux of hylan molecules within articular tissues is slow; the half-life of exogenous elastoviscous or non-elastoviscous hyaluronan and hylan in the rabbit joint is 1–2 days [27]. Therefore, it is not surprising that a relatively long latency time is required for reduction of nerve discharges. The different accessibility of the diffusing hylan molecules to the receptive sites of the nociceptive terminals [42] could explain why some of the fibres were not affected. Most importantly, the non-elastoviscous hylan solutions contain the same concentration of chemically identical but smaller molecules of hylan as the elastoviscous hylan solution. The difference in average molecular weight results in differences in physical (rheological) properties. This strongly indicates that the difference in effect is related to the elastoviscosity of the solution.

The reduction of nociceptive discharges in articular afferent fibres of the cat by elastoviscous hylan solutions provides experimental support to the clinical observation that viscosupplementation with elastoviscous hyaluronan and hylan solution decreases pain and improves the mobility of joints in race horses with traumatic arthritis [28,31], and in osteoarthritic human knee joints [32,35,37, 38,45,46].

Several mechanisms could contribute to the attenuation of nociceptive activity in normal and inflamed joints by elastoviscous hylan solutions. The first mechanism, suggested by Balazs [26,29,30,33], assumes that the hyaluronan of the intercellular matrix surrounding the nerve terminals may act as an elastoviscous filter, reducing the transmission of mechanical stretch to the membrane of nerve endings where mechano-electric transduction takes place [14,18,20]. The fact that non-elastoviscous hylan does not reduce nociceptive activity in joint afferents speaks in favour of a mechanical (rheological) role for the hyaluronan or hylan molecules. This explanation is supported by the finding that hyaluronan, injected into the knee joint cavity of rats, decreases pseudo-effective signs of pain induced by a parallel intra-articular injection of bradykinin [47]. This effect did not occur with non-elastoviscous solutions of low-molecular-weight hyaluronan; hence, it is not attributable to binding of hyaluronan to bradykinin receptors.

Another mechanism that may contribute to the attenuation by hylan of sensory discharges in joint nociceptors, particularly in arthritic joints, could be that the polyanionic polysaccharide molecular network constitutes a diffusion barrier for ions and for certain algogenic molecules released in the injured joint

[48,49]. Proteoglycans present in the capsular space in the muscle spindle appear to act both as a diffusion barrier and as a buffer for cations such as calcium [50,51]. Monovalent cations and calcium play an important role in the modulation of the excitability and sensitization of nociceptors [22,52]. Calcium ions, however, do not show binding affinity to hylan molecules in the presence of physiological concentrations of sodium chloride. Therefore, a reduction by hylan of the availability of extracellular calcium ions is unlikely. The possibility may still exist that hyaluronan or hylan molecules interfere with the access of algogenic molecules to nociceptive endings and contribute to the maintenance of the ionic milieu, thus regulating the sensitivity of nociceptive terminals to mechanical stimuli. However, with this mechanism alone, it is difficult to explain the effect of hylan on intact joints and the lack of effect of the chemically and ionically identical but non-elastoviscous hylan solutions. Thus, a filtering action on the transmission of mechanical forces to the mechanotransducing elements appears as the more presumable explanation for the attenuation by hylan of nociceptive discharges of joint afferent fibres and, consequently, of articular pain.

We thank Dr. J. Denlinger for her assistance and suggestions. Supported by grants PM95-0167 of the DGICYT, Spain and by the Matrix Biology Institute.

References

1. Skoglund, S. (1956) Acta Physiol. Scand. **36** (Suppl. 124),1–101
2. Freeman, M.A.R. and Wyke, B. (1967) J. Anat. **101**, 505–512
3. Langford, L.A. and Schmidt, R.F. (1986) Anat. Rec. **206**, 71–78
4. Burgess, P.R. and Clark, F.J. (1969) J. Physiol. **203**, 317–335
5. Schaible, H.G. and Schmidt, R.F. (1983) J. Neurophysiol. **49**, 35–44
6. Schaible, H.G. and Schmidt, R.F. (1983) J. Neurophysiol. **49**, 1118–1126
7. Schaible, H.G. and Schmidt, R.F. (1985) J. Neurophysiol. **54**,1109–1122
8. Bessou, P. and Perl, E.R. (1969) J. Neurophysiol. **32**,1025–1043
9. Perl, E.R., Kumazawa, T., Lynn, B. and Kenins, P. (1976) in Somatosensory and Visceral Receptor Mechanisms. Progress in Brain Research (Iggo, A. and Ilyinsky, B., eds.), vol. 43, pp. 263–276, Elsevier, Amsterdam
10. Handwerker, H.O. and Reeh, P.W. (1992) in Hyperalgesia and Allodynia (Willis, W.D., Jr., ed.), pp. 107–115, Raven Press, New York
11. Hanesch, U., Heppelmann, B., Messlinger, K. and Schmidt, R.F. (1992) in Hyperalgesia and Allodynia (Willis, W.D., Jr., ed.), pp. 81–106, Raven Press, New York
12. Schaible, H.G. and Schmidt, R.F. (1996) in Neurobiology of Nociceptors (Belmonte, C. and Cervero, F., eds.), pp. 202–219, Oxford University Press, Oxford
13. Shepherd, G.M. (1991) Cell, **67**, 845–851
14. Belmonte, C. (1996) in Neurobiology of Nociceptors (Belmonte, C. and Cervero, F., eds.), pp. 243–257, Oxford University Press, Oxford
15. Heppelmann, B., Pfeffer, A., Schaible, H.G. and Schmidt, R.F. (1986) Pain **26**, 337–351
16. Guharay, F. and Sachs, F. (1984) J. Physiol. **352**, 685–701
17. Sachs, F. (1986) Membrane Biochem. **6**, 173–195
18. Sachs F (1991) in Sensory Transduction (Corey, D.P. and Roper, S.D., eds.), pp. 241–260, The Rockefeller University Press, New York
19. Hudspeth, A.J. and Gillespie, P.G. (1994) Neuron, **12**, 1–9
20. Hamill, O.P. and McBride, D.W. (1995) Am. Sci. **83**, 30–37
21. Belmonte, C., Gallar, J., Lopez-Briones, L.G. and Pozo, M.A. (1994) in Cellular Mechanisms of Sensory Processing (Urban, L., ed.), NATO Series, Vol. H79, pp. 87–117, Springer-Verlag, Berlin
22. Pozo, M.A., Gallego, R., Gallar, J. and Belmonte, C. (1992) J. Physiol. **450**, 179–189
23. Bevan, S. and Geppetti, P. (1994) Trends Neurosci. **17**, 509–512
24. Rang, H.P., Bevan, S. and Dray, A. (1991) Br. Med. Bull. **47**, 534–548
25. Bevan, S. (1996) in Neurobiology of Nociceptors (Belmonte, C. and Cervero, F., eds.), pp. 298–324, Oxford University Press, Oxford

26. Balazs, E.A. (1974) in Disorders of the Knee (Helfet, A.J., ed.), pp. 61–74, J.B. Lippincott, Philadelphia.
27. Denlinger, J.L. (1982) Ph.D. Thesis, Université des Sciences et Techniques de Lille, France
28. Balazs, E.A. and Denlinger J.L. (1984) Equine Vet. Sci. **5**, 217–227
29. Balazs E.A. and Denlinger, J.L. (1984) in Modern Aging Research 4. Comparative Pathobiology of Major Age-related Diseases: Current Status and Research Frontiers (Scarpelli, D.G. and Migaki, G., eds.), pp. 129–143, Alan R. Liss, New York
30. Balazs, E.A. and Denlinger, J.L. (1985) in Osteoarthritis: Current Clinical and Fundamental Problems (Peyron, J.C., ed.), pp. 165–174, Laboratoires Ciba Geigy, Rueil-Malmaison, France
31. Rydell, N.W., Butler, J. and Balazs, E.A. (1970) Acta Vet. Scand. **11**, 139–155
32. Weiss, C., Balazs, E.A., St. Onge, R. and Denlinger, J.L. (1981) in Seminars in Arthritis and Rheumatism (Talbott, J.H., ed.), vol. 11 pp.143–144, Grune & Stratton, New York
33. Balazs, E. A. and Denlinger, J.L. (1989) in The Biology of Hyaluronan (Evered, D. and Whelan, J., eds.), Ciba Foundation Symposium No. 143, pp. 265–280, Wiley, Chichester
34. Balazs, E.A. and Denlinger, J.L. (1993) J. Rheumatol. **20** (Suppl), 3–9
35. Adams, M.E., Atkinson, M.H., Lussier, A., Schutz, J.I., Siminovitch K.A., Wade, J.P. and Zummer, M. (1996) Osteoarthritis Cartilage **3**, 213–226
36. Peyron, J.G. (1993) Osteoarthritis Cartilage **1**, 85–87
37. Peyron, J.G. (1993) J. Rheumatol. **20** (Suppl), 10–15
38. Scale, D., Wobig, M. and Wolpert, W. (1994) Curr. Ther. Res. **55**, 220–232
39. Gardner, E. (1944) J. Comp. Neurol. **80**, 11–32
40. Balazs, E.A. and Leshchiner, A. (1989) in Cellulosics Utilizations: Research and Rewards in Cellulosics (Inagari, H. and Phillips, G.O., eds.), pp. 233–241, Elsevier Applied Science, New York
41. Guilbaud, G., Iggo, A. and Tegner, R. (1985) Exp. Brain Res. **58**, 29–40
42. Heppelmann, B., Messlinger, K., Neiss, W.F. and Schmidt, R.F. (1990) J. Comp. Neurol. **292**, 103–116
43. Brune, K. (1982) Eur. J. Rheumatol. Inflam. **5**, 335–349
44. Mehlisch, D.R. (1983) Am. J. Med. **75**, 47–52
45. Peyron, J.G. and Balazs, E.A. (1974) Pathol. Biol. **22**, 731–736
46. Adams M.E. (1993) J. Rheumatol. **20** (Suppl), 16–18
47. Gotoh, S., Miyazaki, K., Onaya, J., Sakamoto, T., Tokuyasu, K. and Namiki, O. (1988) Folia Pharmacol. Jpn. **92**, 17–27
48. Schaible, H.G. and Schmidt, R.F. (1988) J. Physiol. **403**, 91–104
49. Grubb, B.D., Birrell, G.J., McQueen, D.S. and Iggo, A. (1991) Exp. Brain Res. **84**, 383–392
50. Fukami, Y. (1986) J. Physiol. **376**, 281–297
51. Fukami, Y. (1988) Brain Res. **463**, 140–143
52. Sato, J., Mizumura, K. and Kumazawa, T. (1989) J. Neurophysiol. **62**, 119–125
53. Pozo, M.A., Balazs, E.A. and Belmonte, C. (1997) Exp. Brain Res. **116**, 3–9

23

Induction of inflammatory gene expression by low-molecular-weight hyaluronan fragments in macrophages

Paul W. Noble*, Charlotte M. McKee and Maureen R. Horton

Department of Medicine, Division of Pulmonary and Critical Care Medicine, Johns Hopkins University School of Medicine, 858 Ross Research Building, 720 Rutland Avenue, Baltimore, MD 21205, U.S.A.

Hallmarks of chronic inflammation and tissue fibrosis are the increased synthesis and degradation of components of the extracellular matrix (ECM). The dynamic turnover of ECM results in the net deposition of matrix in the interstitium of affected tissues. An additional consequence of the degradation of ECM components is the generation of fragments from larger precursor molecules. In the cases of collagen and fibronectin, fragments generated as a result of proteolytic cleavage exhibit biological activities not attributable to the intact precursor [1,2]. A major component of the ECM that undergoes dynamic regulation during inflammation is the glycosaminoglycan (GAG) hyaluronan (HA). HA is a non-sulphated, linear GAG consisting of repeating units of (β1-4)-D-glucuronic acid-(β1-3)-N-acetyl-D-glucosamine [3]. In its native state, such as in normal synovial fluid, HA exists as a high-molecular-weight polymer, usually in excess of 10^6 [3]. However, under inflammatory conditions, HA has been shown to be more polydisperse, with a preponderance of lower-molecular-weight forms [4]. The accumulation of lower-molecular-weight forms of HA has been postulated to occur by a variety of mechanisms, including depolymerization by reactive oxygen species, enzymic cleavage, and *de novo* synthesis of lower-molecular-weight species [3,5,6]. Several studies have suggested that high and lower-molecular-weight HA may exhibit different biological effects on cells and in tissues [7,8]. High-molecular-weight HA (10^6 and greater) has been shown to inhibit macrophage phagocytosis; whereas low-molecular-weight HA (10^5 and lower) either has no effect or enhances phagocytosis [7].

HA has been suggested to play an important role in a number of biological processes, including wound healing [9], embryonic development [10] and tumour growth [11] by providing a provisional matrix for supporting cellular migration and adherence [3]. Further evidence has suggested that certain HA functions are mediated by interactions with CD44, a recently described HA receptor [12,13]. Recent work has provided evidence that, in addition to serving as a structural scaffold, HA may function as a cellular signalling molecule under certain circumstances [14]. The observation that low-molecular-weight HA stimulated the expression of interleukin-1β (IL-1β), tumour necrosis factor-α (TNF-α) and insulin-like growth factor-1 (IGF-1) in primary mouse bone

*To whom correspondence should be addressed at Yale School of Medicine, VA Connecticut Healthcare System, Pulmonary Section IIIA, 950 Campbell Avenue, West Haven, CT 06516, U.S.A.

marrow-derived macrophages, raised the possibility that HA may have a role in macrophage activation [15].

Products of activated macrophages play an important role in initiating and maintaining the chronic inflammatory state that is characteristic of disorders such as rheumatoid arthritis and pulmonary fibrosis [16,17]. The mechanisms responsible for macrophage activation in the chronic inflammatory state, however, have not been well characterized. We have investigated the hypothesis that fragments of HA generated at sites of inflammation may signal macrophages to induce a number of genes with functions important in regulating the inflammatory response. Using the technique of differential screening we have identified a subset of genes that are induced by both lipopolysaccharide (LPS) and HA [18,19]. These include four members of the chemokine gene family: macrophage inflammatory protein-1α (MIP-1α), macrophage inflammatory protein-1β (MIP-1β), cytokine-responsive gene-2 [crg-2 (murine inflammatory protein-10)], and murine regulated on activation, normal T-cell expressed and secreted (RANTES) as well as the important transcriptional regulator of inflammatory gene expression I-κBα [19,20]. Chemokines represent a newly described family of small peptides, many of which are derived from macrophages, which are potent modulators of immune cell function [21].

Relationship between HA size and induction of inflammatory genes in macrophages

As shown in Figures 1(a) and 1(b), purified high-molecular-weight HA does not induce chemokine gene expression, but smaller fragments generated in several ways are capable of inducing gene expression in a murine alveolar macrophage cell line (MH-S), primary bone-marrow-derived macrophages and elicited peritoneal macrophages (not shown). HA fragments as small as hexamers are able to induce gene expression (Figure 1c). Disaccharides are unable to induce gene expression. Fluorescein-labelled high-molecular-weight HA, while unable to induce gene expression, does bind to cultured mouse alveolar macrophage cells. This binding is blocked by low-molecular-weight HA, suggesting that the different sized polymers recognize the same cell surface receptor [19]

The role of CD44 in mediating low-molecular-weight HA-induced gene expression in macrophages

Over the last several years, multiple studies have established HA as the principal ligand for CD44 [12,13]. CD44 is a type I transmembrane glycoprotein with multiple isoforms that is expressed on the surface of most leucocytes, fibroblasts, keratinocytes and epithelial cells [22,23]. CD44 and its isoforms (20 have currently been described) have been implicated in binding to ECM, lymphopoiesis, eosinophil maturation, homotypic lymphocyte aggregation, T-cell and monocyte/macrophage activation and tumour development and metastasis [23]. Macrophages express the 85–95 kDa form that is referred to as the 'standard'

Figure 1

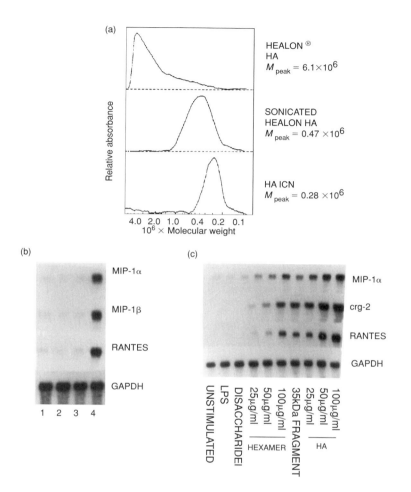

Effects of HA size on chemokine gene expression in MH-S cells

(a) Size distribution of HA preparations. The molecular-weight distribution of purified Healon® HA (top panel), HA fragments generated by sonicating purified Healon® HA for 2 min (middle panel), and commercially obtained purified HA fragments from human umbilical cord (HA-ICN, bottom panel) was analysed by gel electrophoresis as described in [19]. (b) HA fragments but not high-molecular-weight HA induce chemokine gene expression in MH-S cells. Purified high-molecular-weight Healon® HA was sonicated as described in [19]. MH-S cells were stimulated with medium alone (lane 1), LPS (100 ng/ml, lane 2), Healon® HA (500 μg/ml, lane 3) or the sonication-generated HA fragments (500 μg/ml, lane 4) for 4 h at 37 °C. Total RNA was isolated and Northern blotting analysis performed as described in [19]. (c) Effect of HA size on chemokine gene expression. MH-S cells were stimulated for 4 h at 37 °C with medium alone (unstimulated), LPS (100 ng/ml), HA disaccharide (100 μg/ml), increasing concentrations of HA hexamer, a 35 kDa HA fragment (100 μg/ml), and increasing concentrations of purified HA fragments from human umbilical cord (HA). Total RNA was isolated and subjected to Northern blotting analysis as described in [19]. MIP, macrophage inflammatory protein.

or 'haematopoietic' form CD44H [24]. CD44H is synthesized as a 42 kDa polypeptide that undergoes subsequent N- and O-linked glycosylation and is variably phosphorylated [24]. Thus post-translational modification and/or alternative splicing may confer tissue-specific properties [25]. Recent reports have shown that glycosylation alters the HA binding properties of CD44 [26,27]. Several characteristics of CD44 make it an attractive candidate to play an important role in both leucocyte recruitment and activation in inflammation [25]. First, while expressed on resting lymphocytes and macrophages, CD44 can be up-regulated under certain inflammatory conditions. Second, most resting leucocytes express CD44 but do not bind HA. It appears as though a post-translational event and/or additional molecule is required for CD44 to acquire HA binding activity. However, a central unanswered question is what are the physiological stimuli that convert CD44 into an HA-binding state. Thus, constituents of the inflammatory milieu may have the capacity to convert CD44 into a ligand-binding state, thereby allowing specific regulation at the tissue level. It is also possible that the HA-binding capacity of CD44 may depend on the state of cell maturation. Freshly isolated human monocytes do not bind HA, but when allowed to mature into macrophages in tissue culture, binding activity is acquired [28]. Several recent studies have suggested that ligation of CD44 on monocytes/macrophages can induce gene expression. Webb and co-workers showed that human monocytes plated on tissue culture dishes coated with anti-CD44 mAb released IL-1 and TNF into the media [29]. In addition, we provided evidence that HA induced IL-1β, TNF-α and IGF-1 expression in mouse bone-marrow-derived macrophages [15]. This provided the first evidence for a potential link between an extracellular matrix molecule produced as part of the inflammatory response and the production of a growth factor with the ability to promote fibroblast proliferation.

As shown in Figure 2, HA-induced chemokine gene expression is inhibited in the presence of anti-CD44 monoclonal antibody but not control antibodies. However, the inhibition is incomplete, suggesting that there may be a CD44-independent component to low-molecular-weight HA signalling. These results support the hypothesis that HA fragments induce genes whose functions are important in regulating the chronic inflammatory response.

Induction of inflammatory gene expression by low-molecular-weight HA requires an inflammatory state or the presence of stromal milieu

In addition to HA size, the state of macrophage activation appears to play an important role in the low-molecular-weight HA signalling. As shown in Figure 3, normal mouse alveolar macrophages do not respond to low-molecular-weight HA. However, if alveolar macrophages are isolated following the instillation of the inflammatory stimulus bleomycin sulphate, they acquire the ability to respond to low-molecular-weight HA. Figure 3 shows that low-molecular-weight HA induces the expression of inducible nitric oxide synthase (iNOS) mRNA in inflammatory alveolar macrophages [30]. Similar results are found for resident and elicited peritoneal macrophages; only the inflammatory macrophages respond to

Figure 2

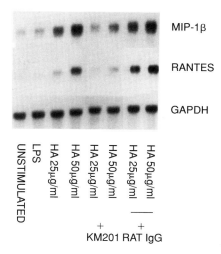

MIP-1β

RANTES

GAPDH

UNSTIMULATED
LPS
HA 25μg/ml
HA 50μg/ml
HA 25μg/ml
HA 50μg/ml
HA 25μg/ml
HA 50μg/ml

+ +
KM201 RAT IgG

Role of CD44 in HA-induced chemokine gene expression by MH-S cells

HA-induced chemokine gene expression is inhibited by mAb to CD44 [19]. MH-S cells were stimulated for 4 h at 37 ° C with medium alone (unstimulated), LPS (100 ng/ml), the indicated concentrations of purified HA fragments alone, HA fragments plus the anti-CD44 monoclonal antibody KM201 (25 μg/ml), or HA fragments plus nonspecific rat IgG (25 μg/ml). Cells that received antibody were preincubated with 25 μg/ml KM201 or rat IgG in 1X PBS for 45 min at 4 ° C prior to stimulation with HA fragments. Total RNA was then isolated and Northern blotting analysis was performed as described in [19].

Figure 3

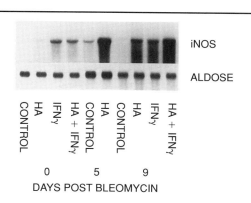

iNOS

ALDOSE

HA CONTROL
IFNγ
HA + IFNγ
HA CONTROL
HA
HA
IFNγ
HA + IFNγ

0   5   9
DAYS POST BLEOMYCIN

HA fragments induce iNOS gene expression in inflammatory alveolar macrophages

Alveolar macrophages were collected by bronchodreolar lavage from rats on the indicated days following intratracheal instillation of bleomycin as described in [30]. Cells were then stimulated for 4 h with medium alone (control), purified HA fragments (100 μg/ml), interferon-γ (IFNγ) (300 U/ml), or HA fragments plus IFNγ. Total RNA was collected and Northern blotting analysis was performed as described in [30]. iNOS, inducible NO synthase.

low-molecular-weight HA [31]. Bone-marrow-derived macrophages also respond to low-molecular-weight HA. These are not 'inflammatory' macrophages, but they are cultured in the bone marrow stroma in the presence of CSF-1. Collectively, these data suggest that in addition to the size of the HA, a priming step is required for macrophages to attain responsiveness to low-molecular-weight HA.

Implications of the generation of low-molecular-weight HA fragments on the resolution of the inflammatory response

High- and low-molecular-weight HA appear to present very different messages to cells. High-molecular-weight HA represents the normal homeostatic state and has 'anti-inflammatory' properties. High-molecular-weight HA provides a suitable matrix for cell adhesion and migration. The generation of low-molecular-weight HA fragments signals a disruption of the normal homoeostatic environment and alerts the immune system that significant tissue damage has occurred at a site of inflammation [32]. If unabated, the generation of low-molecular-weight HA fragments may contribute to the pathophysiology of chronic tissue inflammation. Elucidating the mechanisms that regulate the interactions between HA and inflammatory cells could provide important new therapeutic options for chronic inflammatory diseases such as rheumatoid arthritis and pulmonary fibrosis.

The authors would like to thank Dr. Mary Cowman for performing HA sizing and Drs. Cheryl Knudson and Judy Yannaerello-Brown for providing HA fragments. We also thank Clare Bao for expert technical assistance. This work was supported by the National Institutes of Health (K11HL02880) and the American Lung Association.

References

1. Clark, R., Wikner, N., Doherty, D. and Norris, D. (1988) J. Biol. Chem. **263**, 12115–12123
2. Laskin, D., Soltys, R., Berg, R. and Riley, D. (1994) Am. J. Respir. Cell Mol. Biol. **10**, 58–64
3. Laurent, T. and Fraser, J.R.E. (1992) FASEB J. **6**, 2397–2404
4. Saari, H. (1991) Ann. Rheum. Dis. **50**, 389–396
5. McNeil, J.M., Wiebkin, O.W., Betts, W.H. and Cleland, L.G. (1985) Ann. Rheum. Dis. **44**, 780–789
6. Sampson, P., Rochester, C., Freundlich, B. and Elias, J. (1992) J. Clin. Invest. **90**, 1492–1503
7. Forrester, J. and Balazs, E.A. (1980) Immunology **40**, 435–446
8. West, D., Hampson, I., Arnold, F. and Kumar, S. (1985) Science **228**, 1324–1326
9. Oksala, O., Salo, T., Tammi, R., Hakkinen, L., Jalkanen, M., Inki, P. and Larjava, H. (1995) J. Histochem. Cytochem. **43**, 125–135
10. Underhill, C., Nguyen, H., Shizari, M. and Culty, M. (1993) Dev. Biol. **155**, 324–336
11. Bartolazzi, A., Peach, R., Aruffo, A. and Stamenkovic, I. (1994) J. Exp. Med. **180**, 53–66
12. Aruffo, A., Stamenkovic, M., Melnick, M., Underhill, C.B. and Seed, B. (1990) Cell **61**, 1303–1313
13. Miyake, K., Underhill, C.B., Lesley, J. and Kincade, P.W. (1990) J. Exp. Med. **172**, 69–75
14. Bourguignon, L., Lokeshar, V., Chen, X. and Kerrick, W. (1993) J. Immunol. **151**, 6634–6644
15. Noble, P., Lake, F., Henson, P. and Riches, D. (1993) J. Clin. Invest. **91**, 2368–2377
16. Gauldie, J., Jordana, M. and Cox, G. (1993) Thorax **48**, 931–935.
17. Yanni, G., Whelan, A., Feighery, C. and Bresnihan, B. (1994) Ann. Rheum. Dis. **53**, 39–44
18. Shin, H., Drysdale, B-E., Shin, M., Noble, P., Fisher, S. and Paznekas, W. (1994) Mol. Cell. Biol. **14**, 2914–2925

19. McKee, C., Penno, M., Cowman, M., Burdick, M., Strieter, R., Bao, C. and Noble, P. (1996) J. Clin. Invest. **98**, 2403–2413
20. Noble, P., McKee, C., Cowman, M. and Shin, H. (1996) J. Exp. Med. **183**, 2373–2378
21. Schall, T.J. (1991) Cytokine **3**, 1–32
22. Haynes, B.F., Telen, M.J., Hale, L.P. and Denning, S.M. (1989) Immunol. Today **10**, 423–428
23. Lesley, J., Hyman, R. and Kincade, P. (1993) Adv. Immunol. **54**, 271–335
24. Camp, R.L., Kraus, T.A. and Pure E. (1991) J. Cell Biol. **115**, 1283–1292
25. Lazaar, A.L. and Pure, E. (1995) Immunologist **3**, 19–35
26. Lesley, J., English, N., Perschl, A., Gregoroff, J. and Hyman, R. (1995) J. Exp. Med. **182**, 431–437
27. Katoh, S., Zheng, Z., Oritani, K., Shimozato, T. and Kincade, P.W. (1995) J. Exp. Med. **182**, 419–429
28. Culty, M., O'Mara, T.E., Underhill, C.B., Yeager, Jr., H., and Swartz, R.P. (1994) J. Leukocyte Biol. **56**, 605–611
29. Webb, D., Shimizu, Y., Van Seventer, G.A., Shaw, S. and Gerrard, T.I. (1990) Science **249**, 1295–1297
30. McKee, C., Lowenstein, C., Horton, M., Wu, J., Bao, C., Chin, B., Choi, A. and Noble, P. (1997) J. Biol. Chem. **272**, 8013–8018
31. Hodge-Dufour, J., Noble, P.W., Horton, M.R., Bao, C., Wysocka, M., Burdick, M.D., Strieter, R.M., Trinchieri, G. and Pure, E. (1997) J. Immunol. **159**, 2492–2500
32. Aruffo, A. (1996) J. Clin. Invest. **98**, 2191–2192

24

Tumour hyaluronan in relation to angiogenesis and metastasis

*D.C. West and D.M. Shaw

Department of Immunology, Faculty of Medicine, University of Liverpool, Liverpool L69 3BX, U.K.

It is now widely accepted that the growth of most solid tumours is dependent on their ability to induce and maintain an adequate blood supply, through the stimulation of new capillary vessel growth (angiogenesis), [1,2]. In addition, the level of tumour vascularization has been shown to correlate with the numbers of circulating tumour cells and metastasis, in a transplantable mouse fibrosarcoma model, and tumour metastasis and poor prognosis in many human tumours [3–6]. Although endothelial cells can express all the information necessary to construct a capillary tube and assemble a vascular network [7], it is clear that their growth is tightly controlled by the balance of angiogenic stimulators and inhibitors [8]. Both *in vivo* and *in vitro* studies suggest that tissue hyaluronan (HA) metabolism may be an important regulator of angiogenesis [9].

HA and angiogenesis *in vivo*

The high concentration of HA found in avascular tissues, such as cartilage and vitreus, and at relatively avascular sites, such as the desmoplastic region of invasive tumours, suggests that extracellular matrix HA can inhibit angiogenesis [9,10]. *In vivo* studies indicate that a stroma rich in macromolecular hyaluronan inhibits blood vessel formation in granulation tissue [11,12], in a concentration- and size-dependent manner, and induces regression of the capillary plexus in the developing chick limb bud [13]. In contrast, low-molecular-weight HA-oligosaccharides (4–25 disaccharides in length) stimulate angiogenesis on the chick chorioallantoic membrane [14], in a quantitative rat impaired-graft model [15], in both rat and pig wound healing models ([16], D.C. West, R.L. Smither, D.M. Shaw, M. Joyce, S. Kumar and T.-P. Fan, unpublished work) and after sub-cutaneous implantation or topical application, in rabbits and rats respectively [17]. Hirata and co-workers [18] have independently confirmed the angiogenic activity of this range of HA-oligosaccharides, using the rabbit corneal assay. The size specificity of the angiogenic activity of HA-oligosaccharides *in vivo* has been confirmed in both the chick chorioallantoic membrane (CAM) assay and the impaired-graft model.

The extracellular matrix of healing wounds and the tumour stroma share several important properties, including hyperpermeable blood vessels, extravasation of fibrinogen and extravascular clotting. In both, the deposits of fibrin gel serve as a provisional stroma that is later replaced by granulation tissue. This similarity has prompted Dvorak [19] to compare the invasive edge of a tumour to

---

*To whom correspondence should be addressed.

a 'never healing wound'. Such comparisons encouraged us to investigate the temporal relationship between angiogenesis and tissue HA metabolism in a rat sponge-implant wound healing model and a freeze-injured rat skin-graft model [15]. In both systems angiogenesis coincided with the degradation of matrix hyaluronan, evidenced by a rapid fall in tissue HA size and content, and increasing hyaluronidase levels. The initial stages of HA 'degradation' were apparent immediately before significant angiogenesis, and addition of exogenous HA-oligosaccharides accelerated both the onset and rate of HA degradation, in parallel with angiogenesis, supporting the hypothesis that macromolecular HA is an important inhibitory regulator of angiogenesis. Furthermore, *in vivo* degradation of fetal wound hyaluronan, by the addition of *Streptomyces* hyaluronidase, decreased the wound HA content and increased both fibroplasia and capillary formation [20], i.e. a change from fetal, regenerative healing to an adult fibrotic type of healing. A similar change occurs in late gestation and we have recently shown that this is associated with an increase in hyaluronidase activity and a decrease in both hyaluronan concentration and, to a lesser extent, size [21]. In this context, high-molecular-weight HA has been reported to inhibit the action of transforming growth factor-β_1 [22], a promoter of tissue fibrosis, and to stimulate the secretion of tissue inhibitors of metalloproteinases [23], an inhibitor of angiogenesis, at concentrations similar to those in the early wound. These effects were found to be size dependent, decreasing significantly with the size of the HA.

Such reports imply that HA degradation must precede the onset of vascularization. Recent reports suggest that most HA degradation occurs intracellularly, within the lysosomal system, after binding to a surface receptor, usually CD44, and internalization [24]. Initial extracellular depolymerization of HA or proteoglycan complexes appears to be a necessary precursor to internalization, and is thought to be mediated by either oxygen-derived free radicals or proteinases [25]. As no specific inhibitors of hyaluronidase are available, local hyaluronidase action cannot be ruled out, but the increased hyaluronidase activity associated with hyaluronan degradation in many tissues [26] may reflect increased phagocytic activity.

HA and angiogenesis *in vitro*

In vitro studies on the effects of macromolecular HA on cultured endothelial cells have shown that HA inhibits endothelial proliferation and migration [9,27–30] and disrupts newly formed monolayers, at physiologically relevant concentrations ($\leqslant 100$ μg/ ml), i.e. at concentrations present in avascular tissues and during tissue remodelling. The inhibitory effect of HA decreases with reducing size, and the HA-mediated inhibition cannot be reversed by addition of exogenous growth factors, such as basic fibroblast growth factor (bFGF; [9,29]). Watanabe and co-workers [31] have recently confirmed that HA inhibits the proliferation of cultured bovine adrenal capillary endothelial cells and disrupts endothelial interactions, at physiologically relevant concentrations. Furthermore, high concentrations (1 mg/ml) of HA have been reported to inhibit the migration of human adipose capillary endothelial cells into fibrin gels, *in vitro*, and blood vessel formation in fibrin gels implanted subcutaneously in guinea pigs [12,32].

In our hands, HA-oligosaccharides of 2–8 kDa (4–20 disaccharides) specifically stimulate both proliferation and, to a lesser extent, migration of both bovine and human endothelial cells [9,27,30]. Hirata and co-workers [18] have reported that similar 'angiogenic' HA-oligosaccharides do not stimulate vascular endothelial cell proliferation, migration or plasminogen activator secretion of vascular endothelial cells cultured on 'matrigel'. However, they did find that, in common with several other angiogenic substances, HA-oligosaccharides markedly stimulated endothelial tube formation. Since then two recent studies, and our own preliminary studies, have shown that HA-oligosaccharides (4–20 disaccharides in length) stimulate angiogenesis in three-dimensional collagen and fibrin matrices [33,34]. Trouchon and co-workers [34] also confirmed that the 3-disaccharide oligosaccharide is not angiogenic. Preliminary studies have shown that HA-oligosaccharides also induce proliferation-related proteins, stimulate the phosphorylation of a 32 kDa protein and increase type I and VIII collagen synthesis by 4–6-fold [9,35]. Synthesis of type VIII collagen is thought to be restricted to active, proliferating endothelial cells. These data indicate that both HA-oligosaccharides and macromolecular HA interact directly with endothelial cells, inducing or inhibiting angiogenesis, respectively.

Several workers, including ourselves, have reported that endothelial cells possess high-affinity HA-binding proteins, or receptors, on their cell surface that both bind and internalize HA [27,29,36–38]. Initially, using macromolecular HA (10^6 Da), we detected 2000 receptors/cell, with a K_d of 10^{-12} M. More recently, using a defined 42 kDa HA fraction, we have found approximately 10^5 receptors/cell, with a similar avidity as before; the disparity being due to steric exclusion of large HA molecules from the cell surface [37]. We have identified several putative HA-receptors on human and bovine endothelial cells. Five major I^{125}-labelled cell-surface proteins, between 90 and 125 kDa, were detected on SDS/PAGE, with two minor bands at 78 and 46 kDa [29,30,36]; western blotting with anti-CD44 and anti-ICAM-1 (anti-intercellular adhesion molecule-1) antibodies, together with PCR determination of CD44 splice-variants, indicates that the 78–125 kDa proteins are forms of non-variant CD44, differing in their glycosylation, and ICAM-1 isoforms (D.C. West, unpublished work).

Polyvalent anti-CD44 antibodies have been reported to inhibit the migration and tube formation of bovine and porcine endothelial cells in collagen and fibrin gels [34,39], suggesting a role for CD44 in angiogenesis. It is probable that CD44 is involved in endothelial–endothelial interactions but, as the 3-disaccharide HA-oligosaccharide is not angiogenic and is bound by CD44, it is unlikely that it is the primary 'angiogenic' receptor for the active oligosaccharides. Montesano and co-workers [33] found that angiogenic HA-oligosaccharides acted synergistically with vascular endothelial growth factor (VEGF) but not bFGF. The VEGF up-regulates ICAM-1 expression on endothelial cells, but bFGF has only a marginal effect ([40], D.C. West., unpublished work). In contrast, CD44 expression is not altered. This suggests that ICAM-1 is the HA receptor mediating the angiogenic activity of the HA-oligosaccharides. Recently, Noble and co-workers [41] have reported that HA-oligosaccharides, between the hexasaccharide and 440 kDa, rapidly induce and activate NFκB and reduce IκBα expression in cultured macrophages. In an earlier study, they found that the binding of HA (80 kDa in size) to CD44 induced tumour necrosis factor-α and interleukin-1β expression within 1 h, and insulin growth factor-1 after 12 h [42]. These results suggest that binding of HA-oligosaccharides to CD44 could also up-regulate ICAM-1 and thus predispose the endothelial cell to stimulation by the HA-oligosaccharides.

Tumour HA metabolism and metastasis

The increased levels of hyaluronan, often associated with human and animal tumours [43], are difficult to equate with the increased angiogenesis reported for many tumours, especially metastatic tumours. To shed some light on this conundrum, we have analysed the HA content/size and hyaluronidase activity of syngeneic murine and rat tumours, human xenograft tumours and cultured tumour cell lines. Most of the 22 different tumours had raised levels of HA (38–4938 μg/g of tissue), compared with normal adult tissues. Whilst HA in these tumours showed a wide variation in size (ranging from 200 to 2000 kDa) there was a significant shift to a lower molecular mass, compared with normal tissues. The size of HA in both the syngeneic tumours and xenografts showed a loose relationship to the relative levels of tumour hyaluronidase activity and HA content [43–45], i.e. those tumours with high hyaluronidase activity generally had smaller HA. This suggests that increased tumour hyaluronidase levels may reduce tumour HA content and size and, consequently, should be indicative of increased angiogenesis and a poor tumour prognosis. We have previously reported that a bone metastasizing form of Wilms' tumour, bone metastasising renal tumour of childhood (BMRTC), secretes high levels of low-molecular-weight HA into the patients circulation [46]. Wilms' tumours are rarely metastatic and produce high levels of high-molecular-weight HA. Furthermore, a recent report by Lokeshwar and co-workers [47] shows an association between elevated hyaluronidase levels and prostate cancer progression, giving some support to this hypothesis. Closer examination of the data for the syngeneic rat and murine tumours showed that only three tumours had HA levels below 100 μg/g of tissue, the B16-F10 and B16-BL6 murine melanomas and a transplantable spontaneous rat uterine tumour. Although all three had high-molecular-weight HA ($2–3 \times 10^6$ Da), they were all highly metastatic *in vivo*, suggesting that both HA degradation and low tumour HA content are indicative of tumour metastasis and probably increased tumour angiogenesis.

Tumours are complex mixtures of both tumour and stromal cells, and tumour–stromal cell interactions are reported to increase tumour HA, especially in invasive tumours [43]. Thus we were interested to compare HA synthesis with hyaluronidase activity of cultured tumour cell-lines and have analysed over 30 murine and human cell-lines. Most cell lines produced medium to high levels (1–20 μg/10^6 cells), compared with normal fibroblast cultures, of mainly high-molecular-weight HA (>800 kDa). Only 5–10% of the HA produced remained cell-associated and this fraction tended to be slightly smaller than that in the medium. Hyaluronidase activity was difficult to detect in most cultures and when detected it was only present in the culture medium. However, there were several notable exceptions.

Metastatic mouse melanoma B16-F10 and B16-BL6 cells produced very little HA (58–87 ng/10^6 cells). The B16-F10 cells produced mainly large HA (900 kDa) and low levels of hyaluronidase in the medium. In contrast, the more metastatic B16-BL6 cells produced HA of 100 kDa, or less, and had 10-fold higher levels of hyaluronidase, some of which was cell-associated. Of the human tumour cell-lines, 6 lines were found to produce low amounts of HA (<500 ng/10^6 cells). Three were aggressive cervical carcinoma cell-lines and three (A2058 melanoma, Du145 prostatic carcinoma and HCT116(A+) colon carcinoma cells) are highly metastatic in animal models. Similarly, Timar and co-workers [48] have reported that highly metastatic variants of Lewis lung tumour cell-lines produce significantly lower amounts of HA.

We have recently analysed in more detail several related human colon carcinoma cell-lines for their HA synthesis and angiogenic activity. Two colon carcinoma cell-lines, HCT116 A(+) and HCT116 B(+), were originally isolated from the same colorectal carcinoma by Brattain (Houston). They differ in their ability to invade collagen gels *in vitro* and a subclone of the more invasive HCT116 B(+) line, called 2010-1, was selected by Kinsella (Liverpool) for its increased invasiveness *in vitro* [49]. However, results in an *in vivo* metastasis model contradicted those observed *in vitro* [44]. Cells injected into the ceacum wall of nude mice established a primary tumour and eventually secondary liver metastases. The greatest number of liver metastases were found in animals bearing HCT116 A(+) tumours.

Analysis of these cell lines for HA production and hyaluronidase activity gave some interesting results. HCT116 A(+) cells produced considerably less HA than either of the other two cell lines, 0.5 µg/10^6 cells versus 2–4 µg/10^6 cells, with a similar reduction in both cellular and medium HA levels. Examination of HA size showed that HCT116 A(+) HA was significantly smaller than that synthesized by the other cell lines, 100 kDa and <300 kDa, cell-surface and medium, respectively, compared with 1 800–3 000 kDa and 900–1 500 kDa, respectively. In addition, HCT116 A(+) cells had relatively high levels of hyaluronidase activity, which was mainly cell-associated, whereas none could be detected in the other cell lines, suggesting a causative relationship between high enzyme levels and the reduced HA size.

We have hypothesized that a reduction in HA size and concentration would allow endogenous angiogenic cytokines to stimulate angiogenesis. Furthermore, the range of HA secreted by these cells, <30–300 kDa , suggests that angiogenic HA-oligosaccharides are probably present. Analysis of conditioned serum-free media from the three cell lines in the CAM assay showed that only HCT116 A(+)-conditioned medium was angiogenic. When this was mixed with HCT116 B(+)-conditioned medium the angiogenic activity was marginal, suggesting that HCT116 B(+)-conditioned medium contained an anti-angiogenic activity, the most obvious candidate being the high levels of macromolecular HA. Although our data support the hypothesis that hyaluronidase-mediated breakdown of HA is necessary for increased tumour angiogenesis and metastasis, hyaluronidase is thought to act only in the lysosomal environment. Liu and co-workers [50] have recently found a potential 'neutral' surface-associated hyaluronidase, similar to the glycosylphosphatidylinositol-anchored sperm PH-20 hyaluronidase, expressed by angiogenic tumour cell-lines. This may be the cell-associated activity we find in tumour cell-lines producing low-molecular-mass HA.

Conclusions

In vivo wound healing studies have shown a close temporal relationship between HA breakdown and tissue vascularization. It is probable that in these situations HA breakdown is mediated by either hydroxyl radicals and proteinase activity, followed by phagocytosis and further breakdown by lysosomal hyaluronidase. However, it is obvious from our data that partial breakdown products are present in the extracellular matrix. A number of studies have now confirmed the angiogenic nature of HA-oligosaccharides (4–20 disaccharides), both *in vivo* and *in vitro*. Present evidence suggests that these oligosaccharides bind to both CD44 and ICAM-1, but that ICAM-1-binding is probably the most important for the stimulation of angiogenesis.

Studies on transplantable tumours and cultured tumour cell-lines suggest that tumour metastasis/angiogenesis is associated with increased HA degradation. This may be mediated by the increased expression of a cell-surface 'neutral' hyaluronidase, possibly related to the sperm PH-20 enzyme.

References

1.   Folkman, J. (1990) J. Natl. Cancer Inst. **82**, 4–6
2.   Folkman, J. (1995) Nat. Med. **1**, 27–31
3.   Liotta, L., Kleinerman, J. and Saidel, G. (1974) Cancer Res. **34**, 997–1004
4.   Liotta, L., Saidel, G. and Kleinerman, J. (1976) Cancer Res. **36**, 889–894
5.   Weidner, N., Semple, J.P., Welch, W.R. and Folkman, J. (1991) N. Engl. J. Med. **324**, 1–8
6.   Hart, I.R. and Saini, A. (1992) Lancet **339**, 1453–1457
7.   Folkman, J. and Haudenschild, C.C. (1980) Nature **288**, 551–556
8.   Folkman, J. and Klagsbrun, M. (1987) Science **235**, 442–447
9.   West, D.C. and Kumar, S. (1989) in The Biology of Hyaluronan, (Evered, D. and Whelan, J., eds.), Ciba Foundation Symposium No. 143, pp. 187–207, Wiley, Chichester
10. Barsky, S.H., Nelson, L.L. and Levy, V.A. (1987) Lancet **1**, 13336–13337
11. Balazs, E.A. and Darzynkiewcz, Z. (1973) in Biology of Fibroblast (Kulonen, E. and Pikkarainen, J., eds.), pp. 237–252, Academic Press, New York
12. Dvorak, H.F., Harvey, S., Estralla, P., Brown, L.F., McDonagh, J. and Dvorak, A.M. (1987) Lab. Invest. **57**, 673–686
13. Feinberg, R.N. and Beebe, D.C. (1983) Science **220**, 1177–1179
14. West, D.C., Hampson, I.N., Arnold, F. and Kumar, S. (1985) Science **228**, 1324–1328
15. Lees, V.C., Fan, T-P.D. and West, D.C. (1995) Lab Invest. **73**, 259–266
16. Arnold, F., Jia, C.Y., He, C.F., Cherry, G.W., Carbow, B., Meyer-Ingold, W., Bader, D. and West, D.C. (1995) Wound Rep. Reg. **3**, 10–21
17. Sattar, A., Rooney, P., Kumar, S., Pye, D., West, D.C., Scott, I. and Ledger, P. (1994) J. Invest. Dermatol. **103**, 576–579
18. Hirata, S., Akamarsu, T., Matsubara, T., Mizuno, K. and Ishikawa, H. (1993) Arthritis Rheum. **36**, S247(abstr.)
19. Dvorak, H.F. (1986) N. Engl. J. Med. **315**, 1650–1659
20. Mast, B.A., Haynes, J.H., Krummel, T.M., Diegelman, R.F. and Cohen, I.K. (1992) Plast. Reconstr. Surg. **89**, 503–509
21. West, D.C., Shaw, D.M., Lorenz, P., Adzick, N.S. and Longaker M. (1996) Int. J. Biochem. Cell Biol. **29**, 201–210

22. Locci, P., Marinucci, L., Lilli, C., Martinese, D. and Becchetti, E. (1995) Cell Tissue Res. **281**, 317–324

23. Yasui, T., Akatsuka, M, Tobetto., K, Umemoto, J., Ando, T., Yamashita, K. and Hayakawa, T. (1992) Biomed. Res. **13**, 343–348

24. Laurent, T.C. and Fraser, J.R.E. (1992) FASEB J. **6**, 2397–2404

25. Ng, C.K., Handley, C.J., Preston, B.N. and Robinson, H.C. (1992) Arch. Biochem. Biophys. **298**, 70–79

26. Toole, B.P. (1991) in Cell Biology of Extracellular Matrix (Hay, E.D., ed.), pp. 305–339, Plenum Press. New York

27. West, D.C. and Kumar, S.(1989) Exp. Cell Res. **183**, 179–196

28. West, D.C. and Kumar, S. (1988) Lancet **1**, 715–716

29. West, D.C. and Kumar, S. (1991) Int. J. Radiol. **61/62**, 55–60

30. Sattar, A., Kumar, S. and West, D.C. (1992) Semin. Arthritis Rheum. **21**, 43–49

31. Watanabe, M., Nakayasu, K. and Okisaka, S. (1993) Nippon Ganka Gakkai Zasshi **97**, 1034–1039

32. Fournier, N. and Doillon, C.J. (1992) Cell Biol. Int. Rep. **16**, 1251–1263

33. Montesano, R., Kumar, S., Orci, L. and Pepper, M.S. (1996) Lab. Invest. **75**, 249–262

34. Trouchon, V., Mabilat, C., Bertrand, P., Legrand, Y., Smadia-Joffe, F., Soria, C., Delpech, B. and Lu, H. (1996) Int. J. Cancer **66**, 664–668

35. Rooney, P., Wang, M., Kumar, P. and Kumar, S. (1993) J. Cell Sci. **105**, 213–218

36. West, D.C. (1993) in Vascular Endothelium: Physiological Basis of Clinical Problems II (Catravas, J.D., ed.), pp. 209–210, Plenum Press, New York

37. Smedsrød, B., Pertoft, H., Eriksson, S., Fraser, J.R.E. and Laurent, T. (1984) Biochem. J. **223**, 617–626

38. Madsen, K., Schenholm, M., Jahnke, G. and Tengblad, A. (1989) Invest. Ophthalmol. Vis. Sci. **30**, 2132–2137

39. Banerjee, S.D. and Toole, B.P. (1992) J. Cell Biol. **119**, 643–652

40. Melder, R.J., Koenig, G.C., Witwer, B.P., Safabakhsh, N., Munn, L.L. and Jain, R.K. (1996) Nat. Med. **2**, 992–997

41. Noble, P.W., McKee, C.M., Cowman, M. and Shin, H.S. (1996). J. Exp. Med. **186**, 2373–2378

42. Noble, P.W., Lake, F.R., Henson, P.M. and Riches, D.W.H. (1993) J. Clin. Invest. **91**, 2368–2377

43. Knudson, W., Biswas, C., Li, X-Q., Nemec, R.E. and Toole, B.P. (1989) in The Biology of Hyaluronan (Evered, D. and Whelan, J., eds.), Ciba Foundation Symposium No. 143, pp. 150–169, Wiley, Chichester

44. Shaw, D.M. (1996) Ph.D. Thesis, University of Liverpool

45. Shaw, D.M., West, D.C. and Hamilton, E. (1994) Int. J. Exp. Pathol. **75**, A67

46. Kumar, S., West, D.C., Ponting, J. and Gattamaneni, H.R. (1989) Int. J. Cancer **44**, 445–448.

47. Lokeshwar, V.B., Lokeshwar, B.L., Pham, H.T. and Block, N.L. (1996) Cancer Res. **56**, 651–657

48. Timar, J., Moczar, E., Timar, F., Pal, K., Kopper, L., Lapis, K. and Jeney, A. (1987) Clin. Exp. Metastasis **5**, 79–87

49. Kinsella, A.R., Lepts, G.C., Hill, C.L. and Jones, M. (1994) Clin. Exp. Metastasis **12**, 335–342

50. Liu, D., Pearlman, E., Diaconu, E., Guo, K., Mori, H., Haqqi, T., Markowitz, S., Willson, J. and Sy, M-S. (1996) Proc. Natl. Acad. Sci. U.S.A. **93**, 7832–7837

25

Hyaluronan and its derivatives as viscoelastics in medicine

Janet L. Denlinger

Biomatrix, Inc.,65 Railroad Ave., Ridgefield, NJ 07657, U.S.A.

The first comprehensive publication on the therapeutic use of hyaluronan was published in 1971 by Endre A. Balazs [1], which summarized 15 years of research and development on this subject. There were two main conclusions of this publication. The first was that to use high-molecular-weight hyaluronan for medical purposes it had to be purified in a specific way to remove a fraction that caused acute inflammation in the animal and human body. The purified hyaluronan remaining, prepared from human umbilical cord or rooster combs, was called the non-inflammatory fraction of sodium hyaluronate or, as it is known by its acronym, NIF-NaHA. The second conclusion of this publication was that the medical utility of hyaluronan solutions resides in their physical properties. In other words, the medical benefits, whatever they might be, depend on the rheological or physical properties of the material, rather than its chemical nature. This physical effect can be attributed to the viscosity, elasticity or the physical barrier nature of a highly hydrated, entangled molecular network.

The first NIF-NaHA used in human and equine clinical studies was prepared in the late 1960s from human umbilical cord, and shortly thereafter from rooster comb [2,3]. This NIF-NaHA was tested in three medical areas: in human idiopathic and traumatic osteoarthritis [4,5], equine arthritis (traumatic) [6] and in ophthalmic surgery in the anterior chamber and vitreus [7–11].

During the 1960s, Balazs, with Penti Seppälä, Nils Rydell, and David Gibbs, studied the elastoviscous properties of human synovial fluids [6,12–17]. These studies centred mostly on the effect of aging and arthritis on the elastoviscosity of the hyaluronan in the human knee joint. The information collected formed the experimental basis for the new concept of using NIF-NaHA as a synovial fluid replacement in arthritic joints. The first clinical studies using viscosupplementation with NIF-NaHA were started in 1968, using race horses and human patients with osteoarthritis [4,6]. The NIF-NaHA used for these clinical studies had a weight-average molecular weight of 2–3 million, and had rheological properties nearly as high as those of healthy human synovial fluid. This 1% solution was given the name Healon®. The therapeutic intention was to replace the pathological synovial fluid of low elastoviscosity with a fluid that had an elastoviscosity nearly as high as the normal healthy fluid. Because this new therapeutic modality was based on the supplementation of a body fluid, the name *viscosupplementation* was later suggested for this new therapeutic paradigm [18].

By 1972 it was clear that viscosupplementation would be the new therapy for the treatment of arthritis in horses and humans. That year Pharmacia AB took over the task of bringing this new treatment to patients worldwide. Between 1978 and 1982 Pharmacia succeeded in bringing NIF-NaHA [19], under the trade name of Hylartil® [20], into use in the practice of veterinary medicine as the first viscosupplementation product worldwide. They were not successful, however, in completing this task in the field of human arthritis. It was not until 1987, nearly two decades after it had been established that viscosupplementation was a safe and efficacious treatment for osteoarthritis, that two companies – Seikagaku in Japan and Fidia in Italy, succeeded in preparing a low-molecular-weight NIF-NaHA and bringing it to patients in their respective countries. Because of the lower average molecular weight of these preparations, a 1% solution of this NIF-NaHA has a much lower elastoviscosity than Healon. It was found that such low elastoviscous NIF-NaHA was efficacious only if it was injected more often than the more elastoviscous NIF-NaHA tested decades earlier [21–23].

Dr. Balazs and his co-workers were convinced from the beginning that the most effective viscosupplementation treatment would result from using a solution with a viscosity as close as possible to that of the synovial fluid of a healthy young adult. Since his first attempt with Pharmacia did not succeed in bringing highly viscoelastic NIF-NaHA to patients, he and Ed. Leshchiner at Biomatrix developed two new hyaluronan derivatives, called hylans. Hylan A is a fluid with a weight-average molecular weight of 6–26 million, and hylan B is a hydrated gel with an infinite molecular network [24,25]. An elastoviscous solution containing both hylans became the first hyaluronan-derived product available for viscosupplementation of human and equine arthritis: Synvisc® and Gelvisc®Vet, respectively.

Hyaluronan and hylans are also used in another medical therapeutic paradigm called *viscosurgery* [26]. The surgical use of viscoelastic solutions was initiated with the use of hyaluronan in ophthalmic surgery [27]. After the completion of several clinical trials, Balazs and co-workers concluded that highly elastoviscous NIF-NaHA "should be used as an implant in conjunction with various surgical procedures in vitreoretinal diseases, in cataract surgery and in keratoplasty. In these cases, the elastoviscous jelly, acting as a mechanical buffer, may help in the surgical restoration and maintenance of normal anatomical situations and will serve as a protection against mechanical stress" [10]. This concept was confirmed by later work [11,28,29].

The concept of viscosurgery is based on a very special rheological property of hyaluronan and its derivatives. Solutions made of these biopolymers have a very high viscosity at low shear and a very low viscosity and high elasticity at high shear. At low shear rate, a 1% solution of hyaluronan with weight-average molecular weight of 4 million has a half-million times greater viscosity than physiological salt solution. At high shear this viscosity drops to 100 times that of water. This high shear dependence of viscosity makes high-molecular-mass hyaluronan and hylan A ideal viscosurgical tools. Figure 1 shows schematically that at low shear the elastoviscous solution can be used to make and maintain space and retain tissues in the desired position. When the elastoviscous fluid moves under high shear, the viscosity drops dramatically (pseudoplasticity, shear thinning), and therefore the solution can pass through narrow channels or spaces with ease. But when the elastoviscous fluid is exposed to high frequency forces under dynamic conditions the solution becomes more and more elastic as the

Figure 1

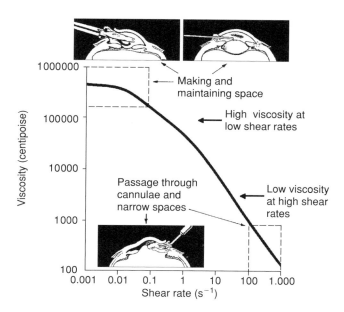

The shear viscosity of a 1% hyaluronan solution (dissolved in physiological buffered salt solution) made of hyaluronan of a weight-average molecular weight of ~4 million

See text for further explanation.

frequency increases. Under these conditions, the elastoviscous solution serves as an elastic shock-absorbing body. This property of the fluid is used for the protection of delicate tissues and cell layers, such as the corneal endothelium, the iris and the retina (Figure 2).

In 1976 Balazs turned over the worldwide commercialization of NIF-NaHA for ophthalmic viscosurgery to Pharmacia. The first clinical data on the use of elastoviscous NIF-NaHA during intraocular lens implantation was reported in 1979 by Balazs, Miller and Stegmann [26]. At this time, the descriptive term *viscosurgery* was also introduced and the utility of NIF-NaHA for viscosurgery was confirmed by Pape and Balazs [29] and Miller and Stegmann [30]. By 1982 Healon® was available for ophthalmic surgeons in most major countries of the world. That year, the first clinical symposium was organized by Professor Meyer-Schwickerath in which 18 clinical investigators described clinical benefits of viscosurgery with Healon® [31]. An ideal viscosurgical tool provides protection from mechanical trauma for instruments and implants, it makes and maintains space for surgical manipulation, is usable as a soft tool for dissecting tissues and tissue adhesion, controls bleeding by its barrier effect, makes removal of tissue debris such as pieces of cataractous lens better controlled and therefore safer, and in general provides elastoviscous protection from mechanical damage by suction or vibrating instruments [32,33]. By the mid-1980s viscosurgery with Healon® had become an increasingly popular therapeutic modality in ophthal-

Figure 2

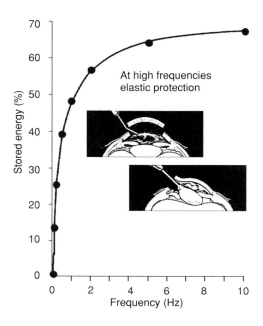

The percentage elasticity is calculated from the phase angle measurement of the dynamic elastic modulus (G′) at various frequencies.

The elastic properties of a 1% hyaluronan solution (dissolved in physiological buffered salt solution) made of hyaluronan of a weight-average molecular weight of ~4 million
The percentage elasticity is calculated from the phase angle measurement of the dynamic elastic modulus (G′) at various frequencies.

mology and was used in the majority of surgical procedures. Now, a decade later, in addition to Healon®, at least 10 different hyaluronan preparations are available for the ophthalmic surgeon. Interestingly, none of them have the high elastic and viscous properties of Healon®, mostly because the molecular weight of hyaluronan in all of these viscoelastics is much lower than that of Healon® [34–36].

Principles of viscosurgery are being extended to other surgical specializations. Ophthalmic surgeons have used Healon® to protect the surface of the cornea from dryness and the heat of operating lamps ever since this material became available. However, it was not until 3 years ago that a hylan A product, HsS™ (Hylan surgical Shield; Biomatrix, Inc.) was developed specifically for this purpose. In the fields of otolaryngology and orthopaedic surgery, viscosurgery with hyaluronan and with hylan A was explored in the 1980s [18,37,38]. In orthopaedic surgery, viscosurgical techniques have been used to protect cartilage surfaces and soft tissue surfaces from mechanical damage during arthroscopy [39,40]. In otology, viscosurgery has been used in middle ear surgery [41,42].

Another application for viscoelastics is their use in combination with drugs to make surgical procedures more effective and to promote post-surgical healing, a process called *viscomodulation.* This use of viscoelastics as surgical tools combined with analgesics, antibiotics, mydriatics, coagulants, anti-growth, anti-tumour or anti-inflammatory agents, has as its primary goal viscosurgery that is aided or supported by the added drugs. These uses of viscoelastics as tools for viscomodulation combined with drugs prior to, during or directly after surgery must be clearly separated from the use of viscoelastics as drug delivery systems or viscoregulation. The aim of viscoregulation is to deliver a drug, and the viscoelastics are used to modulate the dose and time of delivery. In contrast, using viscoelastics as viscomodulators has as its primary goal viscosurgery that is aided or supported by the added drugs.

Balazs, in his 1971 essay, also addressed another issue that today we call *viscoaugmentation,* which is the replacement or augmentation of the intercellular matrix. Viscoaugmentation was first used in ophthalmic surgery to provide a transparent, elastoviscous solution to replace the partially liquified vitreus and to push the detached retina back to its natural position and keep it there. By the end of the 1970s, a multicentre clinical study established that an elastoviscous solution of NIF-NaHA was safe to use for viscoaugmentation of the vitreus, and that it had certain utility in retinal surgery [10,11]. It is now 38 years since the first International Symposium of Retinal Surgery, organized by Charles Schepens in Ipswich, Massachusetts, where Balazs announced that elastoviscous solutions of hyaluronan should be used as vitreus substitutes [43]. Since that time elastoviscous solutions of NIF-NaHA and viscoelastic gels of hylan have been designed and used for this purpose. Today, however, the challenge still remains to find the ideal viscoelastic gel for viscoaugmentation of the vitreus. For the time being, the most frequently used, long-term replacement for the vitreus is silicone oil. The use of silicone oil for filling up the entire vitreus space is approved in the U.S. by the FDA and in Europe by the competent authorities. One may wonder about the logic in medical therapeutic thinking when one considers that silicone oil is regarded as highly toxic and dangerous in a plastic bag implanted into the breast muscle, but is regarded as completely safe and harmless when poured into the space bordered by vascular and sensitive tissues such as the retina, ciliary body and sometimes the lens.

Tissue augmentation with hyaluronan and its derivatives was later extended for use in the skin. Hylan B gel has been used for viscoaugmentation of the dermal intercellular matrix to correct depressed scars and to smooth wrinkles and furrows on the face.

The use of viscoelastics in *viscoseparation* continues to be one of the more interesting applications of these substances in medicine. In 1970, Balazs described the medical use of dry membranes made of the acid form of NIF-NaHA [1]. The membranes were used in several animal experiments to prevent adhesions and excessive scar formation. These studies were based on a very simple concept. When a biologically inert membrane, to which cells, fibrin coagulum or newly formed collagen cannot adhere, is placed between two healing tissue surfaces, the two surfaces will remain separated at the completion of the healing process. Because they can move freely relative to each other, their function is not impaired. In other words, post-surgical adhesion between two tissue surfaces is prevented. This obviously has great importance in preventing adhesions between tissue surfaces in the musculoskeletal system, such as between tendon and tendon sheaths, between surfaces of serous membranes, such as can exist after pleural, abdominal or cardiac surgery, and between surfaces covered with mucous membranes, such as the uterus and in the nasal passages. In the 1970s NIF-NaHA membranes were used in hand tendon surgery in primates [44]. When hylan B gel became available for the first time, in the early 1980s, it was used in human clinical studies to prevent adhesions in reconstructive hand tendon surgery and to prevent excessive scar formation around perineural nerves [45,46].

Briefly then, to summarize the past and present use of viscoelastics in medicine:

• First: the concept of using elastoviscous solutions and membranes made of highly purified hyaluronan called NIF-NaHA was developed during the 1960s by Biotrics in Arlington, MA, U.S.A., and the broad medical uses of NIF-NaHA were firmly established in the 1971 essay by Balazs.

• Second: NIF-NaHA, under the name Healon® in human medicine and under the name Hylartil® in veterinary medicine, became the first hyaluronan available for therapeutic use.

• Third: the first two derivatives of hyaluronan, called hylan A (an elastoviscous fluid) and hylan B (a viscoelastic gel), used as therapeutic agents in medicine were developed by Biomatrix, Inc., Ridgefield, NJ, U.S.A. The process of developing hyaluronan and hylans as viscoelastic medical therapeutic agents has been a long and arduous process. This process began in Boston in the late 1950s, and a great number of scientists and clinical investigators were involved in this development. Without any doubt, Balazs is the leading discoverer and tireless advocate of the use of viscoelastics as therapeutic agents. Elastoviscous solutions of hyaluronan and its derivatives, viscoelastic gels of cross-linked hyaluronan and membranes and tubes made of cross-linked or otherwise derivatized hyaluronan are the new materials that will be used for a wide variety of surgical procedures and innovative medical applications.

Thank you, Torvard, for making it possible for all of us in Bandi's life to be here together, and to you, Bandi, for having and sharing so generously those myriad ideas in many areas of science and medicine. Because of this, all of us have been challenged over the years, and have also had the chance to work on these ideas with you – from your first student – Torvard – to me as your last student.

References

1. Balazs, E.A., ed. (1971) Hyaluronic Acid and Matrix Implantation, Biotrics Inc., Arlington, MA.
2. Balazs, E.A., Mars, P.H. and Szirmai, J. (1955) Am. Chem. Soc., 128th Meeting., p. 46C , Minneapolis, MN
3. Balazs, E.A., Szirmai, J.A. and Bergendahl, G. (1959) J. Biophys. Biochem. Cytol. **5**, 319–326
4. Rydell, N. and Balazs, E.A. (1971) Clin. Orthoped. **80**, 25–32
5. Peyron, J.G. and Balazs, E.A. (1974) Pathol. Biol. **22**, 731–736
6. Rydell, N.W., Butler, J. and Balazs, E.A. (1970) Acta Vet. Scand. **11**, 139–155
7. Balazs, E.A. and Sweeney, D.B. (1965) in Controversial Aspects of the Management of Retinal Detachment (Schepens, C.L. and Regan, C.D.J., eds.), vol. 3, pp.200–202, Little, Brown and Company, Boston
8. Balazs, E.A. and Sweeney, D.B. (1966) in Modern Problems in Ophthalmology (Surgery of Retinal Vascular Diseases, Amersfoort, 1963)(Streiff, E.B., ed.),vol. 4, pp.230–232, S. Karger, Basel
9. Balazs, E.A. and Sweeney, D.B. (1968) in New and Controversial Aspects of Retinal Detachment (McPherson, A., ed.), pp. 371–376, New York

10. Balazs, E.A., Freeman, M.I., Klöti, R., Meyer-Schwickerath, G., Regnault, F. and Sweeney, D.B. (1971) XXI Concilium Ophthalmologicum Mexico ACTA, Part I, pp. 555–558, Excerpta Medica, Amsterdam

11. Balazs, E.A., Freeman, M.I., Klöti, R., Meyer-Schwickerath, G., Regnault, F. and Sweeney, D.B. (1972) in Modern Problems in Ophthalmology (Secondary Detachment of the Retina, Lausanne, 1970)(Streiff, E.B., ed.), Vol. 10, pp. 3–21, S. Karger, Basel

12. Balazs, E.A., Watson, D., Duff, I.F. and Roseman, S. (1967) Arthritis Rheum. 10, 357–375

13. Balazs, E.A. (1968) Symposium: Prognosis for Arthritis: Rheumatology Research Today and Prospects for Tomorrow, Ann Arbor, Michigan, 1967, Univ. Mich. Med. Ctr. J. (Suppl.), 255–259

14. Balazs, E.A. (1968) in Third Nuffield Conference on Rheumatism (Ditchley Park, Oxfordshire, 1967), pp. 13.1–13.10, The Nuffield Foundation, London

15. Seppälä, P. and Balazs, E.A. (1969) J. Gerontol. 24, 309–314

16. Gibbs, D.A., Merrill, E.W. and Smith, K.A. (1968) Biopolymers 6, 777–791

17. Balazs, E.A. and Gibbs, D.A. (1970) in Chemistry and Molecular Biology of the Intercellular Matrix (Balazs, E.A., ed.), pp. 1241–1254, Academic Press, London

18. Weiss, C. and Balazs, E.A. (1988) Third Int. Symposium on Arthroscopy of the Temporomandibular Joint (abstract), New York

19. Balazs, E.A. (1977) U.S. Pat. No. 4141973, 27 February 1979

20. Balazs, E.A. (1981) Presented at the First European Congress on Medicine and Equestrian Sports,18–20 September 1981, Samur, France (E.A. Balazs, personal communication)

21. Namiki, O., Toyoshima, H. and Morisaki, N. (1982) Int. J. Clin. Pharmacol. Ther. Toxicol. 20, 501–507

22. Boni, M. and Cherubino, P. (1984) Clin. Orthoped. Univ. Pavia 23 March 1984

23. Bragantini, A., Cassini, M., De Bastiani, G. and Perbellini, A. (1987). Clin. Trials J. 24, 333–340

24. Balazs, E.A. and Leshchiner, A. (1983) U.S. Pat. No. 4500676, 19 February 1985

25. Balazs, E.A. and Leshchiner, A. (1984) U.S. Pat. No. 4582865, 15 April 1986

26. Balazs, E.A., Miller, D. and Stegmann, R. (1979) Paper presented at the International Congress and First Film Festival on Intraocular Implantation, May 1979, Cannes, France (E.A. Balazs, personal communication)

27. Balazs, E.A. (1989) in Viscoelastic Materials: Basic Science and Clinical Applications (Rosen, E.S., ed.),17–19 July 1986, Second International Symposium of The Northern Eye Institute, 167–183, Pergamon Press, Oxford

28. Graue, E.L., Polack, F.M. and Balazs, E.A. (1980) Exp. Eye Res. 31, 119–127

29. Pape, L.G. and Balazs, E.A. (1980) Ophthalmology 87, 699–705

30. Miller, D. and Stegmann, R. (1980) Am. Intraocul. Imp. Soc. J. 6, 13–15

31. Balazs, E.A. (1984) in Viskochirurgie des Auges (Beiträge des ersten nationalen Healon®/Symposiums, 15–16 October 1982), pp. 1–13, Ferdinand Enke Verlag, Stuttgart

32. Balazs, E.A. (1986) in Ophthalmic Viscosurgery – A Review of Standards, Techniques and Applications (Eisner, G., ed.), pp. 3–19, Medicöpea, Bern

33. Eisner, G., ed. (1986) Ophthalmic Viscosurgery – A Review of Standards, Techniques and Applications, Medicöpea, Bern

34. Arshinoff, S. (1992) in Proceeding of the National Ophthalmic Speakers Program Medicöpea International Inc., Montreal, pp. 7–12

35. Arshinoff, S. (1989) Ophthalm. Pract. 7, 16–19; 36–37

36. Mensitieri, M., Ambrosio, L., Nicolais, L., Balzano, L. and Lepore, D. (1994) J. Mater. Sci.: Mat. Med. 5, 743–747

37. Weiss, C. and Balazs, E.A. (1987) Arthroscopy 3, 138–139

38. McCain, J.P., Balazs, E.A. and de la Rua, H. (1989) J. Oral Maxillofac. Surg. 47, 1161–1168

39. Weiss, C. (1991) in Arthroscopy of the Temporomandibular Joint (Mohan, T. and Bronstein, S., eds.), ch. 22, pp.335–337, W.B. Saunders, Philadelphia

40. Weiss, C., Drucker, M. and Levitt, R. (1991) Robert Jones Lecture Program given at the Hospital for Joint Diseases, 15 November, 1991, New York (C. Weiss, personal communication)

41. Laurent, C., Hellström, S. and Fellenius, E. (1988) Arch. Otolaryngol. Head Neck Surg. 114, 1435–1441

42. Laurent, C., Hellström, S. and Stenfors, L-E. (1986). Am. J. Otolaryngol. 7, 181–186

43. Balazs, E.A. (1960) in Importance of the Vitreus Body in Retina Surgery with Special Emphasis on Reoperations, 30–31 May 1958, The Second Conference of the Retinal Foundation, (Schepens, C.L., ed.), pp. 29–48, C.V. Mosby, St. Louis

44. Weiss, C., Balazs, E.A., St. Onge, R. and Denlinger, J.L. (1981) in Seminars in Arthritis and Rheumatism Vol. 11 (Talbott, J.H., ed.), pp. 143–144, Grune and Stratton, New York

45. Weiss, C., Levy, H.J., Denlinger, J., Suros, J. and Weiss, H. (1986) Bull. Hosp. Joint Dis. **46**, 9–15
46. Weiss, C., Suros, J.M., Michalow, A., Denlinger, J., Moore, M. and Tejeiro, W. (1987) Bull. Hosp. Joint Dis. **47**, 31–39

26

Viscosupplementation as articular therapy

Mark E. Adams

Department of Medicine, and McCaig Centre for Joint Injury and Arthritis Research, The University of Calgary, 3330 Hospital Drive NW, Calgary, AB T2N 4NI Canada

Objectives

The main objective of this chapter is to review the results of the use of hyaluronan (HA) and HA-derived products in the treatment of osteoarthritis (OA) of the knee. (The use of HA and HA derivatives in other joints is covered elsewhere in this volume. See also [1] for another review of the therapeutic uses of HA.) This chapter will also include a brief description of normal joints, normal joint physiology, particularly related to the role of HA, to define and describe briefly OA, to put it in perspective, and to describe briefly its pathogenesis, most specifically as it relates to viscosupplementation. The rationale and development of viscosupplementation will be outlined in order to frame the use of HA and HA derivatives for articular therapy in an historical context. Finally, the results of various HA related products that are marketed for use in OA of the knee will be compared, thus helping to define the features of the product that are important for successful therapy

Background

Roles of HA in normal joint structure

The normal diarthrodial joint is a remarkable structure whose purpose is to allow for movement. The joint is contained within a joint capsule, which is lined with synovium, the joint cavity contains synovial fluid, and the two ends of the bones are covered with articular cartilage. The articular cartilage is the bearing surface, and its structure is crucial for the joint – its disintegration, as occurs in OA, leads to joint dysfunction and failure.

HA has many crucial roles in the structure and function of joints. In the articular cartilage, the HA macromolecules form the backbone for the aggregation of aggrecan into the enormous hydrophilic supramolecular structures that are trapped within the collagenous network. By exerting a swelling pressure that is balanced by the tensile restraints of the collagen the cartilage is endowed with a deformable rigidity. The surface layer of the articular cartilage, called the lamina splendens, is permeated with HA. Furthermore, the articular cartilage surfaces are separated from each other by a thin layer of a viscoelastic fluid, the synovial fluid, which is rich in HA. Finally, the connective tissue matrix of the synovial interstitial space is filled with HA.

Functions of HA in the joint

HA plays a key role at three levels in the homoeostasis of the joint: macro-homoeostasis, or homoeostasis of the rheological environment, mini-homoeostasis, or homoeostasis of the fluid environment, and micro-homoeostasis, or homoeostasis of the chemical environment. See [2,3] for excellent reviews of the functions of HA in the joint, or other chapters in this volume.

Macro-homoeostasis (rheological environment)

The HA in the synovial fluid, on the cartilage surface, and in the soft tissues is essential for rheological properties such as lubrication, protection and shock absorption. The complex viscoelastic properties of the synovial fluid are crucial for joint function, and are entirely due to its HA content [2]. At low impact frequencies of joint movement, as in walking, the fluid is viscous, and at higher frequencies, as in running, the fluid is elastic [2]. The HA in the synovial fluid also participates in the lubrication of the tissue planes, the ligaments and other collagenous structures in the joint tissues, and the surfaces of the articular cartilage. It protects the cells and sensory apparati, including nociceptive receptors of the tissues [4], and absorbs and transmits in an even manner the forces between the two opposing cartilage surfaces.

Mini-homoeostasis (fluid environment)

HA is a major component of the fluid environments of the joint: the synovial fluid phase, the gel phase of the synovium and the gel phase of the cartilage. It regulates the lymphatic flow, including the drainage of the metabolites of all the joint tissues, inclusive of those of HA itself, and also regulates the diffusion through the joint tissues, including that of nutrients, and thus the nutrition of the HA itself. It, therefore, has an intrinsic auto-regulatory function in the homoeostasis of the fluid environment. Normal articular cartilage in the adult obtains all its nutrients via diffusion through the articular surface rather than through the subchondral bone. Thus, the diffusion of all metabolites and waste products of the chondrocytes must pass through the lamina splendens, which is largely HA, and into the synovial fluid, which is a 20–40 mg/l solution of HA [2]. The metabolites of the cartilage emanate from capillary flow in the synovium, and thus must pass through the HA of the synovium and the HA of the synovial fluid. The waste products of the chondrocytes must pass through the HA-enriched matrix of the cartilage, through the lamina splendens, through the synovial fluid and thence through the synovium to the lymphatics.

Micro-homoeostasis (chemical environment)

In addition to affecting the biomechanics, rheology and the fluid flow in the joint, hyaluronan also affects the chemical environment. The HA in all the joint tissues participates actively and importantly in this micro-homoeostasis. HA is an effective free-radical scavenger, and thus provides protection from oxygen-derived free radicals. By nature of its gel filtration properties, it protects both the chondrocytes and the synoviocytes from degradative enzymes, chemical agents and toxins. HA protects the cell membranes by stabilizing them. It desensitizes the sensory receptors, and thus is involved in the perception of pain. Finally, perhaps most importantly, as this is auto-regulatory, HA maintains a proper, stable environment for normal HA synthesis. Thus, a large (and thus properly functional) HA ensures that a large HA is synthesized.

OA
Terminology
OA is also called osteoarthrosis, and sometimes is still known as degenerative joint disease. There has been much debate over which of these terms is correct. Few would consider OA a primarily inflammatory condition (except perhaps the nodal and erosive forms of OA of the fingers), and thus, perhaps, osteoarthrosis is the more appropriate term. However, few would also question that the condition is characterized by episodes of inflammation, and that, in general, these are secondary to the damage in the joint and joint debris ('detritic synovitis'). Maybe those episodes should more precisely be called osteoarthritis, and the whole process osteoarthrosis. However, it is easy to avoid this semantic dilemma simply by referring to the condition as OA.

Definition
OA is often defined on the basis of morphology, stipulating cartilage loss, which in OA, is usually focal, and bony remodelling of two types: subchondral sclerosis, and osteophyte formation. Unfortunately, these changes occur rather late in the course of the natural history of OA and reflect joint damage with repair or remodelling, rather than a specific, aetiological or pathological process. Furthermore, this definition does not reflect fully the extent of the pathological changes in the joint. Most significant for this chapter are the changes in the synovial fluid and the HA, which are outlined below, but OA is usually, eventually, accompanied by tears in the menisci and periarticular ligaments, thickening of and inflammation in the synovium, joint effusions, capsular fibrosis, subchondral cysts, periarticular bursitis and tendonitis, depositions of calcium pyrophosphate dihydrate, hydroxyapatite or other calcium crystals in and around the joints, periarticular muscular atrophy, alterations of periarticular blood flow and many other changes. These changes can develop irrespective of the original insult to the joint because the joint functions as an organ with an interdependence of all of its component parts. These pathological changes affect the function of the joint and contribute to the symptoms; the inflammation and disordered biomechanics may also contribute to the progression of the disease.

Overview – social and economic
OA is the most common disease of synovial joints. Radiographically, its prevalence is strongly correlated with age: fewer than 5% of men and women under 35 years of age show radiographical OA, while over 70% of those over the age of 65 do. In hip and knee OA, which are often disabling, the severity of symptoms generally correlates with the severity of the radiographical changes.

OA also is the most costly form of arthritis: it causes about 20 times more visits to physicians, about 35 times more days bed rest and about 30 times more days lost from work than rheumatoid arthritis [5].

Changes in HA

HA is much smaller in the synovial fluid of OA joints and, because of fluid exudation, the concentration is lower (Table 1) [2,6]. Thus, mechanical protection, the barrier protection, and the chemical protection, provided by the HA is impaired. The articular cartilage is subjected to abnormal impact loading stresses and the synoviocytes and nociceptors in the synovium are less protected from the stretching forces. The tissues, cells and nociceptors are less protected from irritation from noxious substances, such as degradative enzymes, free radicals, tissue breakdown products and debris.

Table 1

Synovial fluid	Complex viscosity (Pa s at 0.02 Hz)	Elasticity (%) (at 3 Hz)
Young	137	94
Old	33	71
Osteoarthritic	0.4–5	36–73

Changes in the rheological properties of human synovial fluid with age and OA [2,6]

Treatment of OA

Conventional therapy of OA

The therapy of OA, not including viscosupplementation, generally starts with mild analgesia, exercises, weight loss and physiotherapy. All of these can be effective and are generally safe. If the patient still has symptoms after these interventions, then non-steroidal anti-inflammatory drugs (NSAIDs) are usually used. Unfortunately, while NSAIDs are often fairly effective in relieving pain, they also have a high incidence of adverse effects, predominantly gastrointestinal (GI) damage, especially erosive gastritis, often including GI bleeding, which can be fatal. For a 'benign' disease like OA, the risk of such an outcome should not be taken without serious consideration. 'Cytoprotective' therapy with prostaglandin antagonists is often effective in blocking these adverse effects, and, thus, can be quite beneficial; however, their use adds extra expense and complications to the therapy of OA.

Rationale for viscosupplementation in OA

Viscosupplementation is a treatment of OA in which HA or a derivative of HA is added to the synovial fluid to restore proper homoeostasis to the diseased fluid [7]. It overcomes some of the limitations of current conventional therapy: it is a physiological therapy, restoring the macro-, mini- and micro-homoeostasis that are altered due to changes in the synovial fluid HA; it is local therapy, which is appropriate, because OA is usually a local disease; and it is a therapy that fills the need for a safe therapy beyond simple analgesia, exercise, weight loss and physiotherapy.

Table 2

Trade name (manufacturer)	Generic name	Molecular weight, M_w (millions)	Elasticity (%) (at 3 Hz)	Complex viscosity (Pa s at 0.02 Hz)	Polysaccharide concentration (mg/ml)	Dosage Schedule (no. of weekly injections)
Hyalgan® (Fidia)*	Hyaluronan	0.5–0.65	26	<0.1	10	5
Artz® (Seikagaku)	Hyaluronan	0.75	33	0.3	10	3
Orthovisc® (Anika)	Hyaluronan	1.5	66	42	15	2
Synvisc® (Biomatrix)	Hylan	6	88	213	8	3

*The product marketed in Canada fomerly as Replasyn® and now as Suplasyn® is very similar if not identical to Hyalgan®.

Rheological properties of products marketed for viscosupplementation

Properties of viscosupplements

There are several hyaluronan products approved for intra-articular use in humans (Table 2). They vary in weight-average molecular weight (M_w), from a low of about 65×10^4 to about 2×10^6.

Any product used for viscosupplementation must be biocompatible with the tissues and with blood, and must be permeable to metabolites and macromolecules. To be ideal as a viscosupplement, it must have better rheological properties than the HA in normal synovial fluid and have a slower export rate, i.e. long half-life within the joint.

HA is biocompatible and is permeable to metabolites and macromolecules, so it can be used for viscosupplementation. However, the M_w of the HA products that are marketed is not higher than that in normal synovial fluid, nor do these products have a long residence time in the joint. Because HA is a linear polysaccharide, its M_w is significantly decreased by only a few breaks in the molecule, and HA is very susceptible to free-radical degradation [8].

It is for this reason that HA was cross-linked to form a new class of molecules, the hylans [9]. If the cross-linked hyaluronan is to be used for viscosupplementation, it is crucial that the cross-linking leaves the chemical exterior of the HA intact, so that the new product is not inflammatory or immunogenic. At present, there is only one hylan, Synvisc® (hylan G-F 20), that is approved for clinical use in humans (Table 2), and its chemical exterior is exactly that of native hyaluronan [9].

Treatment of OA with HA

Clinical trials – methodological concerns

In analysing the results of clinical trials of treatments of OA, it is important to recognize that patients in all treatment groups generally tend to improve. While some of this may be due to a placebo effect, the greater part is likely attributable to a 'regression to the mean'. Because OA is a remitting and relapsing disease, and because the entry criteria for the trial generally specify a minimum pain level, patients tend to be worse on entry and to improve during the trial. So single arm studies, while of some value in assessment of safety, are of little or no value in evaluation of efficacy. Only controlled (with a comparison arm), randomized, blinded trials are reliable for the evaluation of efficacy in OA, although the control does not necessarily need to be a placebo control.

HA used in trials for the treatment of OA

Even in the early studies of HA in the treatment of OA, there were different classes of preparations of HA that were used in the studies. One group (Hyalgan®, Artz®) has a M_w of about $6–7 \times 10^5$, and the other (Healon®) has a M_w of about 2×10^6. (Healon® is not currently marketed for human knee OA.) Furthermore, it is important to note that the $6–7 \times 10^5$ M_w products have been marketed as 'high-molecular-weight HA'. It is also important to note that the HA in normal synovial fluid has a M_w of about $2–3 \times 10^6$, and that HA in osteoarthritic synovial fluid has a M_w of about 6×10^5.) Studies of both the $6–7 \times 10^5$ and 2×10^6 M_w preparations of HA started at about the same time, the late 1960s. There are no published trials of Orthovisc® in human knee OA.

Studies with 6–7 × 10⁵ molecular-weight HA

There are two preparations of HA that are currently marketed for the treatment of human OA: Hyalgan® and Artz®. There have been many studies of Hyalgan® (reviewed in [10]) though many of these have been either single-blind or uncontrolled. Some [11–13], but not all [14–16], of the controlled or double-blind trials show efficacy and safety of this preparation. Several studies show that Hyalgan® is about as effective as corticosteroid injections [17–20], a treatment that, though widely used and advocated by some, has only been shown to have marginal and transient benefit in clinical trials [21–23].

Likewise, there have been many studies of Artz®, but many are in Japanese with only an English abstract. Again, most show efficacy and safety [24], but several recently published large-scale double-blind trials only showed efficacy of this preparation in a small subset of the patients [25,26].

Studies with 2 × 10⁶ molecular-weight HA

The first study to use the higher-molecular-weight HA in human OA was performed by Rydell in 1969–1970 [27]. In this double-blind controlled trial of injections of Healon®, there were 14 patients with OA of the knee in each group. The patients treated with Healon® fared statistically significantly better than the controls. There were no adverse events. Helfet performed an open trial of Healon® in 1969–1971. The patients had OA in 40 knees and 22 hips. Most of the patients with the Healon® injections showed improvement. There were no adverse events reported. Between 1970 and 1971, Peyron and Balazs studied injections of Healon® in a variety of patients, with different conditions and in different joints. Some of the trials were controlled. Most of the patients improved and, once again, there were no adverse events. The results of these trials have been reviewed [28,29].

Summary of treatment of OA with HA

Twenty-five years of clinical trials and clinical practice using HA for the treatment of OA have shown that it is a remarkably safe treatment. The incidence of side effects is approximately 2%. No systemic events have been reported. Joint infections have occurred, but, to the knowledge of this author, none have been attributable to any of the products themselves. These results also show that HA injections are effective, but multiple injections, as many as 5–10, are usually needed, especially for the 6–7 × 10⁵ M_w products. The effect appears to last longer than the residence time of the injected material, which suggests that the injected HA helps restore the homoeostasis of the joint in humans [30]. (Animal studies do show that the M_w of the HA in the fluid is higher, long after the injected material is gone [3], and suggest that HA injections may alter the natural history of some forms of experimental OA [27,31], with better results with the higher-molecular-weight material.) Though there had been, until recently, no direct comparison of HA products of differing M_w in humans, a consideration of the results from all these studies suggests that the efficacy is improved with a higher M_w product. Again, this is shown in animal studies [32].

Treatment of OA with hylan

Results of the clinical trials of hylan

To date there have been seven clinical trials of Synvisc®, six of them double-blind, and controlled. In five of the six double-blind trials, the comparison was with a putatively inactive treatment, sham injections or injections of normal saline; in one trial the comparison was with a definitively active treatment, NSAIDs. The first two trials were double-blind studies comparing two injections in one trial and three injections in the other, each over the period of two weeks, and in each case the Synvisc® was compared with a control injection of an equal volume of normal saline. Despite the fact that there were small numbers of patients in each of these studies (24 per group in one and 15 per group in the other), in both studies the Synvisc® was statistically significantly superior to a saline injection control, and a three-injection regimen was better than a two-injection regimen. There were no serious or systemic adverse events [33]. Following this, a larger-scale double-blind study was performed with 117 patients using a saline control. Again, the Synvisc® was clearly superior to control, and the incidence of adverse events was very low (<2%). In another study, Synvisc® was shown to be more effective in the patients who flared upon withdrawal of their conventional therapy. In another study, Synvisc® was compared with NSAIDs alone or Synvisc® plus NSAIDs. Analysis of variance showed that the Synvisc® was nominally better at 12 weeks, and q-statistical analysis showed that the therapies were at least 60% equivalent [34]. Covariate analysis correcting for the strength of the NSAID (based on a 4-point ordinal scale), body mass index, sex, age, synovial fluid volume at the first visit, degrees of varus, duration of disease, X-ray grade and initial baseline VAS score shows that, in all cases, the patients in either the Synvisc®-alone group or the Synvisc®-plus-NSAID group had significantly better improvement as a result of the treatment than did the patients in the NSAID-alone group. For all the patient-assessed variables, the patients treated with Synvisc® alone had statistically significantly more improvement than did the patients treated with NSAIDs alone.

There is only one clinical study that compares viscosupplementation with hylans and HA. This four-arm study compares the results of a course of three injections of Synvisc® versus Healon® versus Artz®, an HA of about 7.5×10^5-M_w versus degraded hylan. The data shown in Figure 1 show the dependence of the results on the M_w. Data from animal studies also shows this molecular weight dependence of the results [32].

Results from clinical experience with hylan

A retrospective analysis of the results of the use of Synvisc® in clinical practice in Canada has recently been published [35]. Data were presented on 336 patients who received 1537 injections over a two-and-a-half-year period. Some patients had bilateral treatment and some had a second course of Synvisc® treatment. Fully 80% of the patients were improved or much improved, and the median duration

Figure 1

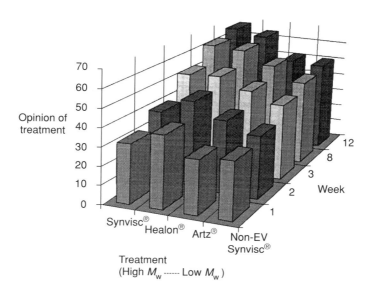

Patients' evaluation of the improvement after treatment

The patients in this trial were given injections of one of the viscosupplements at weeks 0, 1 and 2. The vertical axis represents the patient's overall evaluation of the therapy on a 100 mm visual analogue scale. The viscosupplements are arranged in order of weight-average molecular weight, though this is not linear. The weeks are the times when the evaluations were done. This scale is not linear. This graph shows both that the treatments improve over time, up to 12 weeks, and that the efficacy of the treatment increases with a higher molecular weight. In other studies, Synvisc® has been shown to have a prolonged effect, lasting longer than 12 weeks.

of relief was 8.3 months. The patients also used fewer NSAIDs than they had been using prior to the therapy with Synvisc®. Thus, the results from clinical experience corroborate those from the clinical trials. In addition, data were provided that demonstrate that a second course was equally effective and safe. It is notable that, as in the clinical trials, there were no systemic adverse events (AE), and the local AE rate was essentially the same as in the clinical trials, about 2 to 3%.

One study purports that the incidence of AE is as high as 20% with hylan G-F 20 injections [36]. However, this study was based on a very small number of patients [37], and this high incidence of AE is not borne out by either the clinical trial data, by the results of other clinicians [38] or by the analysis of many more patients in clinical practice that is cited above [35].

Clinical results – summary of trials and practice

The results of all the clinical trials and from clinical practice with hylan show that it is as safe as HA, with about a 2–3% chance of AE per injection. All of the reactions are local and transient. The results of treatment with Synvisc® using fewer injections (3 versus 5 or more) seem to be superior to treatment with HA, and to last longer, with a median response of at least 6 months.

The rate of response and the fact that all the clinical trials show a significantly better result versus control, while those with the HA products have sometimes not differentiated from control, suggest that hylan is superior to hyaluronan; the only trial in humans that has compared these products corroborated this result. Various products have been compared in veterinary use, and the higher-molecular-weight products are clearly superior [32]. Thus, despite a review that suggests that the molecular weight is not of consequence [39], human and veterinary clinical trials and experimental results [30,31,40] suggest that the higher the molecular weight, the better the response.

Summary and conclusion

Viscosupplementation is a significant advance in the therapy of OA. It is the first new treatment paradigm for OA since the introduction of NSAIDs. It is safe, effective and physiological. From a comparison of the results of clinical trials, clinical use and in the one trial that compared various products directly, hylan is better than HA in the treatment of OA.

References

1. Goa, K.L. and Benfield, P. (1994) Drugs. 47, 536–566
2. Balazs, E.A. (1982) in Disorders of the Knee (Helfet, A., ed.), pp. 61–74, J.B. Lippincott Co., Philadelphia
3. Balazs, E.A. and Denlinger, J.L. (1985) J. Equine Vet. Soc. 5, 217–231
4. Pozo, M.A., Balazs, E.A. and Belmonte, C. (1997) Exp. Brain Res. 116, 3–9
5. Kramer, J.S., Yelin, E.H. and Epstein, W.V. (1983) Arthritis Rheum. 26, 901–907
6. Balazs, E.A., Watson, D., Duff, I.F. and Roseman, S. (1967) Arthritis Rheum. 10, 357–376
7. Balazs, E.A. and Denlinger, J.L. (1993) J. Rheumatol. 20 (Suppl. 39), 3–9
8. Al-Assaf, S., Phillips, G.O., Deeble, D.J., Parsons, B., Starnes, H. and von Sontag, C. (1995) Radiat. Phys. Chem. 46, 207–217
9. Balazs, E.A., Band, P.A., Denlinger, J.L., Goldman, A.I., Larsen, N.E., Leshchiner, E.A., Leshchiner, A. and Morales, B. (1991) Blood Coag. Fibrin. 2, 173–178
10. Maheu, E.(1995) Eur. J. Rheumatol. Inflamm. 15, 17–24
11. Grecomoro, G., Martorana, U. and DiMarco, C. (1987) Pharmatherapeutica 5, 137–141
12. Dougados, M., Nguyen, M., Listrat, V. and Amor, B. (1993) Osteoarthritis Cartilage 1, 97–103
13. Carrabba, M., Paresce, E., Angelini, M., Re, K.A., Torchiana, E.E.M. and Perbellini, A. (1995) Eur. J. Rheumatol. Inflamm. 15, 25–31
14. Dixon, A.S.J., Jacoby, R.K., Berry, H. and Hamilton, E.B.D. (1988) Curr. Med. Res. Opin. 11, 205–213
15. Henderson, E.B., Smith, E.C., Pegley, F. and Blake, D.R. (1994) Ann. Rheum. Dis. 53, 529–534
16. Creamer, P., Sharif, M., George, E., Meadows, K., Cushnaghan, J., Shinmei, M. and Dieppe, P. (1994) Osteoarthritis Cartilage 2, 133–140
17. Leardini, G., Franceschini, M., Mattara, L., Bruno, R. and Perbellini, A. (1987) Clin. Trial J. 24, 341–350
18. Pietrogrande, V., Melanotte, P.L., D'Agnolo, B., Ulivi, M., Benigni, G.A., Turchetto, L., Pierfederici, P. and Perbellini, A. (1991) Curr. Ther. Res. 50, 691–701
19. Leardini, G., Mattara, L., Franceschini, M. and Perbellini, A. (1991) Clin. Exp. Rheumatol. 9, 375–381
20. Jones, A.C., Pattrick, M., Doherty, S. and Doherty, M. (1995) Osteoarthritis Cartilage 3, 269–273
21. Dieppe, P.A., Sathapatayavongs, B. and Jones, H.E. (1980) Rheum. Rehab. 19, 212–217
22. Miller, J.H., White, J. and Norton, T.H. (1958) J. Bone Joint Surg. (Am.) 40B, 636–643
23. Friedman, D.M. and Moore, M.E. (1980) J. Rheumatol. 7, 850–856
24. Shichikawa, K., Igarashi, M., Sugawara, S. and Iwasaki, Y. (1983) Jpn. J. Clin. Pharmacol. Ther. 14, 545
25. Dahlberg, L., Lohmander, L.S. and Ryd, L. (1994) Arthritis Rheum. 37, 521–528

26. Lohmander, L.S., Dalén, N., Englund, G., Hämäläinen, M., Jensen, E.M., Karlsson, K., Odensten, M., Ryd, L., Sernbo, I., Suomalainen, O. and Tegnander, A. (1996) Ann. Rheum. Dis. **55**, 424–431

27. Rydell, N. and Balazs, E.A. (1971) Clin. Orthop. **80**, 25–32

28. Weiss, C., Balazs, E.A., St. Onge, R. and Denlinger, J.L. (1981) Osteo. Symp. **81**, 143–144

29. Balazs, E.A. and Denlinger, J.L. (1984) in Osteoarthritis: Current Clinical and Fundamental Problems (Peyron, J.G., ed.), pp. 165–174, Ciba Geigy, Paris

30. Balazs, E.A., Briller, S.O. and Denlinger, J.L. (1981) Semin. Arthritis Rheum. **11**, 141–143

31. Kikuchi, T., Yamada, H. and Shimmei, M. (1996) Osteoarthritis Cartilage **4**, 99–110

32. Phillips, M.W. (1989) J. Equine Vet. Soc. **9**, 39–40

33. Scale, D., Wobig, M. and Wolpert, W. (1994) Curr. Ther. Res. **55**, 220–232

34. Adams, M.E., Atkinson, M.A., Lussier, A., Schulz, J., Siminovitch, K.A., Wade, J.P. and Zummer, M. (1995) Osteoarthritis Cartilage. **3**, 213–226

35. Lussier, A., Cividino, A.A., McFarlane, C.A., Olszynski, W.P., Potashner, W.J. and De Médicis, R. (1996) J. Rheumatol. **23**, 1579–1585

36. Puttick, M.P.E., Wade, J.P., Chalmers, A., Connell, D.G. and Rangno, K.K. (1995) J. Rheumatol. **22**, 1311–1314

37. Adams, M.E. (1996) J. Rheumatol. **23**, 944–945

38. O'Hanlon, D. (1996) J. Rheumatol. **23**, 945

39. Aviad, A.D. and Houpt, J.B. (1994) J. Rheumatol. **21**, 297–301

40. Larsen, N.E., Lombard, K.M., Parent, E.G. and Balazs, E.A. (1992) J. Orthoped. Res. **10**, 23–32

27

Viscoseparation and viscoprotection as therapeutic modalities in the musculoskeletal system

Charles Weiss*

Department of Orthopaedics and Rehabilitation, Mt. Sinai Medical Center, Miami Beach, FL, U.S.A.

Viscoelastics in medicine

Viscoseparation is the use of viscoelastic gels and elastoviscous fluids to separate tissues, prevent adhesions, decrease scar formation and facilitate healing. Viscoprotection is the use of viscoelastic gels and elastoviscous fluids to shield and protect tissue surfaces. These terms describe specific functions of the hyaluronan (HA) molecule and its derivatives, which are useful in the musculoskeletal system to facilitate healing, diminish scar formation, protect surfaces and improve function.

The theoretical basis for the use of these molecules in the musculoskeletal system was first postulated by Endre Balazs and given credence by studies demonstrating that hyaluronan covered the surface of articular cartilage [1], was imbedded in the superficial zone of articular cartilage [2] and could, under dynamic conditions, protect both cells and opposing surfaces owing to its viscoelastic properties [3,4]. The therapeutic use of these molecules in the musculoskeletal system became possible in the late 1960s when Balazs developed the first non-inflammatory, highly purified high-molecular-weight sodium hyaluronate (NIF-NaHA) from umbilical cords and rooster combs [5]. Work done in his laboratory during the late 1960s [6] outlined the major medical applications for this material in the musculoskeletal system. Over the next 25 years experimental laboratory and clinical studies confirmed, explained, refined and only marginally expanded these original observations.

The anti-adhesion properties of the HA molecule (viscoseparation) were demonstrated in several species, utilizing a number of different experimental models [6]. In an abrasion extensor hallucis longus scarring model in rabbits, the tendon sheath on the experimental leg was injected with NIF-NaHA prior to repair. Eighty percent of the injected tendons had scarring of lower tensile strength between the tendon and the sheath than the non-injected controls, 11% were equal and 9% were greater [6,7]. In rabbits, owl monkeys and guinea pigs,

---

* Address for correspondence: 400 Arthur Godfrey Road, Suite 200, Miami Beach, FL 33140, U.S.A.

the fascia separating muscle from subcutaneous tissue was abraded and evaluated 3 to 6 weeks post-surgery. Eighty-two percent of the animals had less scarring on the NIF-NaHA-treated side; in 16% there was no difference and in 2% the control side had less scarring [6]. Another study comparing tissue reaction to subcutaneous foreign bodies (polyethylene tubes) that had been coated with either HA or saline found no rejection on the HA-treated side versus 40% rejection of saline-treated tubes. There was significantly less local tissue reaction and peri-implant fibrous tissue on the HA-treated side [6,7].

The viscoprotective function of the hyaluronan molecule was demonstrated in both dogs and monkeys [6]. NIF-NaHA (molecular weight 1.2 million) was injected every 4 days for 4 weeks into the knee joints of animals who had surgically produced lateral femoral condylar defects resulted in a smoother and less reactive surface on the injected side compared with the non-injected side 6 to 8 weeks post-surgery [6]. The viscoprotective property of this molecule was also studied clinically in race horses [6,8]. A comparison of the functional results of NIF-NaHA plus cortisone versus cortisone injection alone in the treatment of traumatic arthritis in race horses demonstrated a marked improvement in the NIF-NaHA plus cortisone-injected joints compared with the joints injected with cortisone alone (Table 1).

Subsequent studies by Balazs demonstrated that race horses with post-traumatic arthritis had reduced viscosity of the synovial fluid in the affected joint, and that those horses that responded successfully to intra-articular treatment with NIF-NaHA had their synovial fluid viscosity return to normal levels. The synovial fluid in those joints that failed to respond to NIF-NaHA treatment continued to have low viscosity [9]. This data demonstrated the importance of the restoration of joint homeostasis to the functional and clinical recovery of the arthritic or traumatized joint (Table 2).

Table 1

Treatment	Group†	No. of horses	Evaluation‡ Poor	Fair	Good
NIF-NaHA + cortisone injection	I	4	0	0	4
	II	3	0	0	3
	III	5	0	1	4
	Total	12	0	1	11
Cortisone injection alone	I	4	3	0	1
	II	2	2	0	0
	III	2	2	0	0
	Total	8	7	0	1

†Group I, lameness only during racing; Group II, lameness only on movement; Group III, refuses to put full weight on joint.

‡Poor evaluation: unable to race or able to race for less than one month. Fair evaluation: able to race but not as well as before. Good evaluation: able to race as well as before for at least three months

Treatment of arthritic joints in race horses

Table adapted from [6].

Table 2

Result		Before treatment	Weeks after treatment		
			1	2	4-8
Cases with normal viscosity and no arthritis	(n)	5183 ± 12 (6)	5475 ± 201 (4)	6025 ± 440 (2)	—
	p		N.S.	N.S.	
Arthritic cases with good to excellent result	(n)	3083 ± 243 (18)	3941 ± 215 (17)	4736 ± 217 (14)	5280 ± 536 (5)
	p		<0.01	<0.01	<0.01
Arthritic cases with no improvement	(n)	3975 ± 63 (4)	3467 ± 431 (3)	3933 ± 150 (3)	—
	p		NS	NS	

Limiting viscosity number (ml/g) of hyaluronan in horse joints before and after intra-articular NIF-NaHA injection [9]

Viscoseparation

Tendon adhesions following injury or after surgical repair are a significant clinical problem. Numerous attempts have been made, without success, to reduce adhesions and restore function following injury, including interposition membranes, steroids and antimetabolites. As a result of the initial studies by Rydell and Balazs [6], we performed a clinically relevant study in owl monkeys [10]. The superficialis and profundus tendons of the middle and ring fingers were injured at the proximal phalanx by crushing, cutting and then repairing the profundus tendon and placing either a 2% NIF-NaHA solution (molecular weight 1.2 million) or saline solution around the repair. After immobilization with the proximal interphalangeal joint in 90° of flexion for 4 to 5 weeks, a blinded study of range of motion evaluation was performed up to 12 weeks post-surgery. There was a significant ($p < 0.01$) improvement in the post-surgical range of motion with an average of 26.9° less flexion contracture in the NIF-NaHA treated fingers compared with controls. There was no increase in tendon rupture or failure to heal [10]. Subsequent studies using hyaluronans of similar molecular size and concentration in a flexor tendon repair model in dogs demonstrated by gross and histological examination that hyaluronan has a beneficial effect on the architecture of tendon healing by decreasing the paratendon inflammatory response (promoting a contact healing response via epitendon and endotendon cell involvement in the repair process) and decreasing the quality and quantity of adhesions from the wound repair site to the synovial sheath [11,12]. Other studies using a rabbit flexor tendon repair model demonstrated significantly decreased adhesion formation at 8 weeks by graded microscopic evaluation and by tendon sliding test in the hyaluronan-treated tendons compared with saline-treated controls. No difference in the quality of tendon repair or in systemic reactions was seen [13]. A biomechanical evaluation of adhesion formation by simultaneous measurement of tensile load, tendon excursion and joint motion in a rabbit flexor tendon injury model found that at 15 days hyaluronan "significantly limited the strength of adhesions formed without impairment of tensile strength" [14]. Furthermore, the effect of hyaluronan increased with the concentration and molecular weight of the preparation used. Recent studies have confirmed the success of exogenously administrated hyaluronan in limiting adhesion formation in chicken tendon injury and repair [15]. In a digital flexor tendon collagenase-induced adhesion model in horses, hyaluronan was found to decrease adhesion formation and inflammatory cell infiltration with improved tendon structure compared with methylcellulose gel-treated tendons [16]. In a rabbit partial anterior cruciate ligament laceration model, hyaluronan-treated joints showed significantly higher healing grades, more pronounced repair with increased angiogenesis, less inflammatory response and an increased synthesis of Type III collagen when compared with saline controls [17].

 In 1992, the results of a clinical trial showed no significant effect on total active range of motion upon comparing the post-operative results of exogenous hyaluronan and saline after surgical repair of lacerated flexor tendons [18]. However, there was an insufficient number of patients to detect statistical significance (120 digits studied with 540 needed to detect a minimal significant difference of 10°). In addition, this study was characterized by inadequate clinical follow-up, and variations in surgical technique, time of surgery, patient co-operation, type of injury and tissue response to trauma [18]. The brief tissue residence time of the injected hyaluronan was also deemed a significant factor in the lack of positive outcome [18]. The dependence of the inhibition of adhesion formation on the concentration, molecular weight, size and residence time at the local tissue site were factors recognized by our group in the 1970s, and thus when hylans became available in the mid 1980s, their use was investigated [19,20]. A reliable tendon adhesion model, the traumatized extensor digitorum longus tendon of the rabbit, was utilized and hylan B gel was injected into the closed repaired synovial sheath on one leg; the opposite leg served as a control. Animals were sacrificed at three weeks and the sites were studied by gross observation, histological evaluation and pull-out force. The role of viscosity of the hylan B gel was also investigated in this model by utilizing a non-viscous, degraded hylan B gel preparation. The results of these studies revealed a significant ($p < 0.0005$) decrease in tendon adhesions between the hylan-treated and the control group, with more than an 11-fold increase in the number of adhesion-free tendons. Low viscosity hylan B gel failed to reduce adhesion formation ($p < 0.0005$; Table 3). There was no evidence of systemic or local inflammatory reaction as seen histologically at the tendon site on the hylan-treated side, and hylan was found to separate the tendon sheath from the paratendon at 3 weeks (Figure 1). An anti-adhesion study utilizing a fascial abrasion model, separating the subcutaneous tissue from muscle on the dorsum of rabbits and comparing hylan B gel-treated with saline-treated incisions, revealed a significant reduction in the extent and strength of adhesion formation at three weeks on the hylan B gel-treated side (C. Weiss, unpublished work).

In summary, hyaluronan and its derivatives have been shown for over 25 years to be effective in reducing scar and adhesion formation after tendon trauma in several animal models in more than five animal species. These molecules do not hinder and may enhance tendon healing and have no adverse local or systemic effects. Their effectiveness is directly related to the concentration, molecular weight, elasticity, viscosity, pseudoplasticity and residence time of the molecule. To date, hylan fluids and gels have been the only derivatives of this molecule engineered precisely to enhance these properties.

Failed back surgery and its resultant pain and neurological deficits are significant medical problems. Up to 20% of spinal surgery fails to relieve pain or restore function because of adhesions and arachnoiditis, which occurs along the tract of the surgical incision and subsequent disc excision [21]. For more than 50 years, surgeons, clinicians and scientists have utilized unsuccessfully a variety of substances, both biological and non-biological, in an attempt to limit the extent of post-surgical adhesion to the dura. In the late 1980s, recognizing the relatively limited potential of the available hyaluronans because of their short tissue residence time, molecular weight (2–4 million) and the severity of surgical trauma that occurs during disc surgery (muscle, fascial and posterior longitudinal ligament incisions, removal of bone, abrasion of dura and nerve roots, rupture of venous plexus and the extrusion of disc material into areas adjacent to nerve roots), we chose to study the enhanced viscoelastic properties of hylan B gel

Table 3

Rating	Adhesion strength (g)*	Treatment control			Nonviscous hylan			Hylan B gel		
		n	%	Strength (g)	n	%	Strength (g)	n	%	Strength (g)
None (NL)	183 ± 83	12	55	179 ± 47	1	7	129	1	5	150
Mild	<2 × NL	4	18	300 ± 10	1	7	390	3	14	323 ± 18
Moderate	<3 × NL	2	9	480 ± 60	4	29	499 ± 63	4	19	415 ± 20
Severe	>3 × NL	4	18	1075 ± 205	8	57	1182 ± 333	13	62	1064 ± 244

* ± S.D.

Tendon adhesions produced in a moderate trauma model in the rabbit (pull-out force in g)

Significance of hylan B gel versus low viscosity hylan or control: p < 0.0005. Composite table of data from Weiss and co-workers [19,20].

Figure 1

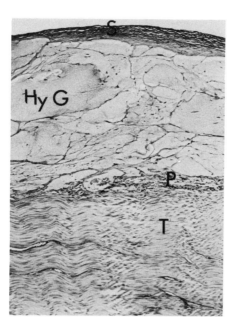

Light micrograph of hylan B-treated extensor hallucis longus tendon 3 weeks post-surgery

Hylan (HyG) separates the synovial sheath (S) from the paratendon (P) and tendon (T).There is no evidence of acute or chronic inflammatory reaction to hylan (Toluidine Blue 67.5 ×).

combined with the distribution characteristics of hylan A fluid in preventing post-laminectomy dural adhesions [22]. In a controlled, blinded evaluation of histology and graded dissection of a laminectomy model in adult rabbits, we found a significant decrease in the amount of post-operative adhesion to the dura (*p* <0.005) at 4, 8 and 12 weeks post-surgery at the hylan-treated site (Table 4). Hylan B gel was found up to 12 weeks post-surgery on histological section (Figure 2), and was biologically inert when placed either extra-durally, intra-durally or within the spinal fluid space. There was no evidence of acute or chronic inflammatory reaction and no systemic adverse events. Subsequent studies in dogs confirmed that the effectiveness of hyaluronan in diminishing adhesions is directly related to molecular weight [21] and that hyaluronan was significantly more effective than fat grafts or Gel Foam® [23]. Gel Foam® appeared to increase fibrosis [23]. Biomechanical studies also confirmed the superiority of hyaluronan-based barriers to scar formation over free fat graft in a lumbar laminectomy and discectomy model [24].

Table 4

Weeks	Mean grade* after treatment	
	Control(n†)	Hylan B gel (n)
4	2.1 ± 0.6 (4)	0.5 ± 0.5 (2)
8	2.0 ± 1.0 (2)	2.0 ± 1.0 (2)
12	3.0 ± 0.0 (2)	1.0 ± 0.0 (2)
Average for combined evaluation period	2.3 ± 0.8 (8)	1.3 ± 1.0 (6)
	$p < 0.005$	

* ± S.D.

†n = Number of cases.

Effect of hylan B gel in a rabbit post-laminectomy adhesion formation model

Graded dissection. Grade 0: no adhesions. Grade 1: thin adhesions easily released by blunt dissection. Grade 2: moderate adhesions released by blunt instrument dissection. Grade 3: severe adherent adhesions released only by sharp instrument dissection. Data from [22].

Viscoprotection

Viscoprotection during surgical procedures and chondral surface protection following injury present additional uses for the viscoelastic derivatives of hyaluronan. The viscoprotective effect of exogenous hyaluronan on the articular cartilage of dogs following trauma as described by Rydell and Balazs provided an impetus for us to study the more viscoelastic hylan fluids and gels in joints undergoing arthroscopy. The use of a pure hylan A fluid with molecular weight of 8 million was studied as a viscoprotective surgical tool to diminish surface scuffing and surface injury that accompany arthroscopic debridement, as the protective layer of hyaluronan is washed from the joint surfaces [25]. A multi-centred,

Figure 2

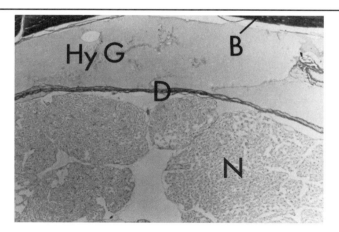

Light micrograph 12 weeks post-laminectomy

Hylan B gel (HyG) is seen separating the bony lamina (B) from the underlying dura (D), which covers the intra-dural neurons (N). There is no evidence of acute or chronic inflammatory response (Masson Trichrome and Toluidine Blue stain 67.5×).

randomized and controlled study, starting with an intra-articular injection of 3 ml of hylan A fluid prior to insertion of the arthroscope and an average of 8 ml used during surgery, showed a statistically highly significant decrease in surface damage accompanying arthroscopic surgery ($p < 0.0005$) (Table 5). In the 80 patients in this study who received hylan fluid there were no local or systemic adverse events, and no increase in post-operative pain, swelling, or stiffness [25–28]. A similar study revealed significant ($p < 0.01$) control of scuffing during arthroscopic surgery of the temporomandibular joint [29].

The decrease in surface injury during surgery as a result of the use of hylan A fluid, and earlier studies [30–32] on the chondral protective effect of injections of hyaluronan to restore joint homoeostasis and diminish the trauma to articular cartilage caused by joint immobilization, encouraged us to pursue the concept of a physical barrier to protect joint surfaces from arthritis-producing trauma. We studied arthritis produced in 40 adult rabbit knees following sectioning of the anterior cruciate ligament, medial collateral ligament and partial resection of the medial meniscus [33]. A single-blinded study, including gross observation and graded histological evaluation, was performed from 1 to 16 weeks post-surgery after a single injection of hylan B gel on the experimental side and saline solution on the control side. Hylan B gel could be found in the perisynovial tissues at 8 to 12 weeks following surgery. There was no evidence of inflammatory response, only progressive migration of gel particles through the synovium and into the pericapsular areas. At all times studied there was a statistically significant decrease in the observable arthritis grade starting at the 3rd and extending to the 16th week post-surgery (Table 6).

Subsequent studies in rabbit models demonstrated significant delays and lessening of surgically induced arthritis that were directly related to molecular weight [34–36] in rabbits following the intra-articular injection of hyaluronan. In a sheep model of early osteoarthritis, intra-articular hyaluronan limited the development of articular cartilage changes and subchondral bone changes characteristic of early osteoarthritis when compared with saline-treated controls [37]. Similar studies revealed that hyaluronans of both low and high viscosity reduced the lameness associated with meniscectomy [38]. Finally, the use of hyaluronan immediately after medial meniscectomy resulted in preservation of the proteoglycans in the lateral meniscus 6 months after surgery [39]. Force-plate analysis

	Control		Hylan A-treated			**Table 5**
	Mean ± SD	n	Mean ± SD	n	Significance	
Grade*	3.6 ± 1.0	49	4.3 ± 0.7	80	$p < 0.0005$	

*Observation grade

Viscosurgery of the knee surface protection with hylan A fluid

Grade 1: Surface easily damaged at start of procedure and damage continues to worsen. 2: Surface becomes progressively more vulnerable to instrument damage than usual. 3: Surface becomes less slippery with time; surface scars and deteriorates on instrument impact. 4: Surface remains slippery; less instrument damage than usual. 5: Surface remains slippery throughout; instruments do not scar or damage surfaces. Data from Weiss and co-workers [27].

Table 6

Weeks	Saline-treated [grade* (n)]	Hylan B gel-treated [grade (n)]	p
1–2	1.78 ± 0.91 (08)	1.44 ± 0.74 (8)	–
3–6	2.83 ± 0.93 (15)	1.86 ± 0.76 (15)	0.005
8–16	2.69 ± 0.81 (17)	2.02 ± 1.06 (15)	0.01

* ± S.D.

Experimental arthritis of the distal femoral condyle in rabbits

Grade 1: early; superficial hypercellularity, loss of staining to middle zone. Grade 2: moderate; loss of staining, chondrocyte clones, clefts to middle zone. Grade 3: moderately advanced; loss of staining, extensive clones, clefts through middle zone. Grade 4: advanced; hypocellular, extensive loss of staining, clefts and erosions to bone.

Data from Weiss and co-workers [33].

suggested that the presence of an extracellular matrix adapted to resist compression and tensional loading in the lateral meniscus of joints treated with hyaluronan as compared with saline controls was probably related to the improved gait seen in meniscectomized sheep after hyaluronan injection [38]. The principal mechanism for this improvement is likely to reside in lessening the discomfort following injury related to the addition of exogenous hyaluronan [40] and modification of the nature of mechanical stresses acting on joint tissues by improving the synovial fluid viscoelasticity [6].

The potential of these molecules and, in particular, the precisely engineered hylans to diminish the untoward effects of impact, scarring and adhesion formation following surgery and trauma in the musculoskeletal system is enormous. The reduction of dural and perineural fibrosis following laminectomy, discectomy, surgical release of trapped nerves, and the repair of nerves following trauma has already been demonstrated in animal models. The significant reduction in tendon adhesions after trauma and improved joint movement has been shown in numerous species. The early restoration of function after prosthetic joint replacement, the separation of adhered surfaces and reduction of pain and soft tissue stiffness following prolonged immobilization of joints, as well as the restoration of motion and decreased pain in bursitis and impingement conditions are still in the early stages of evaluation and hold significant promise. The reduction in scarring and the decrease in granulation tissue response to implanted foreign tissues or tissue debris by molecular exclusion and the direct effects of these high viscosity solutions in limiting migration, mitosis, phagocytosis and prostaglandin release by cells of the lymphomyeloid system (macrophages, lymphocytes and granulocytes) is being studied. Hylans hold significant promise in the mitigation of osteolysis, subsequent prosthetic loosening and bone resorption due to macrophage activation in response to polyethylene debris. The chondral protective function of hyaluronans and, particularly, the more highly viscoelastic hylans suggests their intra- and post-operative utilization in extensive joint surgeries, including cruciate ligament reconstruction and cartilage autograft and allografts. The role of these molecules in the early restoration of joint homoeostasis, movement and function is likely to be significant.

This chapter is dedicated to Dr. Endre Balazs: my mentor, exemplar and friend. His lifelong dedication to the study of the hyaluronan molecule represents the highest ideals of science and medicine. His work has enabled the blind to see, and the lame to walk.

References

1. Balazs, E.A., Bloom, G.D. and Swann, D.A. (1966) Fed. Proc. **25**, 1813–1816
2. Balazs, E.A. (1968) Univ. Mich. Med. Cent. J. (Special Issue), pp. 255–259
3. Gibbs, D.A., Merrill, E.W., Smith, K.A. and Balazs, E.A. (1968) Biopolymers **6**, 777–791
4. Balazs, E.A. (1969) in Thule International Symposium: Aging of Connective and Skeletal Tissue (Engel, A. and Larsson, T., eds.), pp. 107–122, Nordiska Bokhandelns Förlag, Stockholm
5. Balazs, E.A. and Sweeney, D.B. (1968) in New and Controversial Aspects of Retinal Detachment (McPherson, A., ed.), pp. 371–376, Harper and Row, New York
6. Rydell, N.W. and Balazs, E.A. (1971) Clin. Orthop. **80**, 25–32
7. Balazs, E.A., Rydell, N.W. and Freeman, M.I. (1971) in Hyaluronic Acid and Matrix Implantation, 2nd edn, (Balazs, E.A., ed.) Appendix, pp. 1–5, Biotrix Publishers, Arlington
8. Rydell, N.W., Butler, J. and Balazs, E.A. (1970) Acta Vet. Scand. **11**, 139–155
9. Balazs, E.A. and Denlinger, J.L. (1985) Equine Vet. Sci. **5**, 217–229
10. St. Onge, R., Weiss, C., Denlinger, J.L. and Balazs, E.A. (1980) Clin. Orthop. **146**, 269–275
11. Amiel, D., Ishizue, K., Billings, E., Wiig, N., VandeBerg, J., Akeson, W.H. and Gelberman, R. (1989) J. Hand Surgery **14A**, 837–843
12. Amiel, D., VandeBerg, J., Gelberman, F.R., Ishizue, K.K., Sisk, A. and Akeson, W.H. (1988) Trans. Orthop. Res. Soc., 181
13. Thomas, S.C., Jones, L.C. and Hungerford, D.S. (1986) Clin. Orthop. **206**, 281–289
14. Hagberg, L. and Gerdin, B. (1992) J. Hand Surgery **17**, 935–941
15. Seiichi, I., Kazuo, I., Kimura, N., Murakami, Y. and Nisiya, T. (1995) Current Trends in Hand Surgery: Proceedings of the 6th Congress of the International Federation of Societies for Surgery of the Hand (Mvastamaki, ed.), pp. 343–347, Elsevier, NY
16. Gaughan, E.M., Nixon, A.J., Krook, L.P., Yeager, A.E., Mann, K.A., Mohammed, H. and Bartel, D.L. (1991) Am. J. Vet. Res. **52**, 764–773
17. Wiig, M.E., Amiel, D., VandeBerg, J., Kitabayashi, L., Harwood, F.L. and Arfors, K.E. (1990) J. Orthop. Res., 425–434
18. Hagberg, L. (1992) J. Hand Surgery **17**, 132–136
19. Weiss, C. Levy, H.J., Denlinger, J., Suros, J.M. and Weiss, H.E. (1986) Bull. Hosp. Joint Dis. **46**, 9–15
20. Weiss, C., Suros, J.M., Michalow, A., Denlinger, J., Moore, M. and Tejeiro, W. (1987) Bull. Hosp. Joint Dis. **47**, 31–39
21. Songer, M.N., Rauschning, W., Carson, E.W. and Pandit, S.M. (1995) Spine **20**, 571–580
22. Weiss, C., Dennis, J., Suros, J.M., Denlinger, J., Badia, A. and Montane, I. (1989) Trans. Orthop. Res. Soc., 44
23. Songer, M.N., Ghosh, L. and Spencer, D.L. (1990) Spine **15**, 550–554
24. Abitol, J.J., Lincoln, T.L., Lind, B.I., Amiel, D., Akeson, W.H. and Garfin, S.R. (1994) Spine **19**, 1809–1814
25. Weiss, C. and Balazs, E.A. (1987) Arthroscopy **3**, 138–139
26. Weiss, C. and Balazs, E.A. (1988) Viscosurgery and Viscosupplementation in Joints, Third Int. Symp., Arthroscopy of the Temporomandibular Joint, New York (abstract)
27. Weiss, C., Drucker, M. and Levitt, R. (1991) A Multi-center Trial of Hylan as a Surgical Device During Arthroscopy, Robert Jones Lecture, New York
28. Weiss, C. (1991) in Arthroscopy of the Temporomandibular Joint (Thomas, M. and Bernstein, S.L., eds.), pp. 335–337, Saunders, Philadelphia
29. McCain, J.P., Balazs, E.A. and de la Rua, H. (1989) J. Oral Maxillofac. Surg. **47**, 1161–1168
30. Wigren, A., Falk, J. and Wik, O. (1978) Acta Orthop. Scand. **49**, 121–133
31. Amiel, D., Frey, C., Woo, S., Harwood, F. and Akeson, W. (1985) Clin. Orthop. **196**, 306–311
32. Wigren, A. (1981) Acta Orthop. Scand. **52**, 123
33. Weiss, C., Suros, J.M,. Dennis, J., Denlinger, J., Badia, A., Gross, J. and Eremenco, S.I. (1989) Trans. Orthop. Res. Soc. 539
34. Armstrong, S., Read, R. and Ghosh, P. (1994) J. Rheumatol. **21**, 680–688
35. Kikuchi, T., Yamada, H. and Shimmei, M., (1996) Osteoarthritis Cartilage, **4**, 99–110
36. Ghosh, P., Read, R., Numata, Y., Smith, S., Armstrong, S. and Wilson, D. (1993) Semin. Arthritis Rheum. (suppl.) **22**, 31–42

37. Hope, N., Ghosh, P., Taylor, T., Sun, D. and Read, R. (1993) Semin. Arthritis Rheum. (suppl.) **22**, 43–51

38. Kim, N.H., Han, C.D., Lee, H.M. and Yang, I. (1991) Yonsei Med. J. **32**, 139–146

39. Yoshimi, T., Kikuchi, T., Obara, T., Yamaguchi, T., Sakakibara, Y., Itoh, H., Iwata, H. and Miura, T. (1994) Clin. Orthop. **298**, 296–304

40. Miyazuki, K., Goto, S. and Okawara, H. (1984) Pharmacometrics **28**, 1123–1135

28

Management of adhesion formation and soft tissue augmentation with viscoelastics: hyaluronan derivatives

Nancy E. Larsen
Biomatrix, Inc., 65 Railroad Avenue, Ridgefield, NJ 07657, U.S.A.

Introduction

Hyaluronan (HA) and hyaluronan derivatives, hylans, have been investigated for their potential therapeutic use in the management of adhesion formation and in soft tissue augmentation procedures. The rationale for this approach is based on the well-established and extraordinary biocompatibility of the hyaluronan and hylan molecules, their unique physical (rheological) characteristics and the biological effects that result from the combination of these properties.

Hyaluronan in its unmodified form, however, does not provide optimal or adequate effectiveness against adhesion formation or for soft tissue augmentation owing to its relatively short residence time at the site of implantation.

In the mid-1980s, cross-linked hyaluronan derivatives (hylans) were developed and their biological activity and medical usefulness explored [1]. Hylans were found to have the same biocompatibility as the native hyaluronan, but possessed enhanced rheological properties and longer tissue residence times than hyaluronan.

The versatility of the hyaluronan polymer, with regard to modification by cross-linking, has stimulated a broad research effort into the development of cross-linked or otherwise modified forms of hyaluronan for therapeutic applications in the area of adhesion management and soft tissue augmentation.

This chapter will review the current state of research, development and clinical applications of hylans in the management of adhesion formation (abdominal, pelvic, cardiac) and in soft tissue augmentation procedures.

Hyaluronan and hylans: management of abdominal adhesion formation

Adhesion formation

Adhesions are connective tissue fibrous bands that join together adjacent tissues or organs in abnormal configurations. The magnitude of the adhesion may range from a fine wispy film to a dense vascular scar. The prevalence of adhesion formation is tremendous, with a reported 50–100% of abdominal and pelvic surgeries resulting in some degree of adhesion formation [2]. The medical complications resulting from adhesion formation include intestinal obstruction, infection, chronic pain and more difficult subsequent surgery.

Adhesion formation is caused by a cascade of events initiated by protracted inflammatory activity (sequel to trauma, surgery, infection, cancer), accompanied by production of a serofibrinous exudate and reduced plasminogen activator activity (owing to elevated proteolysis). This results in decreased fibrinolysis and enhanced fibroblast proliferation [3]. Decreased fibrinolysis leads to a loss or reduction in the ability of the body to control or modulate fibrous tissue formation, which ultimately leads to unwanted adhesion formation.

Minimization of injury is the single most important component in the prevention or reduction of post-operative adhesions [3]. However, it is not possible to completely eliminate or adequately reduce tissue trauma because medical procedures often involve unavoidable tissue injury. Currently, the methods commonly used to minimize adhesions include either (1) the instillation of various substances such as Ringer's lactate, heparin, Dextran 70, corticosteroids, antibiotics and enzymes directed at mitigating or interfering with the process of fibrous connective tissue formation [2,4–7]; or (2) using barriers that provide mechanical separation between traumatized tissues, i.e. expanded PTFE (polytetrafluorouethylene) films, Gore-Tex® (surgical membrane) [8] or oxidized regenerated cellulose (Interceed® TC7) [9].

The ideal barrier product for prevention of adhesions is biocompatible, self-adherent (not requiring sutures) and resorbable, with an adequate residence time and the ability to moderate fibrous tissue formation without interfering with normal wound healing. The commercial barrier products Gore-Tex® and Interceed®, though reportedly efficacious, are associated with significant disadvantages. Interceed® must be used only when complete haemostasis has been achieved, and the product resorbs quickly, resulting in a short residence time. The use of Gore-Tex membranes requires sutures to keep it at the desired tissue site and the membrane does not biodegrade, i.e. it is permanent and often requires surgical removal.

Recently, the use of hyaluronan and hylans has received widespread attention as potentially effective products for the prevention or reduction of adhesion formation. The scientific basis for this interest and the findings from preclinical and clinical studies are reviewed in this report.

Scientific rationale for the use of hyaluronan and hylans

Over 25 years ago, it was observed that hyaluronan was effective in preventing undesired fibrotic reactions during wound healing [10,11]. Hyaluronan was found to decrease the amount of granulation tissue formation and to reduce the formation of adhesions in subcutaneous wound healing studies in rabbits, owl monkeys and guinea pigs [10].

In 1971, Balazs and co-workers reported on the effect of hyaluronan on adhesion formation between tendon and tendon sheaths and between the conjunctiva and sclera [11]. Solutions of viscoelastic, high-molecular-weight hyaluronan, as well as dry sheets made from soluble hyaluronan, were found to reduce the incidence of adhesion formation without delaying or affecting the healing process. Other investigators have reported mixed results when using soluble hyaluronan to prevent adhesion formation in tendon repair [12–16]. In abdominal and pelvic adhesion prevention, the soluble hyaluronan, however, was found to have significant benefit only when applied before surgical injury, suggesting that the presence of hyaluronan decreases the extent of the injury, thereby decreasing the magnitude of tissue damage and subsequently new adhesion formation [17,18]. Only a 'polymer slab' of cross-linked hyaluronan was found to be effective in a pelvic model, all solutions of hyaluronan were ineffective in reducing adhesions [19]. In abdominal and pelvic procedures, the problem is more difficult owing to the nature and environment of the surgical site [4,19,20].

Biological effects of hyaluronan and hylans: relevance to adhesion formation and prevention

Hyaluronan and hylans have been shown to have effects on biological systems that influence and control adhesion formation.

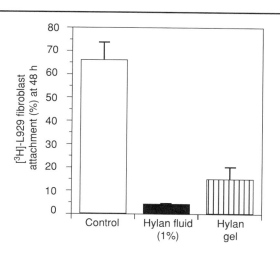

Figure 1

The effect of hylan fluid (hylan A) and hylan gel (hylan B) on migration and attachment of L929 fibroblasts

The bars represent the percentage of cells that have attached to the tissue culture well after 48 h. In the control, more than 65% of the total number of added cells have become attached to the well bottom. In the presence of hylan fluid (1%, 10 mg/ml) and hylan gel (0.5%, 5 mg/ml), cell attachment is markedly reduced. The values are the mean and standard deviation.

Cell migration

In 1973, Balazs reported that high-molecular-weight viscoelastic hyaluronan dramatically effects the movement and migration of cells [21]. This effect is readily demonstrated *in vitro* using viscoelastic hylan matrices. Application of fibroblasts (radiolabelled with tritium) to the surface of a 1% solution of hylan A fluid (viscoelastic, soluble, high molecular weight) or to the surface of 0.5% hylan B gel (viscoelastic, insoluble, cross-linked) results in inhibition of cell migration through the matrix (Figure 1). The hylan molecular network, therefore, forms a barrier to invading cells. Reducing the number of cells that migrate into the wound site will lead to reduced collagen deposition and, hence, reduced collagen maturation, and reduced fibrous tissue. Hyaluronan and hylans can therefore modulate fibrous tissue formation and ultimately adhesion formation.

Fibrinogen diffusion

The molecular network of viscoelastic hyaluronan and hylans form an effective barrier for fibrinogen as well as for cells. Fibrinogen is a 400000 molecular weight blood protein that originates from the tissue exudate after injury and is polymerized to form a fibrin coagulum at the wound site. Fibrinogen cannot penetrate the molecular network of hyaluronan and hylan and, therefore, a fibrous coagulum cannot form between two surfaces separated by hyaluronan or hylan. In the absence of the scaffolding for fibrous tissue formation, tissue adhesions and/or scar tissue formation are prevented [22–24].

Figure 2

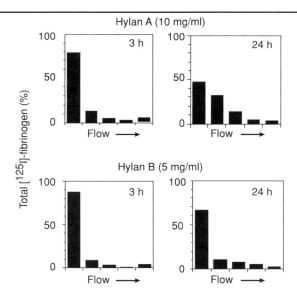

Diffusion of ^{125}I-fibrinogen through hylan matrices

The bars represent the amount (%) of ^{125}I-fibrinogen in a given location in the column of hylan A or hylan B. In each diffusion experiment, a 5 ml vertical column of hylan A or hylan B was prepared in a plastic syringe. A solution of ^{125}I-fibrinogen (300 μl) was applied to the top of the column. After 3 or 24 h, the penetration of fibrinogen was measured by removing 0.5 ml fractions and measuring the amount of radioactivity per fraction using γ-ray spectrometry.

The fibrinogen exclusion effect is demonstrated *in vitro* in diffusion studies using radiolabelled fibrinogen (^{125}I), viscoelastic, soluble hylan fluid (hylan A, 10 mg/ml) and viscoelastic, insoluble hylan gel (hylan B, 5 mg/ml). The results are presented in Figure 2. Fibrinogen diffusion through hylan matrices is markedly inhibited, and after 24 h there is very little penetration of the hylan matrix by the fibrinogen molecules. This effect is completely dependent upon the viscoelastic properties of the hyaluronan and hylan, since non-viscoelastic solutions of hyaluronan do not have the same effect [25].

Fibrin clot formation

The interaction of hyaluronan and hylan with fibrinogen and fibrin has been evaluated by several groups. The most extensive investigations were conducted by LeBoeuf and Weigel and their colleagues [26]. Several interesting findings are reported: (1) hyaluronan (low molecular weight; 32 000) specifically and reversibly binds human fibrinogen; (2) HA decreases the lag time before clotting and increases the rate of clot formation (measured by changes in A); and (3) HA effects the structure of the fibrin gel as assessed by light scattering [27].

Evaluation of viscoelastic hyaluronan and hylans reveals a lack of effect on the kinetics of clot formation in contrast to that reported by Le Boeuf and co-workers [26]. Fibrous clot formation in our laboratories was monitored by measuring the change in rheological properties of the mixture, i.e. by actually measuring the physical changes that occur during gel formation. The results from rheometric kinetic analysis are shown in Figure 3. Viscoelastic soluble hylan fluid (hylan A) does not increase the rate of clot formation. In fact, the rate of 'gel'

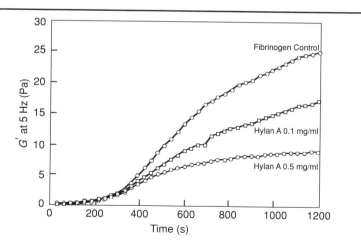

Figure 3

The effect of hylan A on the rheological changes that occur during fibrin formation

Hylan A (0 to 0.5 mg/ml) was incubated (37 °C) in the presence of fibrinogen at 2.5 mg/ml. The control was a solution of fibrinogen in phosphate-buffered saline (2.5 mg/ml). Fibrin formation was initiated by adding 0.14 Units (NIH) of thrombin per ml (11 ml total volume). The elastic modulus G′ measured in pascals, was monitored on a Bohlin rheometer at a frequency of 5 Hz.

Figure 4

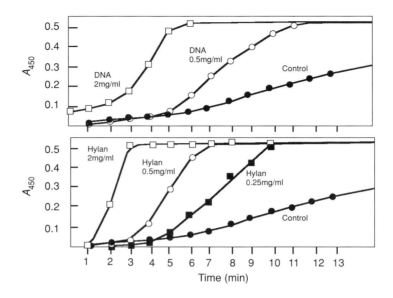

The effect of hylan A and DNA on the absorbance changes during fibrin formation

Hylan A or DNA (0 to 2.0 mg/ml) were incubated (23 °C, room temperature) in the presence of fibrinogen (2.5 mg/ml). The control sample was a solution of fibrinogen in phosphate-buffered saline. Fibrin formation was initiated by adding 0.14 Units (NIH) of thrombin per ml (2.5 ml). The absorbance (A) was monitored on a spectrophotometer at a wavelength of 450 nm.

(clot) formation is reduced slightly by the presence of hylan A. It is expected that a fibrin–hylan A gel has physical properties that differ from those of a pure fibrin gel due to the effect of viscoelastic hylan A on the properties of the composite gel (fibrin–hylan A).

Measurement of fibrin formation (clotting time) using absorbance as described by Le Boeuf and co-workers is problematic owing to the interference of HA on the optical density of fibrin gels through its excluded volume effects. In fact, other high-molecular-weight viscous polymers, for example DNA, have the same effect as hylan under these conditions, i.e. when A changes are measured and, therefore, this is not an effect that is related only to hylan and hyaluronan (Figure 4).

It is concluded that viscoelastic hylan and hyaluronan are not procoagulants and do not interfere with the normal process of thrombin-catalysed fibrin formation.

Polymorphonuclear leucocyte activity (O_2^--generation)

The effect of hylan A fluid on superoxide generation by fresh, human PMN (polymorphonuclear leucocytes, granulocytes) was studied. Hylan A was found to produce dose-dependent and viscoelasticity-dependent inhibition of PMN stimulation when present before the stimulating agent phorbol myristate acetate was added. Hylan B gel was also found to exhibit dose-dependent inhibition of

Figure 5

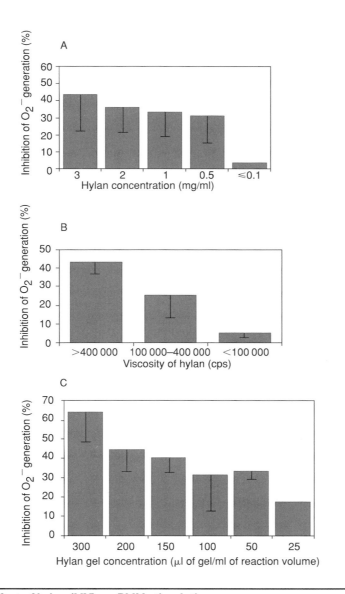

The effect of hylan (HY) on PMN stimulation

The bars represent percentage inhibition by HY of superoxide generation by PMA-stimulated PMN. The positive control was superoxide dismutase (100% inhibition), buffer was used as the negative control (0% inhibition). A, concentration effect: HY (<0.1–3 mg/ml) was incubated with PMA and PMN activation was measured. The viscosity of HY was >400 Pa s at 0.01 s^{-1} shear rate. B, Viscosity effect: the HY concentration was constant (2 mg/ml), whereas the viscosity (centipoise at 0.01 s^{-1} shear rate) of HY was varied, and the superoxide generation was measured. C, The effect of HY gel (insoluble, chemically cross-linked HY) on PMN stimulation: the concentration of gel was increased from 25 to 300 μl/ml and superoxide generation was measured. The values are the means and SD; n ≥ 6.

PMN activation (Figure 5) [28]. This function is indicative of the role as a protective biopolymer in that PMN activation leads to the release/secretion of numerous substances involved in the inflammation response (enzymes, free radicals), which are associated with increased tissue damage and, hence, enhanced adhesion formation.

In summary, hyaluronan and hylan produce diverse effects on biological systems: inhibition of cell migration, restriction of fibrinogen diffusion and inhibition of superoxide release from granulocytes (PMN). These effects are essential to the physical mechanisms of adhesion control by hyaluronan and hylans.

Hyaluronan-based products for management of adhesion formation

The earliest reports of the beneficial effects of HA on the prevention of adhesion formation were made over 25 years ago [10,11]. After numerous investigations using soluble HA, it became clear that hyaluronan with a longer residence time was needed to achieve adequate and consistent adhesion prevention.

Soluble hyaluronan has been reported to be effective in reducing adhesion formation, particularly in tendon repair [23]; however, in 1986, the first use of a cross-linked insoluble hyaluronan gel (hylan gel) was found to provide a consistent and prominent reduction in tendon adhesion formation in a large number of study animals [29,30]. In these studies, the hyaluronan derivative, hylan gel, had a longer residence time at the site of injury, did not elicit an inflammatory response and did not inhibit healing. Hylan gel-based products for orthopaedic applications in adhesion are currently under development by Biomatrix Inc. In post-surgical abdominal, cardiac and pelvic adhesion prevention, it is widely acknowledged that a barrier-type material is most effective in maintaining mechanical separation of injured tissues to prevent fibrin network formation between the tissue surfaces. The first commercially available mechanical barrier using HA as a major component is the product Seprafilm®, produced by Genzyme Corp. (Cambridge, MA). In this product, HA purified from a bacterial fermentation process is chemically modified and combined with carboxylmethylcellulose (CMC) to produce an HA–CMC membrane. The HA is not chemically cross-linked and, therefore, readily hydrates and dissolves in aqueous media. In rat caecal abrasion studies, Seprafilm® dramatically reduced the incidence of adhesion formation following surgical injury (65% incidence in control versus 5% in Seprafilm-treated injuries) [31]. In clinical studies of adhesion formation following abdominal surgery (colectomy and ileal pouch-anal anastomosis with diverting loop ileostomy), 51% of Seprafilm®-treated patients had no adhesions, while only 6% of control, untreated patients were free of adhesions. However, there was no reduction in the incidence of the serious complications of bowel obstruction in the Seprafilm®-treated group. In addition, there was a higher incidence of abscess formation in Seprafilm®-treated patients as compared with control, untreated patients (7.7 versus 2%) [32]. However, this product demonstrates the clinical and commercial feasibility of using hyaluronan-based products in adhesion control.

Figure 6

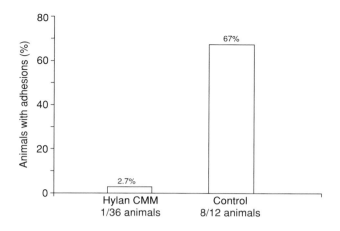

Effect of hylan membranes on rat cecal abrasion

The abdominal cavity of anaesthetized female Sprague–Dawley rats was opened and two defects were created: one in the abdominal wall (removal of peritoneum and muscle) and one on the caecal surface (abrasion injury). In 36 rats, a hylan membrane was applied to the wound surface before closing. A control group of 12 animals with identical injuries did not receive an anti-adhesion agent. After 7 days, all animals (test and control) were killed and evaluated for incidence and strength of adhesion formation.

A resorbable hylan barrier membrane was evaluated in the rat caecal abrasion model. Hylan membranes were found to be tightly adherent to the wound surface, biocompatible and highly effective in reducing and preventing adhesions in this model (Figure 6). Animals were evaluated seven days post-surgery, and of 36 rats treated with hylan membranes, only one developed an adhesion, and this adhesion was much weaker than those formed in untreated controls. The incidence of adhesions was 3% (1/36) in hylan-treated rats, while in the control groups, the incidence of adhesion formation was 67% (8/12) [33].

Hylan membranes made from insoluble, cross-linked hylan (hylan B gel) are currently in development. The advantages of membranes produced from cross-linked insoluble hylan include longer residence time at the site of implantation, as compared with natural (non-cross-linked) HA products, enhanced barrier function and enhanced physical properties. In addition, a hylan membrane composed only of cross-linked hylan and without a cellulosic or other polymer component would have a biocompatibility essentially like that of the native hyaluronan.

Numerous groups have identified the use of HA solutions and membranes as an effective approach to the prevention or reduction of adhesions. Table 1 lists these products in development for adhesion control. Of the eight groups developing new products for adhesion control, five are using an HA-based material for their product. It is clear that the physical and biological properties of hyaluronan and hylan, together with their chemical versatility, make them the 'substances of choice' for many medical procedures, including adhesion management.

Table 1

Company	Product	Description	Indication	Regulatory status
Genzyme	Seprafilm®, Sepracoat® (formerly HAL product line)	Bioresorbable HA-based film	Post-operative adhesions formation following abdominal, pelvic and cardiac surgery	FDA approved PMA. Received CE mark for market in Europe
Biomatrix	Hylan B	Cross-linked hylan membrane	Post-operative indications (abdominal, pelvic)	Preclinical studies, pilot U.S. clinical trials
Gliatech	ADCON-T/N, ADCON-L	Carbohydrate polymer based on research on glial cells	Tendons, nerves and peridural fibrosis following laminectomy and spinal fusions	Received CE mark for market in Europe U.S. clinicals
Life Medical Sciences	Repel	Bioresorbable HA-based film	Post-operative adhesions (abdominal)	Preclinical studies
Lifecore Biomedical	Lubricoat	Ferric HA	Prevent adhesions to tube and ovary to preserve fertility	Clinical trials in U.S. and Europe
Osteotech	PolyActive	Polyethylene glycol Terephthalate/ Polybutylene Terephthalate	Post-operative adhesions	Pilot clinical trials in Europe

Hyaluronan-based and other anti-adhesion products under development, in clinical trials or approved

Table 1 (contd)

Company	Product	Description	Indication	Regulatory status
Anika Research	INCERT	Bioresorbable HA-based film	Post-operative adhesions	Awaiting CE mark approval
Focal Interventional	FocalGel	Polyethylene glycol and polylactic acid photopolymerized by UV light	Post-operative adhesions (abdominal)	Preclinicals in U.S. and Europe

Hyaluronan-based and other anti-adhesion products under development, in clinical trials or approved

Hylans in soft tissue augmentation

Various materials have been used for soft tissue augmentation and correction of soft tissue defects. These materials include paraffin, silicone and collagen [34–36]. Paraffin and silicone have been associated with undesirable tissue reactions, including foreign-body type reactions, and also are reported to migrate from the site of injection. Collagen implants are currently used for the correction of soft tissue defects; however, there is extensive documentation that collagen implants are often short-lived and are known to be associated with some hypersensitivity reaction [37–39]. Prospective patients must undergo skin testing before receiving collagen therapy.

Hylan B gel (Hylaform®, Biomatrix Inc., Ridgefield, NJ) was developed as a soft tissue augmentation material for the treatment of wrinkles and depressed scars [1,40]. Hylan B gel is a viscoelastic, insoluble hyaluronan derivative in which the polysaccharide chains are chemically cross-linked through sulphonyl-bis-ethyl cross-links. Hylan B gel was developed for this purpose because even the very-high-molecular-weight soluble hylan A (4–8 million) does not provide the physical properties and extended residence time required for effective soft tissue augmentation.

Hylan gels are well-suited for use in soft tissue augmentation because of their insolubility and resistance to degradation and migration. Their high water content mimics the natural hydrating properties of hyaluronan. These properties, together with the fact that hylan gel implants do not elicit inflammatory, immuno-logical or foreign body reactions, contribute to their stability and compatibility in the soft tissues [1,41].

Dermal residence time studies in animals indicate that hylan B gel implants have an extended intradermal residence time and produce no significant tissue reaction. In radiotracer studies in guinea pigs, an intradermal half-life of 9 months was estimated by linear regression analysis of the recovery data (recovery of radioactivity) [42] (Figure 7). Histological evaluation of the dermal injection site tissue revealed an absence of inflammation and fibrosis at the site of implantation.

Hylaform® was evaluated clinically in two multi-centre studies in the U.S.A. and Sweden for the treatment of wrinkles and depressed scars. Patients were assessed at regular intervals during a 52-week study period. Independent patient and physician assessments of the degree of correction revealed maintenance of moderate or better correction in more than half of the treated facial sites at the six-month time point. There were no antigenic or immunogenic responses observed and there were no systemic adverse reactions related to the use of Hylaform®. In some patients, transient erythema, swelling and pain were reported, which were anticipated reactions related to the injection procedure. Hylaform® has been approved in Europe as a treatment for correction of wrinkles and depressed scars.

Recently, other HA-based soft tissue augmentation products have been introduced to the medical community in Europe; however, their history of use in animals and humans is extremely limited and, therefore, their effectiveness and safety are not known.

Figure 7

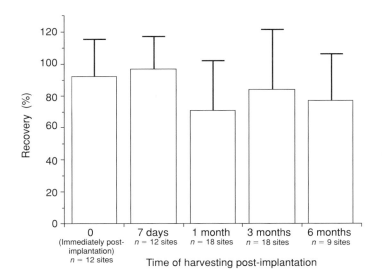

Recovery of [¹⁴C] from guinea pigs injected intradermally with [¹⁴C]hylan B gel

Female guinea pigs (Hartley, n = 14) were injected intradermally with approximately 0.25 µl of [¹⁴C]hylan B gel per each of 8 sites on each animal. The exact amount of radiolabelled hylan B gel was determined for each injection site by the weight of syringe pre- and post-injection. Two animals were killed immediately following the injection procedure. Animals were harvested at intervals of 7 days and 1, 3 and 6 months. Injection site tissue was harvested and analysed for radioactivity by β-particle spectrometry.

Summary: hyaluronan and hylans in adhesion management and soft tissue augmentation

Hyaluronan was identified over 25 years ago by Balazs and co-workers for clinical use as a viscosurgical tool in ophthalmic surgery [11]. Following years of clinical exposure and widespread medical application for hyaluronan, its derivatives, hylans, have now become one of the most interesting and important polymers in medicine. The therapeutic applications of hyaluronan and hylans are included in numerous medical specialties, including adhesion management, soft tissue augmentation (dermal, urological, reconstructive), arthritis therapy, drug delivery, ophthalmology (viscosurgery), neuroradiology (embolization) [43,44] and topical applications. The unique physical, chemical and biological properties of hyaluronan and hylans will ensure their continued and increased use in medicine.

References
1.   Balazs, E.A. and Leshchiner, E.A. (1989) in Cellulosics Utilization: Research and Rewards in Cellulosics, (Inagaki, H. and Phillips, G.O., eds.), pp. 233–241, Elsevier Applied Science, New York
2.   diZerega, G.S. (1994) Fertil. Steril. **61**, 219–235
3.   Gomel, V. (1983) in Microsurgery in Female Infertility, pp. 195, Little Brown and Co., Boston

4. Drollette, C.M. and Badawy, S.Z.A. (1992). J. Reprod. Med. **37**, 107–122
5. Leondires, M.P., Stubblefield, P.G., Tarraza, H.M. and Jones, M.A. (1995). Am. J. Obstet. Gynecol. **172**, 1537–1539
6. Cofer, K.F., Himebaugh, K.S., Gauvin, J.M. and Hurd, W.W. (1994) Fertil. Steril. **62**, 1262–1265
7. Hill-West, J.L., Chowdhury, S.M., Dunn, R.C. and Hubbell, J.A. (1994) Fertil. Steril. **62**, 630–634
8. Boyers, S.D., Diamond, M.D. and DeCherney, A.H. (1988) Fertil. Steril. **49**, 1066
9. Shimanuki, T., Nishimura, K., Montz, F.J., Nakamura, R.M. and diZerega, G.S. (1987) J. Biomed. Mater. Res. **21**, 173–185
10. Rydell, N. (1970) Acta Orthop. Scand. **11**, 307–311
11. Balazs, E.A., Rydell, N.W. and Freeman, M.I. (1971) in Hyaluronic Acid and Matrix Implantation, 2nd edn. (Balazs, E.A., ed.), Appendix 13, Biotrics Inc., Arlington, MA
12. Meyers, S.A., Seaber, A.V., Glisson, R.R. and Nunley, J.A. (1989) J. Orthop. Res. **7**, 683–689
13. Thomas, S.C., Jones, L.C. and Hungerford, D.S. (1986) Clin. Orthop. Relat. Res. **206**, 281–289
14. Amiel, D., Ishizue, K., Billings, Jr., E., Wiig, M., Vande Berg, J., Akeson, W.H. and Gelberman, F.R. (1989) J. Hand Surg. **14A**, 837–843
15. Jones, L.C., Thomas, S.C. and Hungerford, D.S. (1986) in 32nd Ann. ORS, Proceedings of 17–20 February 1986, New Orleans LA, p. 132
16. Green, S., Szabo, R., Langa, V. and Klein, M. (1986) Bull. Hosp. Joint Dis. **46**, 16–21
17. Urman, B., Gomel, V. and Jetha, N. (1991) Fertil. Steril. **56**, 563–567
18. Goldberg, E.P., Burns, J.W. and Yaacobi, Y. (1993) Prog. Clin. Biol. Res. **381**, 191–204
19. Grainger, D.A., Meyer, W.R., DeCherney, A.H. and Diamond, M.P. (1991) J. Gynecol. Surg. **7**, 97–101
20. Goldberg, R.L., Huff, J.P., Lenz, M.E., Glickman, P., Katz, R. and Thonar, E.J-M.A. (1991) Arthritis Rheum. **34**, 799–807
21. Balazs, E.A. and Darzynkiewicz, Z. (1973) in Biology of Fibroblast (Kulonen, E. and Pikkarainen, J., eds.), pp. 237–252, Academic Press, London
22. Rydell, N. and Balazs, E.A. (1971) Clin. Orthop. **80**, 25–32
23. St. Onge, R., Weiss, C., Denlinger, J.L. and Balazs, E.A. (1980) Clin. Orthop. Relat. Res. **146**, 269–275
24. Balazs, E.A. and Denlinger, J.L. (1985) in Osteoarthritis - Current Clinical and Fundamental Problems (Peyron, J.G., ed.), pp. 165–174, Laboratoires Ciba-Geigy, Rueil-Malmaison, France
25. Balazs, E.A. and Denlinger, J.L. (1989) in The Biology of Hyaluronan (Evered, D. and Whelan, J., eds.), Ciba Foundation Symposium No. 143, pp. 265–280, Wiley, Chichester
26. Le Boeuf, R.D., Gregg, R.R., Weigel, P.H. and Fuller, G.M. (1987) Biochemistry **26**, 6052–6057
27. Weigel, P.H., Frost, S.J., Le Boeuf, R.D. and McGary, C.T. (1989) in The Biology of Hyaluronan (Evered, D. and Whelan, J., eds.), Ciba Foundation Symposium No. 143, pp. 248–264, Wiley, Chichester
28. Larsen, N.E., Lombard, K., Parent, E.G. and Balazs, E.A. (1992) J. Orthop. Res. **10**, 23–32
29. Weiss, C., Levy, H.J., Denlinger, J., Suros, J. and Weiss, H. (1986) Bull. Hosp. Joint Dis. **46**, 9–15
30. Weiss, C., Suros, J.M., Michalow, A., Denlinger, J., Moore, M. and Tejeiro, W. (1987) Bull. Hosp. Joint Dis. **47**, 31–39
31. Burns, J.M., Cox, S., Greenawalt, K., Magi, L., Muir, C., Kirk, J. and Colt, J. (1991) Presented at the 17th Annual Meeting of the Society for Biomaterials, 1–5 May 1991, Scottsdale, AZ, 251(abstract)
32. Becker, J.M., Dayton, M.T., Fazio, V.W., Beck, D.E., Stryker, S.J., Wexner, S.D., Wolff, B.G., Roberts, P.L., Smith, L.E., Sweeney, S.A. and Moore, M. (1996) J. Am. Coll. Surg. **183**, 297–306
33. Harris, E.S., Foresman, P.A., Rodeheaver, G.T., Larsen, N.E. and Balazs, E.A. (1996) Fifth World Biomaterials Congress, 29 May to 2 June, 1996, Toronto, Canada, 371(abstract)
34. Burton, J.L. and Cunliff, W.J. (1986) in Textbook of Dermatology (Rook, A., Wilkinson, D.B., Ebling, F.J.G., Champion, R.H. and Burton, J.L., eds.), pp. 1870–1871, Blackwell, Oxford
35. Selmanowitz, V.J. and Orentreich, N. (1977) J. Dermatol. Surg. Oncol. **3**, 597–611
36. Knapp, T.R., Kaplan, E.N. and Daniels, J.R. (1977) Plast. Reconstr. Surg. **60**, 398–405
37. Donald, P.J. (1986) Otolaryngol. Head Neck Surg. **95**, 607–614
38. Arem, A. (1985) Collagen Modif. **12**, 209–220
39. Kamer, F.M. and Churukian, M.M. (1984) Arch. Otolaryngol. **110**, 93–98

40. Larsen, N.E., Kling, M.B., Balazs, E.A. and Leshchiner, E.A. (1990) in Society for Biomaterials 16th Ann. Meeting, Proceedings of 20–23 May 1990, Charleston, SC, Trans. Soc. Biomater. **XIII**, 302
41. Larsen, N.E., Pollack, C.T., Reiner, K., Leshchiner, E. and Balazs, E.A. (1993) J. Biomed. Mater. Res. **27**, 1129–1134
42. Larsen, N.E., Leshchiner, E., Pollak, C.T., Balazs, E.A. and Piacquadio, D. (1995) in Polymers in Medicine and Pharmacy (Mikos, A.G., Leong, K.W., Radomsky, M.L., Tamada, J.A. and Yaszemski, M.J., eds.), pp. 193–197, Materials Research Society, Pittsburgh, PA
43. Hilal, S.K., Leshchiner, E.A., Larsen, N.E., Khandji, A.G., Moser, F.G. and Balazs, E.A. (1987) 25th Ann. Meeting of Am. Soc. Neuroradiol., New York, 80 (abstract)
44. Larsen, N.E., Leshchiner, E.A., Parent, E.G., Hendrikson-Aho, J., Balazs, E.A. and Hilal, S.K. (1991) J. Biomed. Mater. Res. **25**, 699–710

29

The action of hyaluronan on repair processes in the middle ear

Claude Laurent
Department of Oto-Rhino-Laryngology, University Hospital, University of Umeå, S-901 85 Umeå, Sweden

Background

In middle ear surgery, the otosurgeon deals a lot with the consequences of infectious and inflammatory disorders of the middle ear. In children, an acute bacterial infection of the middle ear – an acute otitis media – is one of the most frequent disorders. In this condition, the tympanic membrane is bulging outwards because of collection of pus in the middle ear. The disease is generally treated with antibiotics, and if this is correctly done there are, nowadays, seldom any complications seen. However, in less well-developed countries, acute otitis media is a common cause of chronic tympanic membrane perforations and bad hearing.

In chronic otitis media, the tympanic membrane becomes retracted because of a long-standing negative middle ear pressure and the tympanic membrane may perforate. Sometimes desquamated scales from the ear drum may accumulate in the middle ear and a keratin cyst – a so-called cholesteatoma – is formed. Cholesteatomas have an osteolytic power and in long-standing cases of chronic otitis media, the ear drum and the bony middle ear structures can become severely damaged.

The middle ear

The middle ear is an air filled cavity within the temporal bone. It transmits and amplifies the sound waves on their way from the tympanic membrane, via the ossicular chain to the sensory organ of the inner ear. The tympanic membrane is suspended in air and separates the middle ear cavity from the external auditory canal. Embryologically, the tympanic membrane originates from both the ectodermal and the entodermal germ layers. As a consequence, the tympanic membrane consists of an outer epidermal layer, composed of a keratinizing, stratified, squamous epithelium, followed by a densely packed layer of collagenous fibres arranged in an outer radial and an inner circular layer. Facing the middle ear, there is a simple squamous epithelium.

It is a well-known fact that the subepithelial connective tissue undergoes dynamic changes variously related to different inflammatory conditions in the middle ear [1]. For more than a decade, we have focused our interest on the connective tissue in otitis media and, in particular, on the roles that hyaluronan (HYA, hyaluronic acid) plays in the physiology and pathophysiology of the middle ear [2,3].

Occurrence of HYA in the middle ear

In earlier studies we measured the concentration of HYA in different areas of the rat middle ear using a sensitive radioassay [4]. We found that the mast-cell-rich pars flaccida of the tympanic membrane showed a 20–30-fold higher concentration of HYA than the pars tensa and two mucosal areas of the middle ear.

    More recently we have studied the distribution of HYA in the middle ear with a histochemical method [5]. After use of a microwave oven for fixation and decalcification of tissues, HYA was localized in the specimens by use of a HYA-binding protein as a specific probe. In these studies, we noticed an uneven distribution of HYA in the middle ear, indicating that the polysaccharide may have specific local functions within the tissues. As expected from previous concentration measurements [4], HYA was abundant in the pars flaccida of the tympanic membrane. This is an extremely flexible part of the tympanic membrane and HYA, by virtue of its viscoelasticity, may contribute to this characteristic. The pars flaccida also participates in the generation of middle ear effusions in otitis media and, owing to its osmotic power, HYA may influence such production of effusion. Hyaluronan was also abundant in the subepithelial tissue surrounding the Eustachian tube (Figure 1), where oedematous swellings of the connective tissue can regulate the passage of air and secretions between the middle ear and the nasopharynx. In the middle ear muscles, HYA was found in the connective tissue septa between the individual muscle fibres. In this location, HYA may act as a cushion, separating and lowering the friction between the muscle fibres. The

Figure 1

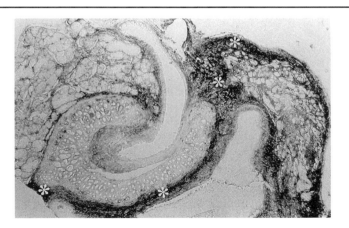

Light micrograph of a transverse section of the Eustachian tube
The subepithelial connective tissue is intensely stained for hyaluronan (white asterisks). Magnification,
×80.

Figure 2

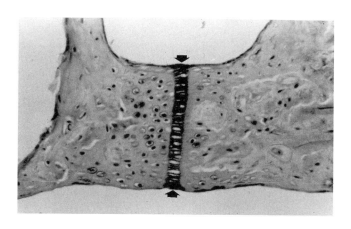

Light micrograph showing the staining pattern for HYA in the annular ligament of the stapes foot-plate (arrows)
Magnification, ×317.

middle ear muscles and the eye muscles, which have to perform quick and precise movements, contain more HYA than skeletal muscles with large fibre dimensions [6]. In a new study, we have discovered an ingenious suspension mechanism for the insertion of the stapes in the oval window facing the inner ear. This small ossicle must be well fixed and still freely mobile to be able to transmit the sound waves to the inner ear. The annular ligament, which is known to surround the stapes, is a more complicated structure than just a ligament. It consists of a cartilaginous joint bridged by elastic and collagenous fibres. Between the fibres, abundant HYA is accumulated (Figure 2).

The tympanic membrane as a wound model

The rat middle ear has many similarities to the human ear. Experimental tympanic membrane perforations in this model are easy to standardize in size and the depth of the penetrating wound is always the same. The healing process occurs in a uniform way and the wound can be topically treated with different substances in order to record various changes in healing pattern [7].

In this model for acute wounds, we found that endogenous HYA accumulated already on the first day in the perforation border. On the fifth day, the HYA staining was at its peak and was localized mainly in the hyperplastic connective tissue of the thickened perforation border around fibroblasts and inflammatory cells. It was also prominent close to dilated vessels at the handle of the malleus ossicle. After closure of the perforation, the HYA staining gradually faded and the healed tympanic membrane showed only patchy HYA staining. Our studies have confirmed that HYA is present only during the first few days of healing and is then diminished, probably because of enzyme activity [8]. In this model, we have also investigated the immunohistochemical staining intensities for fibronectin and various chondroitin sulphates during the course of early healing [8].

Exogenous HYA in middle ear surgery

Exogenous HYA was considered for use in middle ear surgery 10 years ago [2]. In surgery for chronic otitis media, the two major objectives are to eradicate the disease and to preserve or restore the hearing. In this kind of surgery, two of the main problems are: how to repair a chronic tympanic membrane perforation and what material to choose as a supporting substance for the reconstructions in the middle ear.

Effects of exogenous HYA on tympanic membrane repair

Experimental studies

Several experimental studies have shown that exogenous HYA accelerates the healing of acute tympanic membrane perforations [7]. Three days after wounding and application of 1% HYA solution, the perforated area is covered by a sheet of keratin and HYA. Guided by this sheet, the epithelial migration advances within the HYA and keratin cover ahead of a connective tissue layer. It has been shown in several studies [7,9] that about half of the HYA-treated perforations heal before the earliest closures occur in buffer-treated or untreated perforations. We have also shown that, when exogenous HYA solutions in various concentrations and molecular weights are topically applied to experimental perforations, the interval to closure is correlated with the concentrations of the HYA, but not to its molecular weight or viscosity [9].

From a clinical point of view, it is interesting that treatment with exogenous HYA also improves the structural quality of the healed tympanic membrane [9]. As soon as two weeks after closure the tympanic membrane scar is partly transparent. There is also an improved structural organization seen in the histological sections as compared with controls – the HYA-treated tympanic membranes are thinner and show a well-restored connective tissue layer, with few fibroblasts and collagen fibres oriented strictly parallel to the epithelial surface. One may speculate that the exogenous HYA mimics and reinforces the production of endogenous HYA in the wound and that the cell motility is promoted in this HYA-rich environment. Furthermore, it is feasible that a prolonged presence of HYA in the wound can mimic the fetal wound in which the hyaluronan levels remain high throughout the healing process [10]. Fetal wound healing of skin and bony tissues occurs without scar formation. This is in contrast to fetal wounds in muscles, tendon and the gut, which heal by fibrosis and contraction, as in the adult.

Model for chronic wounds

Animal models for studies of chronic wounds are few and unreliable. A breakthrough was made in our laboratory when we created chronic tympanic membrane perforations in our model [11]. Topical application with a 2% hydrocortisone solution, once daily for ten days, on an acute perforation causes a persistent tympanic membrane perforation that will remain open for at least 50 days. After 3 months, 1/3 of the perforations remain open. Glucocorticoids reduce the amount of granulation tissue in the wound area and such tissue is needed to guide and support the migrating epithelium. Another explanation for the retarded healing after application of hydrocortisone is that glucocorticoids suppress the synthesis of HYA. In a recent study on 27 hydrocortisone-induced chronic tympanic membrane perforations, a majority of the perforations treated with HYA alone or in combination with wounding closed within a few days after treatment, whereas the untreated or wounded perforations healed 30–50 days later [11].

Clinical studies

There are a number of clinical studies describing positive effects after application of different preparations of HYA to chronic tympanic membrane perforations [3]. The reported healing rate varies between 30 and 75% and higher concentrated solutions appear to give a better result. In our own controlled clinical study on wounded chronic tympanic membrane perforations [5], we compared the effect of daily applications of 1% HYA solution for 7 days with the application of a rice-paper patch. In both treatment groups, the healing rate after 1 year was about 30%. One visible difference between the groups was that the scar in the HYA-treated group appeared to be more similar to a normal tympanic membrane as compared with the rice-paper treated tympanic membrane, in which the scar was thin and atrophic.

Treatment of dry, small-to-medium sized chronic tympanic membrane perforations with highly concentrated HYA solutions is a cheap, simple and time-saving office procedure that should be an attractive alternative to a tympanoplasty operation. To our knowledge, there is still only one HYA product for this purpose on the market (Otoial®, Fidia S.p.A. – Abano Terme, Italy).

Exogenous HYA as a scaffold in middle ear surgery

Exogenous HYA has been considered as a scaffold in middle ear surgery. So far, gelatine sponge, Gelfoam®, soaked in various solutions has been used for almost 40 years as a supportive substance in otosurgery. It is an artificial material derived from beef protein and converted into an absorbable foam-like material. In our rat model, Gelfoam® applied in the middle ear cavity gives rise to a dense connective tissue with firm adhesions to the surrounding cavity walls [12]. In histological sections, a dense connective tissue with gland-like structures, mast cells and even newly formed bone develops after 3 months. After one year, the histological appearance is somewhat better, with no bone present but still a dense connective tissue. We showed that a mixture of 1% hyaluronan and pieces of Gelfoam® considerably inhibited the formation of unwanted connective tissue that occurred when Gelfoam® was used alone [12]. In histological sections after 3 months, the filling material consisted of a loose connective tissue with few cells and neither gland-like structures nor bone in the stroma. The effect of HYA in inhibiting the development of fibrosis in the gelatine sponge may partly be attributed to its capacity to interact with different elements involved in the inflammatory reaction [3].

One drawback with the commercially available preparations of exogenous HYA is that most of them have too low a viscosity to be retained in the middle ear long enough after a surgical procedure. However, by chemical engineering it is now possible to modify the physicochemical properties of HYA without affecting its properties with respect to biocompatibility and toxicity [13]. In a recent study, we investigated one such product – a 100% ethyl-esterified HYA sponge material that was introduced into the middle ear of 10 rats [14]. After one year all middle ears contained a newly formed connective tissue with a whitish appearance. Histological examination revealed a dense, collagen-rich connective tissue with abundant fibroblasts and various amounts of newly formed bone. The material was non-ototoxic, but the biodegradable features were unacceptable.

The ideal scaffold material for reconstructive surgery of the middle ear has not yet been found. However, chemically modified hyaluronan can be designed according to the needs into, e.g. gels, sponges, films, tubings or threads. In the middle ear, the material should be biocompatible and able to reduce fibrosis and adhesions – it must also be non-toxic and degradable after a limited period of time.

Possible use of exogenous HYA in otosurgery

The use of HYA in otosurgery could possibly be of benefit in tympanic membrane repair, in supporting ossicular chain reconstructions and tympanic membrane grafts. Furthermore, HYA may counteract the formation of connective tissue adhesions in the middle ear and facilitate the re-epithelialization of the middle ear mucosa. When formed into artificial implants, it may be a biocompatible alternative to other materials used in reconstructive middle ear surgery.

The studies reviewed herein were supported by grants from the Swedish Medical Research Council (03X-4; 17X-06578,), the University of Umeå, the Swedish Society of Medicine and the Scientific Council of Dalarna. We thank Cathrine Johansson and Ann-Louise Grehn for excellent technical assistance with the histochemical work. The biotinylated-HABP probe was a kind gift from our friend and collaborator Dr. Anders Tengblad, who died in April 1990.

References

1. Paparella, M.M., Sipilä, P.T., Juhn, S.K. and Jung, T.T.K. (1985) Laryngoscope **95**, 414–420
2. Laurent, C. (1988) in Hyaluronan in the Middle Ear. An Experimental Background to the Evaluation of its Possible Benefits in Otosurgery, New series No. 211, pp. 1–45, Umeå University Medical Dissertations
3. Laurent, C. (1998) Acta Otolaryngol. (Stockh.), in the press
4. Laurent, C., Hellström, S., Tengblad, A. and Lilja, K. (1989) Ann. Otol. Laryngol. **98**, 736–740
5. Hellström, S., Laurent, C. and Yong-Joo, Y. (1994) ORL **56**, 253–256
6. Laurent, C., Johnson-Wells, G., Hellström, S., Engström-Laurent, A. and Wells, A.F. (1991) Cell Tissue Res. **263**, 201–205
7. Hellström, S., Laurent, C., Schmidt, S-H., Spandow, O. and Fellenius, E. (1988) in Cutaneous Development, Aging and Repair (Abatangelo, G. and Davidson, J.M., eds.), Fidia Research Series vol. 18, pp. 179–188, Liviana Press, Padova
8. Laurent, C. and Hellström, S. (1997) Int. J. Biochem. Cell Biol. **29**, 221–229
9. Laurent, C., Hellström, S. and Fellenius, E. (1988) Arch. Otolaryngol. Head Neck Surg. **114**, 1435–1441

10. Longaker, M.T., Chiu, E.S., Adzick, N.S., Stern, M., Harrison, M.R. and Stern, R. (1991) Ann. Surg. **213**, 292–296
11. Spandow, O. and Hellström, S. (1993) Ann. Otol. Laryngol. **102**, 467–472
12. Laurent, C., Hellström, S. and Stenfors, L-E. (1986) Am. J. Otolaryngol. **7**, 181–186
13. Hellström, S., Laurent, C., Söderberg, O. and Spandow, O. (1995) in Novel Biomaterials Based On Hyaluronic Acid and its Derivatives (Williams, D.F., ed.), pp. 38–42, Fifia, Abano Terne, Italy
14. Laurent, C., Söderberg, O. and Hellström, S. (1996) in Transplants and Implants in Otology III (Portmann, M., ed.), pp. 49–52, Kugler Publications, Amsterdam, New York

30

Hyaluronan in drug delivery

Stefan Gustafson

Department of Medical and Physiological Chemistry, University of Uppsala, PO Box 575, S-751 23 Uppsala, Sweden

Introduction

The search for more effective drug treatments with less side effects has led to an interest in macromolecules as drug carriers. The polysaccharide hyaluronan (HA) has many features that make it well-suited for such a purpose. It is a prominent component of soft connective tissues of all higher animals [1], and its simple chemical structure of repeating disaccharide units offers easy chemical modifications, including drug attachment. HA has been used for decades in cosmetics, viscosurgery and viscosupplementation without immunological reactions or any other side effect due to the polysaccharide [2]. In this chapter I will discuss the plausibility of using HA as a drug carrier, review recent published work where HA has been used as a vehicle to deliver pharmaceutically active substances, and also describe a novel way to steer systemically administered HA away from sites of normal uptake without affecting its targeting to pathological sites such as tumours and injured vessels.

Site-directed drug delivery

Most drugs have very little inherent targeting to the pathological site and in many cases they are rapidly metabolized or excreted from the body. Given locally, most drugs rapidly diffuse away from the site in need of treatment so that to have the desired effect on the target tissue, repeated dosing, impractical for the patient and with possible unwanted systemic effects, is required. It is therefore clear that a more specific, selective and prolonged delivery is beneficial in many therapeutic situations. Such a controlled delivery would not only increase the efficiency of the treatment but also reduce unwanted side effects.

Many attempts to alter the pharmacokinetics of drugs have been made and include antibody or liposome drug carriers, sustained release implants, minipumps, transdermal delivery devices, etc. (for review see [3]). Synthetic polymers and macromolecules have been extensively used and studied in drug delivery for many years and their use has expanded to include drugs in cancer chemotherapy, infection, respiratory disorders, diseases of the circulatory system and more. Their use is predominantly based on an increased residential time of the macromolecular–drug complex, resulting in a slow release of the drug. The complexes can be implanted and injected locally, and a controlled release achieved by, for example, polymer degradation or erosion. Natural polymers for use in drug delivery is an active area of research, mainly owing to the better biocompatability offered by these compounds. Proteins such as albumin and collagen, and carbohydrates such as starch, chitin, cellulose and dextran, are some examples. However, immunological responses and biodegradation that vary depending on administration site, amount, molecular weight etc. limit the use of most of these carriers in drug delivery [3,4]. The use of HA or HA derivatives as drug carriers is of relatively new vintage despite the fact that this polysaccharide of naturally high molecular weight is almost ideally suited for the purpose. Apart from being biodegradable and non-immunogenic it is also non-toxic, even at very high concentrations, and lends itself to many ways of chemical modification for the preparation of gels, films, drug complexes, etc. The identification of HA binding sites and receptors has also opened up the perspective of drug targeting via systemic administration of drugs carried by HA to sites of HA binding. As the nature and specificity of the binding sites vary between tissues, some receptor sites can be blocked by ligands other than HA, allowing the HA–drug complex to bypass those sites if desired.

HA

HA is a polysaccharide consisting of repeating units of glucuronic acid and *N*-acetylglucosamine. The properties and functions of HA have recently been reviewed by Laurent and Fraser [1]. In nature, HA is predominantly found as a high-molecular-weight compound at high concentrations in connective tissues such as skin and cartilage, in the vitreus body of the eye and in synovial fluid. Very high concentrations of the polysaccharide are found in specialized structures such as the umbilical cord and rooster comb. In tissues, most of the HA is associated with cells or binding proteins, but some exists as 'free' polysaccharide in the interstitial fluid. Some HA enters the blood via the lymph, after 80–90% has been removed in lymph nodes [5], and is rapidly taken up by liver endothelial cells (LEC) via high affinity receptors [6]. Blood levels are thus normally very low, in humans only 10–50 μg/l, but increased in inflammatory conditions, liver cirrhosis and various malignancies [7]. HA is appreciated in ophthalmic surgery and for treatment of joint diseases because of its beneficial viscoelastic properties at high concentrations [2].

HA as a drug carrier

Drobnik has published excellent reviews of the advantages of biodegradable soluble macromolecules as drug carriers [8] and the unique qualities of HA in this respect [9]. It is pointed out that the safest way to clear a carrier polymer is via biological degradation and that use of macromolecular carriers not only brings prolongation of drug action, but also that targeting, based on non-specific factors or via specific ligand–receptor interactions, can be achieved [8,9]. As HA is non-immunogenic, effectively degraded by many cells [10], exists naturally as high-molecular-weight and viscous solutions, is chemically suitable for drug attachment and targets to specific receptors, primarily LEC and lymph nodes [5,6], it qualifies as a prospective carrier of drugs [9].

The use of HA in drug delivery has until recently primarily been based on the polysaccharide playing a passive role as an immunologically 'invisible' carrier with beneficial viscoelastic properties, without making deliberate use of its interactions with cellular HA binding sites. The use in skin care products is based on a decline in skin HA with age and the formation of viscoelastic, smooth and lubricating HA films when the polysaccharide is applied to the surface of the skin [11]. In such compositions, a drug can be provided with a retard effect and better bioavailability due to the polysaccharide's viscoelastic properties. The HA can form a salt with one or more pharmacologically active substances, in some cases it can be esterified on the carboxylic groups, but it can also be the free acid or cross-linked HA [12–16].

Low-molecular-weight substances dissolved in an HA solution behave as if in pure solvent if there are no interactions between the molecules. The electrostatic interaction between gentamicin and HA results in slow release of the drug from cross-linked gels of HA (hylans), but high salt concentration reverts the release to control levels [17]. The HA network, even in cross-linked gels, can only retard diffusion of very large molecules out of the gel. However, the network will of course be tighter when the HA concentration is increased, e.g. by evaporation of water when HA gels are allowed to dry on the skin, and this could then cause entrapment of smaller molecules as well.

The chemical structure of HA, with carboxy, hydroxy and acetamido groups available, gives several ways by which drugs can be attached to the polymer to ensure the delivery to sites of HA uptake, or to prolong the release from HA until the polysaccharide is degraded enzymically or otherwise. HA-coating of, for example, artificial implants with included drugs, can also provide a biocompatible drug delivery system with slow release characteristics [17].

In addition to the HA–drug salts [15], cross-linked gels of HA with covalently bound drugs [13], conjugation of amino-group-containing ligands after activation of hydroxy groups by cyanogen bromide [18], binding via aldehydes produced by oxidation of hydroxy groups [19] or to free amino groups after deacetylation of the acetamido groups by, for example, hydrazinolysis [20], are some examples of direct coupling of active substances to the polysaccharide. HA can also be derivatized with a spacer carrying, for example, a terminal hydrazido group used for cross-linking or coupling of drugs [21].

An interesting finding related to drug delivery, but not much studied in recent years, is that partially depolymerized (PD) HA facilitates the spreading and absorption of injected drugs in living tissue [22]. An example is the intradermal injection of Trypan Blue in rabbits, where the HA, depolymerized for 15–20 min by hyaluronidase, gave increased spreading (4 times saline control). Shorter as well as longer incubations of HA with hyaluronidase gave less spreading, as did injection of hyaluronidase itself (although about double that of saline control). The PDHA spreading effect was also tested in humans injected with a local anaesthetic where the duration was halved by PDHA. The patent describing this effect [22] also claims the use of PDHA as a hypolipidemic or lipemia clearing agent if given intravenously, subcutaneously or orally. A proposed mechanism is that of a transport agent and/or protective agent, aiding the dispersion of materials in the tissues.

HA in ocular drug delivery

Ocular absorption of drugs is low owing to their rapid clearance from the corneal surface by the tear fluid. Viscous vehicles give increased corneal contact time and better therapeutic results.

Several recent studies, directly related to drug delivery and using HA, are combinations mainly for ocular use and include, in addition to drugs used in ophthalmology, antibiotics, proteins and peptides for local as well as systemic therapy. The use of HA in ocular drug delivery is probably an offspring of the widespread use of HA in ophthalmic surgery. It is an excellent, biocompatible substitute for tear fluid in dry eyes [23].

In a rabbit endophthalmitis model with bacteria inoculated into the anterior vitreus, HA in combination with gentamicin was administered in the anterior chamber and lowered the incidence of endophthalmitis from 10 of 15 eyes for gentamicin alone to 4 of 15 eyes in combination with HA. The half-life of aqueous gentamicin was increased from 0.9 to 2.2 h using HA [24].

A study on human volunteers showed that after topical administration, the gentamicin concentration in the tear fluid was higher when given in combination with HA instead of saline, indicating increased availability of gentamicin at the ocular surface [25].

Gentamicin interacts with HA and this retards and sustains the release of the antibiotic from the gel. The interactions between the cationic gentamicin and anionic HA are predominantly electrostatic as high salt concentration reverts the release to control levels [17].

Tobramycin in 1% HA of molecular weight 500 000, gave highly statistically significant shorter healing time (re-epithelialization) in a double blind clinical study of human bacterial corneal ulcers when compared with tobramycin in saline control (3.5 versus 5.9 days). The authors discuss possible mechanisms such as prolonged antibiotic/corneal contact time but do not rule out direct effects of HA on the repair process, possibly by epithelial lubrication [26].

Pilocarpine in combination with HA, given topically to the eye in glaucoma, gives significantly increased corneal residence time and drug response [27]. Using hylan gel, Larsen and Balazs showed, *in vitro*, that incorporation of pilocarpine into the cross-linked HA gel gave a threefold increase in the time required for 50% of the drug to be released [17].

Betaxolol is a β-adrenoreceptor antagonist used in the treatment of glaucoma and it interacts with hylan gels. Increasing the net negative charge of the polymer by introduction of sulphate groups gives prolonged release rates for this drug, suggesting that electrostatic forces also play a role in this interaction. [17].

Cyclopentolate base in combination with HA gives better mydriatic response in rabbits when compared with the standard hydrochloride salt. The ionic complex of this drug and another polyanionic polymer, polygalacturonic acid, also gives better response than a control preparation [28]

A mucoadhesive property of HA, not related to viscosity and leading to increased bioavailability of pilocarpine in rabbit eyes, has been described by Saettone and co-workers [29]. This property could explain why low-viscosity HA in combination with pilocarpine as a salt gives a better miotic effect on rabbits than pilocarpine alone [30]. The effect of HA disappears after removal of precorneal mucin by acetylcysteine [30].

Anti-inflammatory agents in combination with HA

HA seems to accumulate at sites of inflammation [1], indicating increased synthesis and expression of HA binding sites at these pathological sites. Systemic treatment with HA of animals with experimental inflammatory conditions [31] and of animals subject to balloon angioplasty [32] has been shown to reduce inflammation and occurrence of restenosis, respectively, further indicating that HA targets these pathological sites, thereby possibly inhibiting leucocyte recruitment. Intravenous HA has been shown to reduce leucocyte adherence to endothelial cells both *in vitro*, using human umbilical vein endothelial cells, and *in vivo*, as studied by intra-vital microscopy of hamster post-capillary venules [33,34]. Labelled HA, given intravenously, also accumulates in balloon-catheter-damaged rat carotid arteries [35].

Treatment of arthritis with HA has not only been by intra-articular viscosupplementation therapy that started around 1970 (for review see [2] and [17]), but also by intravenous, intramuscular, subcutaneous or topical (with DMSO to give penetration) application of HA of high molecular weight. Such remote administration gave beneficial results on horses with experimental arthritis, and also on joint and muscle pain in man [36]. HA has also been used to reduce post-surgical adhesions (for review see [37]). These seem to be partially owing to an inflammatory reaction as corticosteroids and non-steroid anti-inflammatories (NSAIDs) also have positive effects on this disorder [37]. It is therefore logical to try combinations of anti-inflammatory drugs and HA against these problems.

The NSAID tolmetin is one of the most widely studied drugs for the prevention of surgical adhesions [37]. In post-surgical peritoneal lavage fluids from rabbits, the protease activity was suppressed after administration of HA and tolmetin in comparison with control [38].

Diclofenac is an NSAID that has been shown to reduce tumour growth and vascularization in rat hepatoma. When labelled diclofenac was given intravenously to mice carrying subcutaneous implanted Colon-26 tumours, the co-administration of HA caused an increase in the amount of diclofenac found in the tumour, while pretreatment with HA inhibited the accumulation of the NSAID in the tumours [39]. In a study of chronic granulomatous inflammation, induced by injection of Freunds' complete adjuvants and croton oil into subcutaneous air pouches in mice, daily topical administration of diclofenac in combination with 2.5% HA significantly decreased the vascularity from day 7 to 21 in relation to all controls, including diclofenac in aqueous cream [40].

The glucocorticoid dexamethasone together with HA gives a synergistic therapeutic effect in the intra-articular treatment of osteoarthritis [41].

The combination of HA and the immunosuppressant cyclosporin was tested in two immune-driven inflammatory conditions, avridine-induced arthritis and cotton pellet granuloma formation, in rats. Whereas the suboptimal dose of cyclosporin and HA given alone had no significant effect on the conditions, the combination caused statistically significant reductions of both granuloma formation and arthritis score [42].

HA in the delivery of proteins and peptides

By mixing 5% HA with 5% polyvinyl alcohol in different proportions, and then adding growth hormone (GH) and repeatedly freeze–thawing the mixtures, hydrogels can be produced [43]. When the amount of GH release was monitored by ELISA, it was found that initially all gels released the same amount of GH, but the gel containing HA/PVA (1:9) ceased to release hormone whereas gels with HA/PVA ratios of 2:8 and 3:7 continued to release GH in therapeutic doses for up to 8 days [43].

Using pancreatectomized dogs the effects on blood glucose and presence of immunoreactive insulin was studied after administration of eye-drops containing insulin with or without HA [44]. In the conscious state, only very low insulin levels were found in the blood and no significant drop in blood glucose was seen when HA was omitted. However, if the insulin was in viscous 0.36% HA, immunoreactive insulin in blood increased 25-fold and the blood glucose fell significantly [44].

Thrombin administered intravascularly in hylan gel can be used for embolization in the treatment of, for example, tumours [17].

Interferon in cross-linked HA shows reduced *in vitro* release, probably caused by HA–protein interactions [17].

So-called bioadhesive liposomes were made by linking HA via carboxyl residues to phospholipid amine residues at the liposomal surface using carbodiimide. The resulting HA-liposome was found to bind in a saturable manner to a human tumour cell line (A431), whereas unmodified liposomes showed no such binding [45]. This indicates binding of HA to a cell surface binding site on the tumour cells. The HA modification did not significantly change the release kinetics of epidermal growth factor, which was continuously released from the liposomes over 24 h [45].

Systemic delivery with altered biodistribution

The diaqua form of the cytotoxic agent cyclohexanediaminoplatinum was reacted with the sodium salt of acid polysaccharides, including HA, giving complexes with slow release kinetics, probably owing to interactions by charge effects. When the complexes were tested for anti-tumour activity in mice with intraperitoneal tumours, it was found that the low-molecular-weight (10 000) HA–Pt complex was the most effective, with all mice surviving. Almost as effective were complexes of chondroitin sulphate (CS) and polysulphated CS, while the heparin group showed very few survivors [46]. If cisplatin or cyclohexanediaminoplatinum were given directly as a combination, the CS increased the anti-tumour effect. CS had no effect on its own. The authors state that the complexes, given intraperitoneally, show selectivity in organ distribution and accumulates more in the liver than in other organs. This is probably owing to binding of the complexes to the endothelial cells of the liver, the natural target, beside sinusoidal cells in lymph nodes, for drugs carried by HA. There is an efficient uptake and degradation of HA by this cell type [1,9]. HA has also been reported to enhance fluorouracil uptake in liver-implanted rat mammary carcinoma and subcutaneous-implanted Fischer bladder carcinoma, possibly via targeting to HA receptors on the tumour cells [47]. A polylysine–graft–HA co-polymer as a gene carrier to liver endothelial cells has also been recently described [48].

Targeting via cell surface HA-binding sites and receptors

Beside HA-binding proteins of the extracellular matrix [1,49], several cell surface HA-binding proteins have been described and include CD44, RHAMM (receptor for HA-mediated motility) [49,50] and ICAM-1(intercellular adhesion molecule-1) [51]. Scavenger receptors on reticuloendothelial cells have also been suggested to clear the body fluids from HA by rapid receptor-mediated endocytosis [6,52]. Targeting of exogenous HA/drug complexes to cells carrying one or more of these proteins should be possible but is restricted for several reasons. As all of the described cell surface binding proteins bind other ligands beside HA, and these as well as HA are endogenously synthesized and turned over in considerable quantities, they may be occupied by endogenous ligands when the exogenous material is introduced, resulting in no or minute binding of the HA–drug complex. This makes it difficult, if not impossible, to calculate which tissues or cell-type will be targeted by HA by virtue of the mere presence or increased expression of HA binding sites. In addition, several isoforms of CD44, ICAM-1 and RHAMM have been described and these may vary in their affinity for HA, and it is at present not possible to easily distinguish between all these variants in biological samples. It is also likely that differences in glycosylation, contacts with the cytoskeleton, lipid environment and other neighbouring molecules in the cell membrane will affect the affinity. A reflection of the complexity is the virtual absence of HA binding to cells in the blood expressing the form of CD44 earlier described to bind HA and be the principal cell surface receptor for hyaluronan [50,53]. This form of CD44 is also absent from the most efficient cells in receptor-mediated endocytosis of HA, the LFC [50,54].

When HA is injected intravenously it is rapidly taken up in tissues rich in reticuloendothelial cells by receptor-mediated endocytosis, primarily by the liver in LEC, but also by spleen, kidney and bone marrow ([1], Figure 1). The HA LEC-receptors recirculate, do not seem to be down-regulated after ligand binding and recognize also other negatively charged polysaccharides [6,55], indicating strong similarity with scavenger receptors, known for rapid turnover, binding of

Figure 1

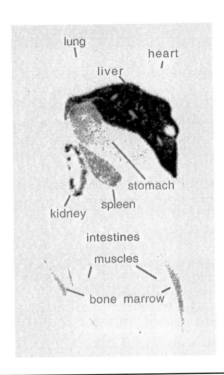

Whole body autoradiography of a 20 μm cryosection of a rat that received a tracer dose of intravenous ^{125}I-HA 20 h prior to being killed

The autoradiograph was produced using the Bio-Rad GS-525 phospho-imager and the Molecular analyst® software on a Macintosh 7200/90 computer. Labelling of HA was performed as previously described [63].

negatively charged polysaccharides, active primarily on LEC and not down-regulated to any great extent after ligand binding [55,56].

Similar to the LEC receptors, the receptors in lymph nodes are on sinusoidal lining cells, have the ability to endocytose and degrade HA in high quantities and can also recognize CS [5]. HA and HA–drug complexes introduced in peripheral tissues would probably be quickly transported to the lymph nodes where much of the material would be degraded and the drug released [9].

The molecular nature of the endocytosing HA receptors is, despite the strong circumstantial evidence in favour of a scavenger receptor [55], not completely clear. The receptors are obviously not related to the CD44 family of lymphocyte homing receptors, as these molecules are found on a variety of cell-types but are absent from LEC [50,54,57]. RHAMM does not seem to be localized to LEC under normal circumstances, and differences in affinity, specificity and other characteristics also suggest that the LEC receptors are different from CD44 and RHAMM [6,50,55,57]. Although ICAM-1 is found on normal liver endothelium in the sinusoids and on the endothelium of lymph nodes, spleen, some capillaries of the kidney [58] and cornea [59], and its localization corresponds well to tissues where HA binding and uptake have been found, e.g.

LEC, macrophages, lymph nodes, corneal endothelial cells and some tumours (reviewed in [1,6,55]), the efficient endocytosis in LEC and the very low uptake in many other endothelial cells expressing ICAM-1, such as corneal endothelial cells [60,61], suggests that normal ICAM-1 is not directly involved in endocytosis.

Despite all efforts in trying to find and characterize the molecules involved in HA binding, only *in vivo* studies of HA–drug turnover and biodistribution in pathological conditions will give clear answers on the suitability of HA as a drug carrier in any given situation.

Influences of CS and other negatively charged polysaccharides on HA metabolism

To determine which HA binding sites are accessible to exogenously introduced HA in physiological and pathological conditions, we have performed turnover and biodistribution studies of labelled HA in animal, mainly rat, models [35,62–67], including those with tumours [64,65] and vascular injury [35]. To reduce liver uptake of intravenous HA we tried to block the liver receptors with other negatively charged polysaccharides, hoping that the receptors at the pathological sites would be more specific for HA and not affected by blocking of the liver. We have then seen that CS and dextran sulphate can inhibit the liver uptake *in vivo*, while heparin is ineffective [67]. The inhibition of liver uptake caused the HA to remain in the circulation and be continuously cleaved by a saturable mechanism to lower-molecular-weight species that, when cleaved to a mean of less than about 40 000, escapes into the tissues and is filtered by the kidneys out into the urine [67].

That the molecular weight of HA in plasma is lower than that of the HA entering the general circulation was known earlier [68] and also that a part of the reduction was due to a process in the circulation [69]. However, as the endogenous HA found in the blood has a weight-average molecular weight of about 150 000 [68] and not much HA is found in urine even when serum levels are moderately increased [70], the intravascular degradation was believed to be rather limited and the major pathway of degradation of circulating HA at all times to be via complete intracellular degradation in the liver after receptor-mediated endocytosis in LEC. Our recent findings suggest that a major part of circulating HA can be excreted in the urine when liver uptake is compromised or saturated by high doses given therapeutically [67]. In studies where the liver is blocked surgically the HA levels in serum rapidly increase. A similar increase, 5–10-fold in 45 min, is seen by blocking the kidneys surgically [71]. The latter increase has not been possible to explain by influences on previously known mechanisms in HA metabolism, where the kidneys are known to normally be responsible for only a small amount of uptake and degradation [1]. The phenomenon was tentatively explained by a previously unknown degradation process in the kidney, important in removal of HA from the circulation [71]. However, the influx of CS into the blood is fairly substantial and this polysaccharide binds to the LEC HA receptors with high affinity [1]. Normally, CS is predominantly excreted in the urine, but this metabolic route is stopped when the kidneys are ligated. Therefore the plasma

Figure 2

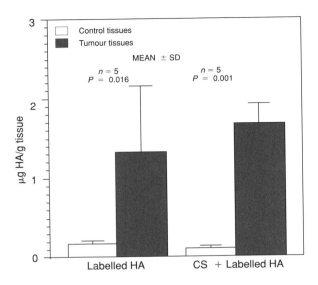

Effect of CS on the accumulation of ^{125}I-HA in rat colon carcinomas

The indicated number of rats carried an implanted rat colon carcinoma in a hind leg; 18–20 h before killing the animals received intravenous injections of 1 mg of ^{125}I-HA or 5 mg of CS and 30 s later 1 mg of ^{125}I-HA. After killing, tumour tissue and control muscle tissue from the healthy leg were weighed and checked for radioactivity. Labelling of HA and measurement of radioactivity were as described earlier [63,65].

level of CS increases and the HA receptors become saturated. A plausible explanation, based on our data [67], is that the increase of plasma HA is secondary to an increase in the plasma level of CS. This could also be the explanation behind the increased blood levels of HA seen in uraemia [1,7]. Interference with HA metabolism in peripheral tissues could also occur and may be an explanation for the increased levels and molecular weight of HA seen in the synovial fluid of osteoarthritic patients treated with oral CS [72].

From the above discussion it should be clear that not only influences on HA turnover and biodistribution by physiological and/or pathological processes, but also influences by other negatively charged polysaccharides, endogenous or exogenous, must be taken into account when using HA in a therapeutic situation or when studying and discussing fluctuations in HA levels in biological samples.

Targeting of intravenous HA to pathological tissues

From our studies on turnover and biodistribution of systemic HA in tumour-bearing animals and animals with experimental vascular injury, it is clear that the normal sites of uptake dominate also in disease states, but that a specific accumulation of HA can be seen in pathological tissues [35,64,65]. This targeting could be used to deliver drugs to these sites using HA. However, the scavenging of HA by

Figure 3

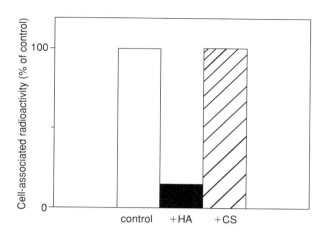

Effect of unlabelled HA and CS on the association of ^{125}I-HA to rat colon cancer cells in culture

All cells received ^{125}I-HA at 1 µg/ml and were incubated at 37 °C for 1 h. Control cells received no unlabelled polysaccharide, +HA and +CS indicate cells receiving the indicated unlabelled polysaccharides at 12.5 µg/ml at the same time as labelled HA. Results are mean values of three determinations. HA and CS, labelling of HA and measurement of radioactivity were as described earlier [63,67]

the liver would target most of the drug to this organ if a HA–drug combination was given intravenously. By using preinjections of CS, the liver uptake of HA could be reduced in experiments on rats with damaged carotid arteries [35]. The reduction of liver uptake did not reduce the specific accumulation of HA at the damaged carotid relative to uninjured vessels, indicating that the accumulation is specific for HA. Likewise, the targeting of HA to subcutaneously implanted rat tumours, derived from a chemically induced rat colorectal carcinoma, was unaffected by preinjections of high doses of CS given to block the liver ([73], Figure 2). The mechanism behind the targeting is presently under investigation. In the arterial damage study, non-specific passive accumulation effects are unlikely to play a role because labelled CS was found not to accumulate at the damaged site [35]. The tumour targeting could in part be owing to a passive mechanism due to the leaky nature of tumour vasculature [74]. However, the very high tumour/non-tumour ratio and the presence of saturable HA binding sites that do not bind CS on the tumour cells in culture (Figure 3) speaks in favour of a more specific receptor-like mechanism.

Conclusions

HA has proven to be a good slow release vehicle for several drugs interacting with the polymer. The biocompatibility of the carrier is excellent and, as a component of the normal body fluids, immunological reactions to the polysaccharide are virtually out of the question. HA can easily be cross-linked into gels of various shapes and also chemically modified for drug attachment. Providing care is taken so that drug attachment does not interfere with the biological properties of the polymer, targeting of the complex to accessible unoccupied HA binding sites can be used. Differences in affinity and specificity of these binding sites can also be exploited to reduce uptake of drugs by, for example, the liver, causing further improved efficiency and reduction of side effects. I believe that as we learn more about cellular HA binding sites and the fate of exogenous HA in pathological conditions, HA and HA derivatives will be used in an increasing number of clinical situations.

I wish to thank Torvard C. Laurent for valuable discussions, and Tomas Björkman and Carolina Samuelsson for excellent technical assistance. This work was supported by Konung Gustaf V:s 80-årsfond, Hyal Pharmaceutical Corporation and Pharmacia & Upjohn.

References

1. Laurent, T.C. and Fraser, J.R.E. (1992) FASEB J. **6**, 2397–2404
2. Balazs, E.A. and Denlinger J.L. (1989) in The Biology of Hyaluronan (Evered, D. and Whelan, J., eds.), Ciba Foundation Symposium No. 143, pp. 265–275 Wiley, Chichester
3. Poznansky, M.J. and Juliano, R.L. (1984) Pharmacol. Rev. **36**, 277–336
4. Domb, A. (1994) in Polymeric Site-specific Pharmacotherapy (Domb, A.J., ed.), pp. 1–26 Wiley, Chichester
5. Fraser, J.R.E. and Laurent, T.C. (1989) in The Biology of Hyaluronan (Evered, D. and Whelan, J., eds.), Ciba Foundation Symposium No. 143, pp. 41–52, Wiley, Chichester
6. Smedsröd, B., Pertoft, H., Gustafson, S. and Laurent, T.C. (1990) Biochem. J. **266**, 313–327
7. Laurent T.C., Laurent, U.B.G. and Fraser, J.R.E. (1996) Ann. Med. **28**, 241–253
8. Drobnik, J. (1989) Adv. Drug Deliv. Rev. **3**, 229–245
9. Drobnik, J. (1991) Adv. Drug Deliv. Rev. **7**, 295–308
10. Smedsröd, B. (1991) Adv. Drug Deliv. Rev. **7**, 265–278
11. Balazs, E.A. (1981) U.S. Pat. No. 4303676
12. Della Valle, F. and Romeo, A. (1982) Eur. Pat. No. 216453
13. Balazs, E.A. and Leschiner, A. (1986) U.S. Pat. No. 4582865
14. Balazs, E.A. and Leschiner, A. (1987) U.S. Pat. No. 4636524
15. Della Valle, F., Romeo A and Lorenzi, S. (1988) U.S. Pat. No. 4736024
16. Della Valle, F. and Romeo, A. (1989) U.S. Pat. No. 4851521
17. Larsen, N.E. and Balazs, E.A. (1991) Adv. Drug Deliv. Rev. **7**, 279–293
18. Glabe, C.G., Harty, P.K. and Rosen, S.D. (1983) Anal. Biochem. **130**, 287–294
19. Raja, R.H., Leboeuf, R.D., Stone, G.W. and Weigel, P.H. (1984) Anal. Biochem. **139**, 168–177
20. DahL, B., Laurent, T.C. and Smedsröd, B. (1988) Anal. Biochem. **175**, 397–407
21. Pouyani, T. and Prestwich, G.D. (1994) Bioconj. Chem **5**, 339–347
22. Seifter, J. and Baeder, D.H. (1957) U.S. Pat. No. 769287
23. Balazs, E.A. (1983) in Healon®:A guide to its use in ophthalmic surgery, (Miller, D. and Stegmann, R., eds.), p. 5–23, Wiley, New York
24. Moreira, C.A., Moreira, A.T., Armstrong, D.K., Jelliffe, R.W., Woodford, C.C., Liggett, P.E. and Trousdale, M.D. (1991) Acta Ophthalmol. **69**, 50–56
25. Bernatchez, S.F., Tabatabay, C. and Gurny, R. (1993) Graefe's Arch. Clin. Exp. Ophthalmol. **231**, 157–171
26. Gandolfi, S.A., Massari, A. and Orsoni, J.G. (1992) Graefe's Arch. Clin. Exp. Ophthalmol. **230**, 20–23
27. Camber, O. and Edman, P. (1989) Curr. Eye Res. **8**, 563–567
28. Huupponen, R., Kaila, T., Saettone, M.F., Monti, D., Iisalo, E., Salminen, L. and Oksala, O. (1992) J. Ocul. Pharmacol. **8**, 59–67
29. Saettone, M.F., Chetoni, P., Torracca, M.T., Burgalassi, S. and Giannaccini, B. (1989) Int. J. Pharm. **51**, 203–212
30. Saettone, M.F., Monti, D , Torracca, M.T and Chetoni, P. (1994) J. Ocul. Pharmacol. **10**, 83–92
31. Ialenti, A. and Di Rosa, M. (1994) Agents Actions **43**, 44–47
32. Ferns, G.A.A., Konneh, M., Rutherford, C., Woolaghan, E. and Änggård, E.E. (1995) Atherosclerosis **114**, 157–164

33. Seed, M.P., Freemantle, C., Brown, J., Alam, C.A.S. and Willoughby, D.A. (1995) in Third International Workshop on Hyaluronan in Drug Delivery (Willoughby, D.A., ed.), Round Table Series, vol. 40, pp. 82–87, Royal Society of Medicine Press, London

34. Seed, M.P., Freemantle, C., Gustafson, S., Brown, J., Alam, C., Peretti, M., Newbold, P., Dwivedi, A., Carrier, M. and Willoughby, D.A. (1995) Inflammation Res. **44**(S3), S245

35. Gustafson, S., Sangster, K., Björkman, T. and Turley, E.A. (1996) in 4th International Workshop on Hyaluronan in Drug Delivery (Willoughby, D.A.,ed.), Round Table Series vol. 45, pp. 27–33, Royal Society of Medicine Press

36. Schultz, R.H., Wollen, T.H., Greene, N.D., Brown, K.K. and Mozier, J.O. (1989) U.S. Pat. No. 4 808 576

37. Wiseman, D. (1994) in Polymeric site-specific pharmacotherapy (Domb, A.J., ed.), pp. 369–421, Wiley, Chichester

38. Abe, H., Campeau, J.D., Rodgers, K.E., Ellefson, D.D., Girgis, W. and diZerga, G.S. (1993) J. Surg. Res. **55**, 451–456

39. Seed, M.P., Alam, C.A.S., Brown, J., Freemantle, C. and Willoughby, D.A. (1995) in Third International Workshop on Hyaluronan in Drug Delivery (Willoughby D.A., ed.), Round Table Series, vol. 40, pp. 74–81, Royal Society of Medicine Press, London

40. Alam, C.A.S., Seed, M.P. and Willoughby, D.A. (1995) J. Pharm. Pharmacol. **47**, 407–411

41. Grecomoro, G., Piccione, F. and Letiziz, G. (1992) Curr. Med. Res. Opin. **13**, 49–55

42. Moore, A.R., Gowland, G., Willis, D. and Willoughby, D.A. (1995) in Third International Workshop on Hyaluronan in Drug Delivery (Willoughby, D.A., ed.), Round Table Series, vol. 40, pp. 106–109, Royal Society of Medicine Press, London

43. Gascone, M.G., Sim, B. and Downes, S. (1995) Biomaterials **16**, 569–574

44. Nomura, M., Kubota, M.A., Kawamori, R., Yamasaki, Y., Kamada, T. and Abe, H. (1994) J. Pharm. Pharmacol. **46**, 768–770

45. Yerushalmi, N., Arad, A. and Margalit, R. (1994) Arch. Biochem. Biophys. **313**, 267–273

46. Maeda, M., Takasuka, N., Suga, T., Uehara, N. and Hoshi, A. (1993) Anti-Cancer Drugs **4**, 167–171

47. Klein, E.S., He, W., Shmizu, S., Asculai, S. and Falk, R.E. (1993) in First International Workshop on Hyaluronan in Drug Delivery (Willoughby, D.A., ed.), Round Table Series, vol. 33, pp. 11–15, Royal Society of Medicine Press, London

48. Maruyama, A., Asayama, S., Nogawa, M., Akaike, T. and Takei, Y.(1996) Abstract to the 5th World Biomaterials Congress, Toronto, Canada

49. Toole, B.P. (1990) Curr. Biol. **2**, 839–844

50. Knudson, C.B. and Knudson, W. (1993) FASEB J. **7** 1233–1241

51. McCourt, P.A.G., Ek, B., Forsberg, N. and Gustafson, S. (1994) J. Biol. Chem. **269**, 30081–30084

52. Eskild, W., Smedsröd, B. and Berg, T. (1986) Int. J. Biochem. **18**, 647–651

53. Aruffo, A., Stamenkovic, I., Melnick, M., Underhill, C.B. and Seed, B. (1990) Cell **61**, 1303–1313.

54. Underhill, C.B.(1989) in The Biology of Hyaluronan (Evered, D. and Whelan, J., eds.), Ciba Foundation Symposium No. 143, pp. 87–99, Wiley, Chichester

55. Gustafson, S. (1996) Trends Glycosci. Glycotechnol. **8**, 13–21

56. Freeman, M.W. (1994) Curr. Opin. Lipidology **5**, 143–148

57. Yannariello-Brown, J., Frost, S.J., Weigel, P.H. (1992) J. Biol. Chem. **267**, 20451–20456

58. Dustin, M.L., Rothlein, R., Bhan, A.K., Dinarello, C.A. and Springer, T.A. (1986) J. Immunol. **37**, 245–254

59. Foets, B.J., van den Oord, J.J., Volpes, R. and Missotten, L. (1992) Br. J. Ophthalmol. **76**, 205–209

60. Madsen, K., Schenholm, M., Jahnke, G., and Tengblad, A. (1989) Invest. Ophthalmol. Vis. Sci. **30**, 2132–2137

61. Forsberg, N., von Malmborg, A., Madsen, K., Rolfsen, W., and Gustafson, S. (1994) Exp. Eye Res. **59**, 689–696

62. Gustafson, S., Forsberg, N., McCourt, P., Wikström, T., Björkman, B. and Lilja, K. (1994) in First International Workshop on Hyaluronan in Drug Delivery (Willoughby, D.A., ed.), Round Table Series vol. 33, pp. 43–59, Royal Society of Medicine Press, London

63. Gustafson, S., Björkman, T. and Westlin, J-E. (1994) Glycoconjugate J. **11**, 608–613

64. Gustafson, S., Björkman, T., Forsberg, N., Lind, T., Wikström, T. and Lidholt, K. (1995) Glycoconjugate J. **12**, 350–355

65. Gustafson, S., Björkman, T., Wikström, T., McCourt, P., Forsberg, N. and Graf, W. (1995) in 3rd International Workshop on Hyaluroan in Drug Delivery (Willoughby, D.A., ed.), Round Table Series, vol. 40, pp. 1–9, Royal Society Medicine Press, London

66. Westerberg, G., Bergström, M., Gustafson, S., Lindqvist, U., Sundin, A. and Långström, B. (1995) Nucl. Med. Biol. **22**, 251–256

67. Gustafson, S. and Björkman, T. (1997) Glycoconjugate J. **14**, 561–568
68. Tengblad, A., Laurent, U.B.G., Lilja K., Cahill, R.N., Engström-Laurent, A., Fraser J.R.E., Hansson, H.E. and Laurent, T.C. (1986) Biochem. J. **236**, 521–525
69. Fraser, J.R.E (1995) in Second International Workshop on Hyaluronan in Drug Delivery (Willoughby, D.A., ed.), Round Table Series, vol. 36, pp. 8–10, Royal Society of Medicine Press, London
70. Laurent T.C., Lilja, K., Brunnberg, L., Engstöm-Laurent A., Laurent, U,B.G, Lindqvist, U., Murata, K. and Ytterberg, D. (1987) Scand. J. Clin. Lab. Invest. **47**, 793–799
71. Engström-Laurent, A. and Hellström, S. (1990) Conn. Tissue Res. **24**, 219–224
72. Conte, A., Volp, I.N., Palmieri, L., Bahous, I. and Ronca, G. (1995) Arzneimittelforschung **45**, 918–925
73. Gustafson, S. (1995) U.S. Pat. Application No. 08 568 489
74. Maeda, H. (1994) in Polymeric site-specific Pharmacotherapy (Domb, A.J., ed.), pp. 95–116, Wiley, Chichester

31

Hyaluronan as a clinical marker of pathological processes

Torvard C. Laurent

Department of Medical and Physiological Chemistry, University of Uppsala, Box 575, S-751 23 Uppsala, Sweden

Introduction

The use of *exogenous* hyaluronan in medical treatment has been dealt with repeatedly during this symposium. This field was pioneered by Endre Balazs and has grown in importance and diversity during the last 25 years. However, during the last 10 years clinical applications based on *endogenous* hyaluronan have also attracted attention owing to our new knowledge of its role in pathological processes.

After the introduction of sensitive techniques for analysis of hyaluronan in body fluids [1] as well as specific staining methods for hyaluronan in tissue sections [2] it became apparent in the middle of the 1980s that hyaluronan is often accumulated in tissues and fluids during disease. Hyaluronan has, therefore, to an increasing extent been utilized as a diagnostic marker. The present paper reviews some conditions in which clinical analyses of hyaluronan have been utilized or in which they can be predicted to be of value. Rheumatic disorders are dealt with in Chapter 32 by Anna Engström-Laurent and recent data on hyaluronan in cerebrospinal fluid are discussed by Ulla Laurent in Chapter 33.

Background

The concentration of hyaluronan (or any component) in a body compartment is a complex function of influx and removal. A variation in volume of the compartment will further increase the complexity. Figure 1 demonstrates schematically that the concentration of hyaluronan depends on the various fluxes and that steady state is only obtained when the influxes of hyaluronan and water are equal to their outfluxes. A perturbation can easily be created by an imbalance between supply and removal of the two components and this actually frequently occurs *in vivo*.

The main factors that determine influx and removal of hyaluronan in a tissue [3–6] are: (1) synthesis within the compartment, which depends both on the presence of cells able to produce the polysaccharide and various regulatory factors such as inflammatory mediators and growth factors; (2) lymphatic influx from surrounding tissues; (3) cellular uptake and degradation of the polysaccharide,

Figure 1

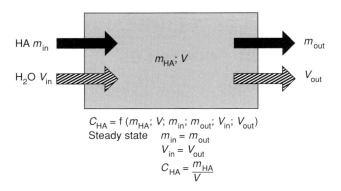

HA m_{in}

$H_2O\ V_{in}$

$m_{HA};\ V$

m_{out}

V_{out}

$$C_{HA} = f\ (m_{HA};\ V;\ m_{in};\ m_{out};\ V_{in};\ V_{out})$$
Steady state $m_{in} = m_{out}$
$$V_{in} = V_{out}$$
$$C_{HA} = \frac{m_{HA}}{V}$$

Schematic description of factors that influence the concentration of hyaluronan (C_{HA}) in a tissue compartment

The total amount of hyaluronan in the compartment at a given time point is m_{HA} and the volume of the compartment V. The fluxes of hyaluronan mass and water volume in and out of the compartment per time unit have ben labelled m_{in}, V_{in}, m_{out} and V_{out}, respectively. Steady-state concentration of hyaluronan is obtained when in- and out-fluxes of hyaluronan and water, respectively, balance each other.

which requires cells with specific endocytosing receptors; and (4) removal by diffusion or bulk flow.

The vascular system is an example of a compartment with a roughly constant volume. We have until now only observed *elevated* levels of serum hyaluronan in disease, which means that we have had either a pathologically increased influx into the circulation or an impaired removal of the polysaccharide from blood. To my knowledge no condition has been connected with the opposite situation, i.e. *lowered* serum level due to a decreased influx of hyaluronan or an increased removal rate.

An example of a compartment with a variable volume is the joint cavity. We commonly observe a lowering of the hyaluronan concentration in synovial fluid in joint disease but this is presumably due to the increased volume of the cavity and a dilution of the polysaccharide. It has been shown that in an arthritic joint there is both an increased total amount of hyaluronan and an increased removal rate [7].

Hyaluronan in blood

The vascular compartment is by far the most important from a clinical point of view owing to the ease of analysing blood. In a normal person the serum level varies considerably with age, as shown by data given in [8] (Figure 2). The concentration in middle-aged adults is 30–40 µg/l. High values are found in the newborn. In old people the level is in the order of 100 µg/l.

Daily, 30–40 mg of hyaluronan is transported by lymph to the circulation. Its normal half-life in serum is 2–5 min [3,4]. One can estimate from animal experiments that 90% is taken up by the liver in the sinusoidal endothelial

Figure 2

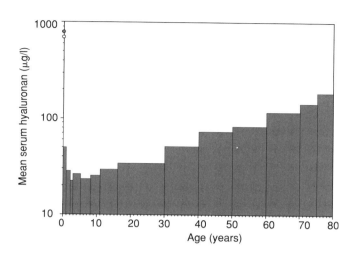

The variation of serum hyaluronan with age

Measurements were made on 585 healthy individuals as published in [8].The mean values in defined age groups have been plotted.The two individual points represent newborn (○) 0–24 h and 7 day old children (●). Note the logarithmic plot on the ordinate. Reproduced from [11] with permission.

cells and about 10% by the kidneys. There is also a small uptake by the spleen. It should be pointed out that the clearance of hyaluronan by various organs not only depends on their capacity of uptake but also on the fractional blood flow through the organ. For example, a lowered circulation in the splanchnicus region will decrease the perfusion of the liver.

However, we can also conceive of a situation when the concentration in serum is so high that the maximal capacity of the liver (350 mg/day) is surpassed. Under these circumstances an increasing part of the hyaluronan must find other escape routes and it is quite possible that recirculation into tissues occurs. When a large dose of hyaluronan is given intravenously some can be traced to bone marrow and lymph nodes.

Elevated serum levels of hyaluronan have been observed in a number of diseases [9–11].

Inborn errors

There are a few rare hereditary conditions that are known to exhibit high levels in blood and urine. To these belong the Werner syndrome and the Hutchinson–Gilford progeria syndrome, two metabolic diseases that are characterized by an apparently accelerated ageing. We have analysed one patient with Werner's syndrome and he exhibited a 10-fold increase both in serum level and urinary excretion of hyaluronan [12]. It is of interest that the serum hyaluronan is increased also in normal ageing [8]. Another disease recently discovered by Ramsden, Fraser and colleagues is congenital cutaneous hyaluronosis (see [11]). A child was born with an excessively folded skin due to an overproduction of hyaluronan in this tissue. The abnormal synthesis also revealed itself by an elevated serum level.

Malignancies

That certain tumours exhibit high serum levels of hyaluronan, causing hyperviscosity of blood, was early recognized. The first cases of human tumours with this behaviour (one neuroblastoma and one reticulum cell sarcoma) were reported by Deutsch in 1957 [13]. In subsequent papers it has mainly been nephroblastomas (Wilms' tumour in children) that have been described [11]. Concentrations of hyaluronan as high as 8.5 g per litre of serum have been observed in one patient. Such high concentrations interfere in the platelet function and cause disturbances of blood coagulation. Among other things the patients often get thrombosis in the hepatic veins. When the tumours are removed the rise in serum hyaluronan disappears. Interestingly, Longaker and co-workers [14] have found that serum and urine of patients with Wilms' tumour contains a factor that stimulates cells to synthesize hyaluronan.

Mesothelioma is another tumour that is known to cause a production of hyaluronan. There are numerous reports on hyaluronan accumulation in pleural fluid and ascites in these patients [15]. Dahl and Laurent have studied patients with mesothelioma as well as other malignancies and their data show a large individual variation in serum hyaluronan of mesothelioma patients, extending from normal level to 1000-folds higher [16]. Therefore hyaluronan analyses cannot be used for a definite diagnosis of the disease. However, from a statistical point of view the mesotheliomas differ significantly from other tumours and assaying hyaluronan has been a valuable tool for evaluating treatment and for prognostic purposes. Rising serum level of hyaluronan is a sign of bad prognosis. Interestingly, Asplund and co-workers [17] have shown that mesothelioma cells do not themselves produce hyaluronan but produce factors that enhance hyaluronan production in other cells.

Liver disease

One of the more successful uses of serum hyaluronan has turned out to be in grading liver conditions. About 40 clinical reports have now been published in this area [11]. As mentioned above, the major part of serum hyaluronan is rapidly taken up by the liver endothelial cells. An impairment of the function of these cells will automatically lead to a rise in serum hyaluronan. However, there could also be other mechanisms for a rise such as a decreased blood flow through the liver and thereby decreased uptake. An increased influx of hyaluronan into the circulation could be caused either by an increased lymph flow from the splanchnicus region due to portal hypertension or an increased hyaluronan production in a fibrotic liver.

Liver injuries of different aetiology have been followed with regard to hyaluronan [11] and the results are the same, i.e. the more severe the condition the higher is the serum level. Figure 3 contains data collected from the literature on patients in different stages of hepatitis, alcoholic liver injury, primary biliary cirrhosis and haemochromatosis. The elevated levels in the cirrhotic end-stage are apparent and patients often have levels 10- or even 100-fold higher than the

Figure 3

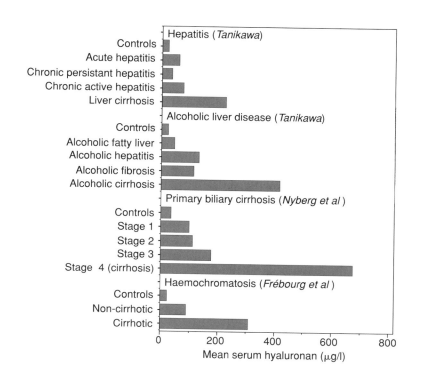

Serum hyaluronan in various liver diseases

Mean concentrations measured in groups of patients at various stages of virus hepatitis, alcoholic liver disease, primary biliary cirrhosis and idiopathic haemochromatosis are displayed. Note the general increase in the cirrhotic stage. Reproduced from [11] with permission. See [11] for full details of references in Figure.

physiological level. As a matter of fact, in studies on primary biliary cirrhosis no other laboratory test correlated better to morphologically diagnosed cirrhosis in liver biopsies. Hyaluronan analyses can obviously be used with advantage to follow the progress of a liver disease.

It is of special interest to note that the sinusoidal liver endothelial cells change character in cirrhosis, a process called capillarization. Normally the cells are not supported by basement membrane and have a very characteristic fenestration. During capillarization the cells become similar to normal capillary endothelium. They lose the fenestrae, a basement membrane is formed and they express Factor VIII-related antigen and Weibel–Palade bodies. Regular vascular endothelial cells are not known to metabolize hyaluronan. Babbs and co-workers [18] and Ueno and co-workers [19] have shown a significant correlation between serum hyaluronan and capillarization and at present this change in cell character is the most probable explanation of decreased clearance of the polysaccharide. Therefore assay of serum hyaluronan becomes a very sensitive test of the function of liver endothelial cells. All other commonly used liver function tests measure the activity of hepatocytes.

In cases of acute liver damage due to intoxication [20] extremely high serum hyaluronan levels were reported – in the order of 1000-fold the normal level. Another major disturbance is the rejection of a transplanted liver. Increased serum hyaluronan is the first sign of rejection and appears about 1 day before the rise of bilirubin [21].

Inflammatory disorders and septicaemia

A number of inflammatory mediators activate hyaluronan synthesis. Therefore increased serum levels of hyaluronan are often seen in inflammatory diseases, e.g. in rheumatoid arthritis. This is treated in Chapter 32 of this volume by Engström-Laurent.

Sepsis displays many molecular mechanisms common to inflammation. Berg and co-workers [22] followed patients that had been admitted to an intensive care unit for septic conditions and found that those who exhibited the highest serum hyaluronan levels, when entering the hospital, usually had a fatal outcome. No other test had the same predictive value. The reason for the increase in serum hyaluronan in sepsis is unknown but animal experiments have shown that injection of tumour necrosis factor-α, endotoxin or live *Escherichia coli* bacteria all lead to hyaluronaemia. Patients with adult respiratory distress syndrome, usually connected with septicaemia, exhibit increased hyaluronan both in blood and bronchoalveolar lavage (see below) [23].

Another group of seriously debilitated patients are uraemic patients treated by dialysis. They have moderately elevated serum hyaluronan but the values increase when amyloid-associated arthropathy develops [24]. We have no clear explanation for the increase of serum hyaluronan in uraemic patients.

Urine

Urine is a typical fluid compartment that varies in volume and it is, therefore, not possible to use urinary hyaluronan concentrations for diagnosis. It becomes necessary to collect 24 h samples and determine daily excretion. In the few experiments published the urinary excretions have been related to the serum concentrations [12]. They were increased in Werner's syndrome, liver cirrhosis and rheumatoid arthritis, but the elevations were much smaller in urine than in blood. Until now urinary hyaluronan has, with a few exceptions, not been used for clinical diagnosis.

Tissue fluids

A number of body fluids have been analysed for hyaluronan. Although hyaluronan in lymph has been thoroughly investigated in animals [6] there are only a few studies of human material and no clinical investigations have been reported. Studies on cerebrospinal fluid are described in Chapter 3 of this volume.

Aqueus has been analysed in patients undergoing surgery but no apparently pathological values were observed [25]. The hyaluronan content of amniotic fluid shows a dramatic drop from about 20 to about 1 mg/l after the 20th week of pregnancy. This coincides with the time when the fetus begins to swallow the amniotic fluid [26]. So far, analyses of amniotic fluid have not served diagnostic purposes. Skin blister fluid has been collected from patients with various dermatological conditions and the hyaluronan level is often increased in various inflammatory conditions [27]. Joint fluid is frequently tapped from patients with joint disease, especially with effusions in the joints, and, as mentioned previously, the hyaluronan concentrations are often lower than normal [7]. Fluids from the pleural, pericardial and peritoneal cavities have been collected from tumour patients with effusions and the clinical rationale has been to diagnose for mesothelioma [15]. A large amount of material may therefore be found in the literature on pleural and peritoneal effusions of various origins. In addition, inflammatory conditions lead to increased levels of hyaluronan, e.g. in the peritoneal fluid after genital intraperitoneal inflammation in women [28].

A second type of fluids, which have received wider clinical use, comes from organ lavage. Bronchoalveolar lavage fluid contains highly increased levels of hyaluronan in various pulmonary disorders such as adult respiratory distress syndrome, sarcoidosis and farmer's lung [23,29,30], and hyaluronan can be used as a parameter of the progress of the condition. Intestinal lavage fluid has similarly been used and hyaluronan has been found to increase in inflammatory diseases in the gut, e.g. in coeliac disease, Crohn's disease and ulcerative colitis [31,32]. Hyaluronan has also been followed in peritoneal dialysis fluid and a 10-fold increase was observed in peritonitis [33].

Impairment of organ functions by hyaluronan accumulation

The increased production of hyaluronan in various diseases, which can be recorded as increased hyaluronan levels in serum or other body fluids, often leads to accumulation of hyaluronan in individual organs and impairment of their normal functions. The first example was accumulation in joint tissue in rheumatoid arthritis, which leads to water retention and 'morning stiffness' [34]. Subsequently, Hällgren and collaborators showed that pulmonary disease causes hyaluronan accumulation in lungs and the resulting interstitial oedema impairs gas exchange [23,35]. Similarly, Waldenström and co-workers have experimentally induced myocardial infarction in rats and described hyaluronan accumulation in parallel with oedema formation [36]. It is clear that such an oedema can be deleterious to the function of the heart. A fourth example of hyaluronan accumulation is the rejected organ transplant. In this case Hällgren and collaborators have described experiments with transplanted hearts [37], kidneys [38] and intestines [39] that, when rejected, show intensive inflammatory reactions, incorporation of hyaluronan and swelling. Wells and co-workers [40] confirmed the same reaction in rejected human kidneys. It is quite possible that increased intracapsular pressure caused by the swelling of the kidney prevents the normal function of the organ.

Future developments

The use of hyaluronan analysis in clinical practice has already given important and useful information. However, so far only single values of the hyaluronan concentration have been measured in the tissue fluids. As pointed out in the introduction, large fluctuations could be due to perturbations from the steady state. Much more information would therefore be available if,for example, serum hyaluronan could be followed by continuous recording during a longer time sequence. In addition, there should be important information to be gained from kinetic studies in which the clearance of hyaluronan from the bloodstream is determined. It should be possible to distinguish an elevated hyaluronan level in blood due to increased influx of the polysaccharide from a decreased liver function. Experiments along these lines are in progress [41,42].

The other aspect of great potential interest will be to find means of avoiding hyaluronan accumulation in diseased organs and subsequent impairment of functions. In principle one should be able to find solutions along two lines: (1) inhibition of the hyaluronan synthesis; (2) removal of hyaluronan, e.g. by hyaluronidase. Great progress is at present being made in elucidating the biosynthetic mechanism, which is discussed elsewhere in this volume. Hyaluronidase was, long ago, used in the treatment of myocardial infarction. With our new knowledge of the role of hyaluronan in pathological processes and our increasing knowledge of hyaluronidases such clinical experiments could, with advantage, be performed again.

This work was supported by a grant from the Swedish Medical Research Council.

References

1. Lindqvist, U., Chichibu, K., Delpech, B., Goldberg, R.L., Knudson, W., Poole, A.R. and Laurent, T.C. (1992) Clin. Chem. **38**, 127–132
2. Rippelino, J.A., Klinger, M.M., Margolis, R.U. and Margolis, R.K. (1985) J. Histochem. Cytochem. **33**, 1060–1066
3. Fraser, J.R.E. and Laurent, T.C. (1989) in The Biology of Hyaluronan (Evered, D. and Whelan, J., eds.), Ciba Foundation Syposium No. 143, pp. 41–59, Wiley, Chichester
4. Laurent, T.C. and Fraser, J.R.E. (1991) in Degradation of Bioactive Substances: Physiology and Pathophysiology (Henriksen, J.H., ed.), pp. 249–265, CRC Press, Boca Raton, FL
5. Laurent, T.C. and Fraser, J.R.E. (1992) FASEB J. **6**, 2397–2404
6. Fraser, J.R.E., Cahill, R.N.P., Kimpton, W.G. and Laurent, T.C. (1996) in Extracellular Matrix: Tissue Function (Comper, W.D., ed.), vol. 1, pp. 110–131, Harwood Academic, Amsterdam
7. Laurent, T.C., Fraser, J.R.E., Laurent, U.B.G. and Engström-Laurent, A. (1995) Acta Orthop. Scand. **66** (Suppl. 266), 116–120
8. Lindqvist, U. and Laurent, T.C. (1992) Scand. J. Clin. Lab. Invest. **52**, 613–621
9. Engström-Laurent, A. and Laurent, T.C. (1988) in Clinical Impact of Bone and Connective Tissue Markers (Lindh, E. and Thorell, J.I., eds.), pp. 235–252, Academic Press, London
10. Engström Laurent, A. (1989) in The Biology of Hyaluronan (Evered, D. and Whelan, J., eds.), Ciba Foundation Symposium No. 143, pp. 233–247, Wiley, Chichester
11. Laurent, T.C., Laurent, U.B.G. and Fraser, J.R.E. (1996) Ann. Med. **28**, 241–253
12. Laurent, T.C., Lilja, K., Brunnberg, L., Engström-Laurent, A., Laurent, U.B.G., Lindqvist, U., Murata, K. and Ytterberg, D. (1987) Scand. J. Clin. Lab. Invest. **47**, 793–799
13. Deutsch, H.F. (1957) J. Biol. Chem. **224**, 767–774
14. Longaker, M.T., Adzick, N.S., Sadigh, D., Hendin, B., Stair, S.E., Duncan, B.W., Harrison, M.R., Spendlove, R. and Stern, R. (1990) J. Natl. Cancer Inst. **82**, 135–139
15. Roboz, J., Greaves, J., Silides, D., Chahinian, A.P. and Holland, J.F. (1985) Cancer Res. **45**, 1850–1854
16. Dahl, I.M.S. and Laurent, T.C. (1988) Cancer **62**, 326–330

17. Asplund, T., Versnel, M.A., Laurent, T.C. and Heldin, P. (1993) Cancer Res. **53**, 388–392
18. Babbs, C., Haboubi, N.Y., Mellor, J.M., Smith, A., Rowan, B.P. and Warnes, T.W. (1990) Hepatology **11**, 723–729
19. Ueno, T., Inuzuka, S., Torimura, T., Tamaki, S., Koh, H., Kin, M., Minetoma, T., Kimura, Y., Ohira, H., Sata, M., Yoshida, H. and Tanikawa, K. (1993) Gastroenterology **105**, 475–481
20. Bramley, P.N., Rathbone, B.J., Forbes, M.A., Cooper, E.H. and Losowsky, M.S. (1991) J. Hepatol. **13**, 8–13
21. Adams, D.H., Wang, L., Hubscher, S.G. and Neuberger, J.M. (1989) Transplantation **47**, 479–482
22. Berg, S., Brodin, B., Hesselvik, F., Laurent, T.C. and Maller, R. (1988) Scand. J. Clin. Lab. Invest. **48**, 727–732
23. Hällgren, R., Samuelsson, T., Laurent, T.C.and Modig, J. (1989) Am. Rev. Respir. Dis. **139**, 682–687
24. Ozasa, H., Chichibu, K., Tanaka, Y., Kondo, T., Kitajima, K. and Ota, K. (1992) Nephron **61**, 187–191
25. Laurent, U.B.G. (1983) Arch. Ophthalmol. **101**, 129–130
26. Dahl, L., Hopwood, J.J., Laurent, U.B.G., Lilja, K. and Tengblad, A. (1983) Biochem. Med. **30**, 280–283
27. Juhlin, L., Tengblad, A., Ortonne, J.P. and Lacour, J.Ph. (1986) Acta Dermatol. Venerol. (Stockholm) **66**, 409–413
28. Edelstam, G.A.B., Lundkvist, Ö., Venge, P. and Laurent, T.C. (1994) Inflammation **18**, 13–21
29. Hällgren, R., Eklund, A., Engström-Laurent, A. and Schmekel, B. (1985) Br. Med. J. **290**, 1778–1781
30. Bjermer, L., Engström-Laurent, A., Lundgren, R., Rosenhall, L. and Hällgren, R. (1987) Br. Med. J. **295**, 803–806
31. Lavö, B., Knutson, L., Lööf, L., Odlind, B. and Hällgren, R. (1990) Gut **31**,153–157
32. Raab, Y., Hällgren, R., Knutson, L., Krog, M. and Gerdin, B. (1992) Am. J. Gastroenterol. **87**, 1453–1459
33. Yung, S., Coles, G.A., Williams, J.D. and Davies, M. (1994) Kidney Intern. **46**, 527–533
34. Engström-Laurent, A. and Hällgren R. (1987) Arthritis Rheum. **30**, 1333–1338
35. Nettelbladt, O., Tengblad, A. and Hällgren, R. (1989) Am. J. Physiol. **257**, L379–L384
36. Waldenström, A., Martinussen, H.J., Gerdin, B. and Hällgren R. (1991) J. Clin. Invest. **88**, 1622–1628
37. Hällgren, R., Gerdin, B., Tengblad, A. and Tufveson, G. (1990) J. Clin. Invest. **85**, 668–673
38. Hällgren R., Gerdin, B., Tufveson, G. (1990) J. Exp. Med. **171**, 2063–2076
39. Wallander, J., Hällgren, R., Scheynius, A., Gerdin, B. and Tufveson, G. (1993) Transpl. Int. **6**, 133–137
40. Wells, A.F., Larsson, E., Tengblad, A., Fellström, B., Tufveson, G., Klareskog, L. and Laurent, T.C. (1990) Transplantation **50**, 240–243
41. Lindqvist, U., Groth, T., Lööf, L. and Hellsing, K. (1992) Clin. Chim. Acta **210**, 119–132
42. Lebel, L.,Gabrielsson, J., Laurent, T.C. and Gerdin, B. (1994) Eur. J. Clin. Invest. **24**, 621–626

32

Hyaluronan analysis as a tool in evaluating rheumatic diseases

Anna Engström-Laurent
Department of Internal Medicine, University Hospital of Umeå, S-901 85 Umeå, Sweden

Introduction

All the rheumatic disorders are true connective tissue diseases, implicating that supportive tissues like bone, cartilage, tendons, muscles and synovial membranes are primarily affected. These tissue types all have a high content and production of hyaluronan (HA) as well as other connective tissue components.

The rheumatic diseases can be roughly divided into two categories; the inflammatory diseases such as rheumatoid arthritis (RA), systemic lupus erythematosus, psoriatic arthritis, gout, etc. and the degenerative or primarily non-inflammatory diseases such as osteoarthritis (OA) and spondylosis. Both groups are characterized by prominent changes in the structure of the supportive tissues.

Serum HA levels and RA

In 1985 the first report on elevated serum A concentrations in RA patients was published [1]. A few years later it was proposed that HA accumulates in and around the joints during the night and in this location it acts as a mechanical hinder and gives rise to the typical morning 'stiffness' in the RA patient. When the patient gets out of bed in the morning and starts moving, HA will be removed from the tissues by muscular pumping. The removal of HA from the tissues is mirrored in the blood as a peak in the serum HA concentration shortly after the patient has risen from bed [2]. The effect of physical exercise on serum HA levels has subsequently been reported in other studies [3,4].

The outflow of HA from the tissues into the circulation of RA patients can be correlated to the number of inflamed joints (synovitis mass) [2,5,6]. It has also been shown that the serum HA levels are more pronounced in RA patients with rapid progression of joint destruction compared with patients with a slow progression [7]. In children with juvenile chronic arthritis serum HA levels have also been analysed and found to be high compared with controls. Increased serum levels were mainly found in children with polyarticular disease [8].

Serum HA analysis in other rheumatic disorders, including osteoarthritis

High serum HA levels have also been recorded in other inflammatory rheumatic diseases, e.g. psoriatic arthritis and systemic scleroderma. In psoriatic arthritis the levels are correlated to the number of inflamed joints and in systemic sclerosis the serum levels are correlated to overall disease severity [9,10]. Ankylosing spondylitis, which mainly affects the spine, is another common inflammatory rheumatic disease. In spite of the inflammatory nature of this condition, serum-HA levels remain within normal ranges. This is probably owing to limited synovial involvement in this disease.

In OA the synovial involvement is often less prominent. However, also in this disease, and especially in patients with OA of the knee, elevated serum HA levels have been observed and correlated to the progression of radiological destruction of the joint [11]. In OA of the knee, not only the synovial membrane, but also the diseased cartilage will contribute to the increased serum concentration.

Analysis of serum HA levels in clinical routine

If serum HA analysis is to be included in the clinical test battery it is important to remember the following:

(1) Serum HA levels are elevated in many of the rheumatic diseases, which means it is not a specific marker for any specific disease.

(2) The circulating levels are not only dependent on inflammatory involvement of the tissues but also on the age, liver function and physical activity of the patient and also at what time of the day blood sampling is performed.

(3) The serum levels do not seem to mirror the inflammatory activity as measured with C-reactive protein or electron spin resonance but rather the extent of tissue involvement or synovitis mass. This hypothesis is supported by the finding that both an inflammatory disease like RA and a degenerative disease like OA exhibit elevated serum-HA levels.

The source of elevated serum HA levels

The elevated serum HA levels measured in the different rheumatic diseases could theoretically depend on an increased local production of HA or a decreased breakdown in lymph nodes or in the liver. Another explanation is an impaired local degradation and/or uptake of the molecule in the tissue itself. In RA there is evidence that supports the theory that the elevated serum levels are a result of an increased production and total amount of HA in the affected joints [12]. There is also an increased turnover rate in arthritic joints and the inflammatory process leads to an increased lymph flux of hyaluronan [12].

Serum HA analysis provides new information

The most common inflammatory rheumatic disease is RA. This disease is characterized by chronic inflammation associated with considerable damage to the musculoskeletal system, particularly in and around the joints. By analysing cartilage, bone and synovium-derived connective tissue molecules, e.g. hyaluronan, new information is provided about the damage caused to the tissues by inflammation and degenerative processes. As increased serum HA levels in both RA and OA have been shown to indicate a progressive erosive or destructive disease there is also valuable prognostic information in the analysis. Until now, there has been a lack of good prognostic markers evaluating tissue engagement both in inflammatory diseases and non-inflammatory diseases. Hopefully analysis of serum HA and other connective tissue components will provide such a tool for early evaluation of pathological processes in the connective tissues [13].

References

1. Engström-Laurent, A. and Hällgren, R. (1985) Ann. Rheum. Dis. **44**, 83–88
2. Engström-Laurent, A. and Hällgren, R. (1987) Arthritis Rheum. **30**, 1333–1338
3. Lindqvist, U., Engström-Laurent, A., Laurent, U., Nyberg, A., Björklund, U., Eriksson, H., Pettersson, R. and Tengblad, A. (1988) Scand. J. Clin. Lab. Invest. **48**, 765–770
4. Saari, H., Konttinen, Y.T. and Nordström, D. (1991) Ann. Med. **23**, 29–32
5. Poole, A.R., Witter, J., Roberts, N., Piccolo, F., Brandt, R., Paquin, J. and Baron, M. (1990) Arthritis Rheum. **33**, 790–799
6. Paimela, L., Heiskanen, A., Kurki, P., Helve, T. and Leirisalo-Repo, M. (1991) Arthritis Rheum. **34**, 815–821
7. Månsson, B., Carey, D., Alini, M., Ionescu, M., Rosenberg, L.C., Poole, A.R., Heinegård, D. and Saxne, T. (1995) J. Clin. Invest. **95**, 1071–1077
8. Andersson Gäre, B. and Fasth, A. (1994) Scand. J. Rheumatol. **23**, 183–190
9. Lundin, Å., Engström-Laurent, A., Michaëlsson, G. and Tengblad, A. (1987) Br. J. Dermatol. **116**, 335–340
10. Levesque, H., Baudot, N. and Delpech, B., Vayssairat, M., Gancel, A., Lauret, P. and Courtois, H. (1991) Br. J. Dermatol. **124**, 423–428
11. Sharif, M., George, E., Shepstone, L., Knudson, W., Thonar, E.J., Cushnaghan, J. and Dieppe, P. (1995) Arthritis Rheum. **38**, 760–767
12. Laurent, T.C., Fraser, J.R.E., Laurent, U.B.G. and Engström-Laurent, A. (1995) Acta Orthop. Scand. **66** (Suppl. 266), 116–120
13. Poole, A.R. and Dieppe, P. (1994) Semin. Arthritis Rheum. **23** (Suppl. 2), 17–31

33

Hyaluronan in cerebrospinal fluid

Ulla B.G. Laurent

Departments of Ophthalmology and *Medical and Physiological Chemistry, University of Uppsala, Box 575, S-751 23 Uppsala, Sweden

Introduction

As this symposium is dedicated to E.A. Balazs it seems appropriate first to mention his pioneering work on the concentration of hyaluronan in two body fluids, namely synovial fluid and vitreus. Balazs had already published in the 1940s a paper on synovial fluid and in the 1950s several papers on the vitreus. Analyses of hyaluronan concentration were then based on determinations of hexuronic acid of the fluids (see, for example, [1,2]). For a review on the vitreus and on much of their work see the chapter by Balazs and Denlinger in Davson's *The Eye* [3].

It would be several years before the concentration could be specifically determined in other body fluids that all have, lower content of hyaluronan. With a sensitive, specific technique introduced by Tengblad in our laboratory [4–6] we analysed several body fluids for hyaluronan. The concentrations in supposedly healthy humans are listed in Table 1. These figures can serve as comparison with the data on hyaluronan in human lumbar and ventricular cerebrospinal fluid (CSF) presented in this chapter.

Hyaluronan concentration in lumbar CSF

The concentration of hyaluronan was determined in 200 specimens of human lumbar CSF. Our original assay [5] or the Pharmacia kit [14], which is based on the same principle [4–6], were used. In most of the specimens protein analyses were also performed. For details see our recently published paper [12].

When we divided the specimens of lumbar CSF according to age of the patients it could easily be seen that the lowest concentration (median 33 μg/l) is found in the age group 1–9 years (Table 2). The concentration increases in puberty to reach a level of around 300 μg/l in adults. In a group of 9 healthy adults, 19–61 years of age, we found a median of 147 μg/l. The mean ± SD was 166 ± 77 μg/l, and the range 52–320 μg/l.

The material can also be grouped according to diagnoses, as seen in Figure 1. Both the reference group and the patient group are small but it can still be seen, especially in inflammatory conditions, that the hyaluronan values are high (in head trauma, vascular insults, encephalitis and notably in meningitis, as well as in medical inflammatory disorders). In some cases with brain tumour, and especially in CSF obtained in connection with neurosurgery, high values were

---

*Department to which correspondence should be sent.

Table 1

Body fluid	n	Hyaluronan concentration range (mg/l)	Reference
Synovial fluid	13	1500–3000	[1,7]
Vitreus body	>1000	100–400	[3]
Thoracic lymph	1	8.5 [a]	[8]
Amniotic fluid			
(gest. w. 16–18)	48	6–45	[9]
(gest. w. 34–39)	58	0.5–2.5	[9]
Aqueous humour	27	0.46–2.2	[10]
Urine	22	av. 0.3	[11]
Cerebrospinal fluid	9	0.05–0.32	[12]
Blood plasma/serum	585	0.03–0.18[b]	[13]

[a] *Rift in the thoracic duct after oesophagus resection due to a malignant tumour.*
[b] *Increasing concentration with age.*

Hyaluronan in body fluids of human 'healthy' adults

found. Note also that hydrocephalus and spinal stenosis seem to be associated with increased values.

Hyaluronan concentration in ventricular CSF

We were able to analyse 27 specimens of ventricular CSF from 16 adults and four children. The concentration was significantly lower than in lumbar CSF (Table 3). In 16 adults we got a mean of 53 µg/l. Of these 16 adults, four had hydrocephalus with values between 85 and 298 µg/l. The four children all had brain tumours. Two of the patients had hydrocephalus when admitted to the hospital and their values were higher than those of the other two.

Table 2

Age	n	Hyaluronan concentration (µg/l)	
		Median	Range
0–28 days	9	86	49–180
29 days– <1year	10	40	<10–320
1–9 years	37	33	15–503
10–16 years	16	75	24–223
19–29 years	11	147	63–522
30–39 years	9	315	118–1300
40–49 years	14	305	74–530
50–59 years	25	260	52–8200
60–69 years	15	345	30–2650
70–79 years	10	305	215–1120
85 years	1	355	

Hyaluronan in lumbar CSF; variation with age

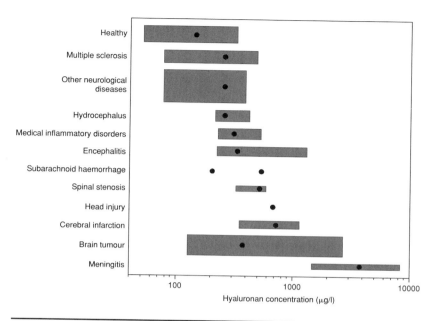

Figure 1

Hyaluronan concentration (μg/l) of lumbar cerebrospinal fluid in twelve groups of patients

The width of the rectangles shows the range, and the height the number of patients, e.g. meningitis (n = 3), other neurological diseases (n = 18). (●) Median or single values.

Molecular weight of hyaluronan

We analysed the molecular weight distributions of hyaluronan from three different sources: (1) ventricular CSF from a child with ventricular drainage before surgery for medulloblastoma; (2) pooled lumbar CSF from 23 adults, none with tumour or meningitis; (3) cyst fluid from a cystic brain tumour. Calibrated gel chromatography was performed on a mixed bed of cross-linked gels [15,16].

The weight-average molecular weight was 2.9×10^5 in ventricular CSF and 3.0×10^5 in lumbar CSF. However, in brain tumour cyst fluid it was higher,

Table 3

| | | Hyaluronan concentration (μg/l) | |
Diagnosis, Adults	n	Mean	Range
Hydrocephalus	4	148	85–298
Brain tumour	8	19	<10–32
Cerebellar infarction	1	41	
Head injury	1	28	
Intracranial infection	1	27	
Subarachnoid haemorrhage	1	<10	
Diagnosis, Children			
Brain tumour	4	63	<10–140

Hyaluronan in ventricular CSF

namely 2.4×10^6. These molecular weights can be compared with earlier published values; for bovine brain 1.4×10^5 [17], rat brain 1.54×10^5 [18] and human brain tumour (glioma) 4×10^6 [19].

Turnover of hyaluronan in CSF

[^{3}H]Hyaluronan was injected through a lumbar or cisternal puncture in the rabbit to study the rate of turnover of hyaluronan in CSF. The technique was the same as used in a study on the turnover of hyaluronan in the anterior chamber [20] and has since been used in several other tissues or cavities. With the lumbar injection we found a half-life of the same order as in the knee-joint, about half a day. With the cisternal injection it was shorter (S.J. Allen, J.R.E Fraser and U.B.G. Laurent, unpublished work; [21]).

Localization of hyaluronan in brain tissues and meninges

Determinations of hyaluronan [5] and histochemistry [22] were carried out on specimens from eight locations in the brain and meninges during autopsy of two elderly females.

We found a higher concentration in white than in grey matter both in frontal cortex and in cerebellum; around 100 μg/g in white matter and around 40 μg/g in grey matter. The values obtained in grey matter agree very well with data by Delpech and co-workers [23]. Much less was observed in plexus choroideus, 9 μg/g, and especially in the leptomeninges, 2.5 μg/g.

Histochemistry verified our biochemical analyses. We could in addition detect a heavier staining for hyaluronan in the superficial layer of the frontal cortex.

Discussion

In our investigation on human lumbar CSF we have found a variation of hyaluronan concentration from <10 to >8000 μg/l. In several neurological diseases we could not see any significant difference compared with the reference values. The results have raised several questions:

How can we explain high hyaluronan concentrations?

Our highest value, 8100 μg/l, was found in a patient with meningitis; generally patients with inflammatory disorders had elevated levels. From other investigations we know that high serum concentrations are found in inflammatory conditions [24], e.g. in rheumatoid arthritis, where hyaluronan comes from the inflamed joint [25]. Inflammatory mediators activate the synthesis of hyaluronan [26]. The reasonable explanation is, therefore, that inflammatory mediators increase hyaluronan production also in the brain.

Brain tumour is another condition with elevated hyaluronan levels. Some tumours found elsewhere are known to synthesize increased amounts of hyaluronan, such as Wilms´ tumour and mesothelioma [27]. This may also be the case with brain tumours. Glioma cells in culture produce hyaluronan [28].

Why does the hyaluronan concentration increase with age?

The pattern found in CSF, low concentration in children (except for neonatals) and then an increase with age, follows the pattern found in serum [13]. Paradoxically, studies both in man and in animals have shown that the concentration of hyaluronan in the brain is higher in young individuals than in adults [23,29]. There is at present no obvious explanation for the age dependence.

Where does hyaluronan in CSF originate?

Is blood a source of hyaluronan in CSF?

CSF is formed in the choroid plexus by filtration and secretion in a volume of 500 ml/24 h. The protein concentration of CSF is about 0.5% of that of plasma while the hyaluronan concentration of CSF is often many times higher than that of plasma/serum. This fact contradicts the hypothesis that hyaluronan originates from blood by ultrafiltration, especially since the polysaccharide has a molecular weight of 300 000 in CSF and hardly passes an ultrafilter, which retains most of the plasma proteins. Furthermore, only minimal amounts of hyaluronan were found in the choroid plexus. Therefore hyaluronan in CSF is presumably of local origin. It should be mentioned, however, that the molecular weight of hyaluronan in CSF is of the same order as that of hyaluronan in blood [8].

Does hyaluronan originate in the meninges?

The CSF is surrounded by the meninges which partly consists of connective tissue. It could therefore be expected that hyaluronan is synthesized by the meninges and secreted over their vast surfaces into CSF. However, it came as a surprise that the hyaluronan content both of dura mater and of leptomeninges was low compared with brain tissue. This was obvious both by direct analysis and by histochemistry. However, the observation does not exclude a production in the meninges.

Is nervous tissue a source of hyaluronan in CSF?

The presence of hyaluronan in the central nervous system has been documented repeatedly [5,23,29–32]. The number of hyaluronan-binding proteins discovered in the brain is increasing. Most of these are found in other tissues as well, but some are confined to the brain. This would indicate that hyaluronan has important structural and cell biological functions in nervous tissues. Other major components of the extracellular matrix found in most tissues, such as laminin, fibronectin and collagen, are absent between or among neurons or glia in the central nervous system [32]. At present it is most probable that hyaluronan in CSF originates from the brain tissue but more investigations are required to explain why the CSF level of hyaluronan increases with age while the extracellular space between the nerve cells, where hyaluronan is situated, is known to contract with age [33,34]. This should make it more difficult for hyaluronan to escape from the tissues.

References

1. Balazs, E.A., Watson, D., Duff, I.F. and Roseman, S. (1967) Arthritis Rheum. **10**, 357–376
2. Balazs, E.A., Laurent, T.C., Laurent, U.B.G., DeRoche, M.H. and Bunney, D.M. (1959) Arch. Biochem. Biophys. **81**, 464–479
3. Balazs, E.A. and Denlinger, J.L. (1984) in The Eye, (Davson, H., ed.), vol. 1a, pp. 533–589, Academic Press, London
4. Tengblad, A. (1980) Biochem. J. **185**, 101–105
5. Laurent, U.B.G. and Tengblad, A. (1980) Anal. Biochem. **109**, 386–394
6. Engström-Laurent A., Laurent, U.B.G., Lilja, K. and Laurent, T.C. (1985) Scand. J. Clin. Lab. Invest. **45**, 497–504
7. Dahl, L.B., Dahl, I.M.S., Engström-Laurent, A. and Granath, K. (1985) Ann. Rheum. Dis. **44**, 817–822
8. Tengblad, A., Laurent, U.B.G., Lilja, K., Cahill, R.N.P., Engström-Laurent, A., Fraser, J.R.E., Hansson, H.E. and Laurent, T.C. (1986) Biochem. J. **236**, 521–525
9. Dahl, L., Hopwood, J.J., Laurent, U.B.G., Lilja, K. and Tengblad, A. (1983) Biochem. Med. **30**, 280–283
10. Laurent, U.B.G. (1983) Arch. Ophthalmol. **101**, 129–130
11. Laurent, T.C., Lilja, K., Brunnberg, L., Engström-Laurent, A., Laurent, U.B.G., Lindqvist, U., Murata, K. and Ytterberg, D. (1987) Scand. J. Clin. Lab. Invest. **47**, 793–799
12. Laurent, U.B.G., Laurent, T.C., Hellsing, L.K., Persson, L., Hartman, M. and Lilja, K. (1996) Acta Neurol. Scand. **94**, 194–206
13. Lindqvist, U. and Laurent, T.C. (1992) Scand. J. Clin. Lab. Invest. **52**, 613–621
14. Brandt, R., Hedlöf, E., Åsman, I., Bucht, A. and Tengblad, A. (1987) Acta Oto-Laryngol. (Stockholm) **442** (Suppl.), 31–35
15. Laurent, U.B.G. and Granath, K.A. (1983) Exp. Eye Res. **36**, 481–492
16. Lebel, L., Smith, L., Risberg, B., Laurent, T.C. and Gerdin, B. (1989) Am. J. Physiol. **256**, H1524–H1531
17. Margolis, R.U. (1967) Biochim. Biophys. Acta **141**, 91–102
18. Iwata, M., Wight, T.N. and Carlson, S.S. (1993) J. Biol. Chem. **268**, 15061–15069
19. Delpech, B., Maingonnat, C., Girard, N., Chauzy, C., Maunoury, R., Olivier, A., Tayot, J. and Creissard, P. (1993) Eur. J. Cancer **29A**, 1012–1017
20. Laurent, U.B.G., Fraser, J.R.E. and Laurent, T.C. (1988) Exp. Eye Res. **46**, 49–58
21. Brown, T.J., Laurent, U.B.G. and Fraser, J.R.E. (1991) Exp. Physiol. **76**, 125–134
22. Hellström, S., Tengblad, A., Johansson, C., Hedlund, U. and Axelsson, E. (1990) Histochem. J. **22**, 677–682
23. Delpech, B., Delpech, A., Brückner, G., Girard, N. and Maingonnat, C. (1989) in The Biology of Hyaluronan (Evered, D. and Whelan, J., eds.), Ciba Foundation Symposium No. 143, pp. 208–232, Wiley, Chichester
24. Laurent, T.C., Laurent, U.B.G. and Fraser, J.R.E. (1996) Ann. Med. **28**, 241–253
25. Engström-Laurent, A. (1989) in The Biology of Hyaluronan (Evered, D. and Whelan, J., eds.), Ciba Foundation Symposium No. 143, pp. 233–247, Wiley, Chichester
26. Laurent, T.C. and Fraser, J.R.E. (1989) in The Functions of Proteoglycans (Evered, D. and Whelan, J., eds.), Ciba Foundation Symposium No. 124, pp. 9–29, Wiley, Chichester
27. Knudson, W., Biswas, C, Li, X-Q., Nemec, R.E. and Toole, B.P. (1989) in The Biology of Hyaluronan (Evered, D. and Whelan, J., eds.), Ciba Foundation Symposium No. 143, pp. 150–169, Wiley, Chichester
28. Glimelius, B., Norling, B., Westermark, B. and Wasteson, Å. (1979) J. Cell. Physiol. **98**, 527–538
29 Margolis, R.U. , Margolis, R.K., Chang, L.B. and Preti, C. (1975) Biochemistry **14**, 85–88
30. Bignami, A. and Asher, R. (1992) Int. J. Dev. Neurosci. **10**, 45–57
31. Bignami, A., Hosley, M. and Dahl, D. (1993) Anat. Embryol. **188**, 419–433
32. Carlson, S.S. and Hockfield, S. (1996) in Extracellular Matrix, vol.1, Tissue Function (Comper, W.D., ed.), pp. 1–23, Harwood, Amsterdam
33. Nicholson, C. and Rice, M.E. (1986) Ann. New York Acad. Sci. **481**, 55–71
34. Bachelard, H., Carrera, D. and Jenkins, H. (1982) in The Ageing Brain (Hoyer, S., ed.), pp. 67–75, Springer, Berlin

Round table discussion: new applications for hyaluronan

Moderators: E.A. Balazs and T.C. Laurent

A two hour discussion was held on the possibilities of finding new applications for hyaluronan in medical practice. The following is a summary of the discussion.

Applications

A list of possible applications based on suggestions found in the literature was presented in the beginning of the session. This list was enlarged during the discussion and a final version is shown as Table 1. Some comments about the items made by discussants (names in brackets) are listed in abbreviated form below.

Viscosurgery and other surgical applications

The classical case of viscosurgery is the use of viscous hyaluronan in the anterior segment during eye operations [1,15]. The hyaluronan both facilitates surgery and protects surrounding tissues. In the middle ear, hyaluronan facilitates the insertion of electrodes during cochlear implants [7,8]. Hyaluronan has also been proposed as an osmotically active compound to extract water from the inner ear in cases of endolymphatic hydrops [9] (*C. Laurent*). Hyaluronan has been useful in arthroscopy to increase visibility in the joint as well as in other endoscopic techniques (*Weiss, Goldman*). It may be used to treat bursitis (*Weiss*). For further references see also Chapter 25 by J.L. Denlinger, and Chapter 27 by C. Weiss.

Viscosupplementation and tissue implants

The use of hyaluronan in joints [12–14] and as vitreus implants [15] has a long history. In mammary implants, silicone can be replaced with cross-linked hyaluronan (hylan) that has the same rheological properties as silicone but without the biocompatibility problems of silicone [17]. It is also believed that surface coating of the bags would increase biocompatibility (*Balazs*). Injection of hyaluronan derivatives is being used in the skin for tissue augmentation and to correct wrinkles and depressed scars (*Juhlin, Denlinger*). Vesico-ureteral implants of dextran–hyaluronan mixtures are made in children with vesico-ureteral reflux [18] and similar implants made in paralysed vocal cords (*C. Laurent*). Microencapsulation of pancreatic islet cells in hyaluronan has been tried for implants [23,24]. However, it has been more successful to make the implants in alginate. Even in implants of alginate the inclusion of hyaluronan may improve the cellular microenvironment for optimal function of the cells (*W. Knudson*). It was also suggested that other cells could be encapsulated in hyaluronan and implanted (e.g. cartilage and bone cells) [25]. For further references see chapters by M.E. Adams (26); P.A. Band (6); J.L. Denlinger (25); N. E. Larsen (28); and C. Weiss (27).

Table 1	**Viscosurgery and other surgical applications [1–11]**

Eyes (anterior segment surgery, strabismus surgery) [1–5]
Ears (middle ear surgery, cochlear implants, Ménière's disease) [6–9]
Microvascular surgery (facilitates sutures) [10]
Clotting haemangiomas
Endoscopies [11]

Viscosupplementation and tissue implants [12–25]

Joints [12–14]
Vitreus replacements [15]
Skin augmentation [16]
Mammary implants [17]
Vocal cord implants
Vesico-ureteral implants [18]
Dentine and bone repair [19,20]
Surface coating (biocompatibility) [21,22]
Microencapsulation of cells for implantation (islet cells, cartilage, bone) [23–25]

Anti-adhesion and wound healing [26–64]

Tendon surgery and tendon healing [26–32]
Prevention of joint contracture [33]
Lumbar laminectomy, discectomy [34–38]
Anti-adhesion in peritoneum and pericardium [39–50]
Prevention of restenosis of vessels [51]
Wound healing (skin, cornea, tympanic membrane, promotion of fetal wound healing) [52–56]
Skin grafting (especially stored skin) and artificial skin [57–63]
Guidance of cells and nerve growth [64]

Peritoneal dialysis [65,66]

Protection of peritoneum
Reduction of water retention
Acceleration of dialysis

Moisturizer. Protection of surfaces [67–72]

Skin moisturizers [67]
Eye drops [68–71]
Protection of mucous surfaces
Protection of skin in chemical warfare
Protection of lung from elastase-induced emphysema [72]

Drug carrier or drug modifier [73–92]

Topical application on skin [73–76]
Application on cornea [77,78]
Formation of depots [79–86]
Receptor-directed uptake [87]
Enhancement of drug action [75]
Gene transfer [88,89]
Enhancement of local anaesthesia [90,91]
Liposome coating [92]
Aerosols

Possible applications of hyaluronan and its derivatives

Some recent references are given in the table. More references are given in the individual chapters of the book.

(contd)	Table 1

Interaction with the immune system and other cells [94–104]

Anti-inflammatory [94,95]

Protection from oxygen radicals [97]

Angiogenic/anti-angiogenic [98–101]

Stimulation of megakaryocytopoiesis [102]

Inhibition of platelet aggregation

Prevents intimal proliferation in vascular injury [103,104]

Treatment of certain diseases [105–111]

Treatment of bronchitis [105,106]

Treatment of insterstitial cystitis [107]

Barrier to cancer metastasis [108]

Hepatoprotection [109]

Pain relief. Treatment of neuromas [110,111]

Technical applications [112–122]

Sperm isolation and tests of sperm performance [112–120]

Isolation of hyaluronan binding cells [121]

Tissue banking [122]

Use of endogenous hyaluronan [123,124]

Diagnostic use of hyaluronan levels in organs and tissue fluids.

Diagnostic use of hyaluronan receptors [123]

Intervention in biosynthesis and removal of hyaluronan [124]

Possible applications of hyaluronan and its derivatives

Some recent references are given in the table. More references are given in the individual chapters of the book.

Anti-adhesion and wound healing

There are many investigations, both positive and negative, in which hyaluronan has been used to prevent adhesions in various compartments. One example is the incorporation of hyaluronan in scaffold material after middle ear operations to decrease fibrosis [6] (*C. Laurent*). Hyaluronan, as well as artificial skins containing hyaluronan, have been used in preclinical wound healing studies [57–63] (*Band*). Hyaluronan has had a positive effect on the healing of lacerated anterior cruciate ligaments in the knee joint [29] (*Wiig*). Recent experiments indicate that circulating hyaluronan prevents restenosis in vessels after balloon injury [51]. Intravenous labelled hyaluronan has been found to accumulate at the damaged site (*Gustafson, Turley*). It was suggested that deposits of hyaluronan could be important in guidance of cell migration (*Heinegård*) or guidance of neural regeneration in polyethylene tubing [64] (*Prestwich*). Injection of hyaluronan oligomers inhibits tumour growth, possibly due to preventing cellular attachment (C. Zeng, J. Kuo, S. Kinney, B.P. Toole and I. Stamencovic, unpublished work) (*Toole*). In addition, hyaluronan fragments (4–20 disaccharides) increase angiogenesis in many systems [98,99] (*West*). For further references see chapters by P.A. Band (6); N.E. Larsen (28); C. Laurent (29); C. Weiss (27); and D.C. West and D.M. Shaw (24).

Peritoneal dialysis

Peritoneal dialysis has been performed on rats in which 0.005 or 0.01% hyaluronan was included in the dialysis fluid. This led to a significant decrease in retention of fluid and also a trend to more rapid transport of urea into the dialysis fluid. The results can be explained in terms of increased resistance to bulk flow across the peritoneal wall, which facilitates diffusion of urea in the opposite direction [65,66] (*Wang*).

Moisturizer; protection of surfaces

Hyaluronan is already used in various cosmetic creams as a moisturizer [67]. It protects dry corneas [68–71] but it has also been used on mucous membranes that are healing, e.g. in the sinus cavities of the nose (*Balazs*). Other mucous surfaces where hyaluronan could be useful as a protection during and after surgery are in the female sexual organs (*Balazs*). The possibility of using hyaluronan or hyaluronan derivatives as protection in chemical warfare was raised (*Bloom*).

Drug carrier and drug modifier

Hyaluronan may be used as a drug depot topically on skin. It was discussed whether hyaluronan itself could penetrate skin. Evidence for this was found in nude mice [93] (*Fraser*).

The feasibility of transcellular or intercellular transport of high-molecular-weight hyaluronan in the epidermis was discussed (*Band, West, Prestwich*). Hyaluronan facilitates diclofenac transport through epidermis [74–76] (*Turley*). The basement membrane between epidermis and dermis should be an impenetrable barrier for high-molecular-weight hyaluronan according to *Hascall*.

Release of drugs covalently or electrostatically attached to depots of hyaluronan has been tried (*Balazs, Prestwich*). Hyaluronan has been described as a gene carrier to liver endothelial and retinal pigment epithelial cells [88,89]. The increase in duration of local anaesthesia by hyaluronan [90,91] is presumably due to a depot effect; the anaesthetic cannot be transported by convection, only by slow diffusion, through viscous hyaluronan (*T. Laurent*). Drug uptake mediated by hyaluronan receptors on specific cells can be utilized (*Gustafson, Prestwich*). Mitomycin and epirubicin have been covalently linked to hyaluronan and encouraging results on tumour cells in cell culture have been described [87]. Hyaluronan–taxol prodrugs are in development at the University of Utah for cancer treatment (*Prestwich*). Enhancement of drug effects has been observed when hyaluronan has been given intravenously simultaneously with the drug [128] but the mechanism is unknown (*T. Laurent*). Is it possible to incorporate hyaluronan in an aerosol for pulmonary use (*Noble*)? For further references see Chapter 30 by S. Gustafson; and Chapter 7 by G.D. Prestwich and co-workers

Interaction with the immune system and other cells

Hyaluronan influences on inflammatory cells, oxygen radicals and angiogenesis have been dealt with in previous chapters [see contributions by E.A. Balazs (21); C. Belmonte and co-workers (22); J. Lesley (13); Z. Lin and co-workers (14); P.W. Noble and co-workers (23); G.O. Philips (11); and D.C. West & D.M. Shaw (24)]. Hyaluronan prevents intimal proliferation in vascular injuries [103,104] (*Turley*).

Treatment of certain diseases

Following on from some old work [105], a recent publication [106] provided evidence that subcutaneous administration of high-molecular-weight hyaluronan reduced the incidence of bronchitis during the winter months. The data are interesting but the mechanism for the reduction in episodes of bronchitis is unknown (*Noble*). The amount injected is minute compared with the normal turnover of hyaluronan in the body. Melanomas have been treated with hyaluronan fragments with positive effects (C. Zeng, J. Kuo, S. Kinney, B.P. Toole and I. Stamencovic, unpublished work) (*Toole*). Injection of ^{125}I-labelled hyaluronan directly into transplanted colon carcinomas caused the hyaluronan to stay localized at these sites and the tumour to regress if injection was with ^{125}I of high specific activity. Tumour growth progressed as usual when the specific radioactivity was low (S. Gustafson and T. Björkman, unpublished work) (*Gustafson*). Hyaluronan may alleviate pain in the injured cornea and when deposited around injured nerves (neuromas) (*Belmonte*).

Technical applications

Hyaluronan has been used for a long time for isolation of viable sperms, i.e. for *in vitro* fertilization [112–120]. Hyaluronan gels can be used for the isolation of hyaluronan binding cells (*T. Laurent*; unpublished work and [121]). Hyaluronan is useful for the preservation of organs to be transplanted, such as tissue banking of corneas [122] (*Denlinger*).

Use of endogenous hyaluronan and hyaluronan receptors

The use of hyaluronan for diagnostic purposes has been described in previous chapters [contributions by A. Engström-Laurent (32); T.C. Laurent (31); and U.B.G. Laurent (33)].

Serum analyses could be used in critical evaluation of therapy, such as for drugs used in portal hypertension (*Fraser*). Analysis of hyaluronan receptors may be used in the diagnosis of malignant cells, e.g. mesothelioma [123] (*Heldin*). Means of regulating synthesis and degradation of hyaluronan will be important in medical treatment, such as when accumulation of hyaluronan impairs organ functions [124] (*T.Laurent*).

Is hyaluronan a single substance?

Armand raised the question whether hyaluronan is a homogenous substance or if we should talk about hyaluronans. *Scott* added that hyaluronan may occur in differently aggregated states and therefore cannot be regarded as homogenous. According to *Jeanloz*, only 80–85% of hyaluronan can be described with certainty as having the same structure [125]. Data from Meyer (cited by *Balazs* and *Fraser*) indicate that close to 100% is degradable by β-glucuronidase and β-*N*-acetylhex-osaminidase, although a certain portion of the products are a result of transglyco-sylation. The data show that hyaluronan is a linear polysaccharide containing equal amounts of glucuronic acid and *N*-acetylglucosamine [126]. It should be added that there is no indication that another structure other than the commonly accepted one exists, except that some deacetylation has been reported [127] (*T.Laurent*).

Is hyaluronan toxic?

A number of observations indicate that hyaluronan is non-toxic even when large doses are administered. *Scott* cited a patient (name, Lucas, St. Mary's Hospital, London, 1959) with mesothelioma, who produced large amounts of intraperitoneal hyaluronan for some years without any special symptoms. *Lindblad* injected large doses intraperitoneally in animals without any apparent effects. Patients have been given large doses of hyaluronan intravenously without any disturbances [128] (*T. Laurent*). When repeated high doses of hyaluronan are given intra-articularly in animals, lymphocyte infiltration is seen in the synovial tissue without any noticeable symptoms [129] (*T. Laurent*). *Toole* asked if the toxicity of hyaluronan fragments is known, but no one could answer.

Is hyaluronan inert?

A discussion arose concerning the issue of how hyaluronan exerts its therapeutic influence when used as a medical product. All of the currently approved indications for hyaluronan (viscosurgery, viscosupplementation, viscoprotection, viscoseparation) are based on concentrated viscoelastic solutions, gels and solids of hyaluronan. These indications rely on the mechanical properties of hyaluronan to accomplish their therapeutic goals, and they are largely regulated as medical devices rather than drugs. Conversely, researchers working with hyaluronan receptors or hyaluronan fragments have observed biological actions, particularly in *in vitro* model systems. It was pointed out that the response of cells to hyaluronan *in vitro* is very different from the *in vivo* situation. *In vitro*, cells are exposed to small amounts of hyaluronan in a relatively hyaluronan-free environment. By contrast, when hyaluronan is applied *in vivo*, into a synovial joint for example, the cells are already in an environment saturated with respect to hyaluronan.

The question was raised whether hyaluronan is properly considered as an inert biomaterial. 'Inert' was subsequently defined as referring to the tissue response to injected hyaluronan. Non-inflammatory hyaluronan (NIF-NaHA) is indeed inert compared with other biomaterials used in medicine in that it does not elicit inflammatory, foreign body or immune reactions when implanted. This does not mean that hyaluronan is inert in the biological sense; only in the medical sense.

The opinion was expressed that medical products based on hyaluronan and its derivatives can be designed to function via mechanical or biological modes, but that currently available hyaluronan products have been designed to function as devices.

References

1. Miller, D. and Stegmann, R., eds. (1983) Healon (sodium hyaluronate). A Guide to its use in Ophthalmic Surgery, Wiley, New York
2. Searl, S.S., Metz, H.S. and Lindahl, K.J. (1987) The use of sodium hyaluronate as a biologic sleeve in strabismus surgery. Ann. Ophthalmol. **19**, 259–268
3. Maselli, E., Talatin, C., Gherardi, G. and Gaddi, D. (1988) Use of Healon in the surgery of strabismus. Implant Refractive Surg. **6**, 98–99
4. Granet, D.B., Hertle, R.W. and Ziylan, S. (1994) The use of hyaluronic acid during adjustable suture surgery. J. Pediatr. Ophthalmol. Strabismus **31**, 287–289
5. Ferreira, R.C., Lamberts, M., Moreira, J.B. and Campos, M.S. (1995) Hydroxypropyl-methylcellulose and sodium hyaluronate in adjustable strabismus surgery. J. Pediatr. Ophthalmol. Strabismus **32**, 239–242
6. Laurent, C., Hellström, S. and Stenfors, L-E. (1986) Hyaluronic acid reduces connective tissue formation in middle ears filled with absorbable gelatin sponge: an experimental study. Am. J. Otolaryngol. **7**, 181–186
7. Lehnhardt, E. (1992) Intrakochleäre Elektrodenplazierung mittels Healon. HNO **40**, 86–89
8. Donelly, M.J., Cohen, L.T. and Clark, G.M. (1995) Initial investigation of the efficacy and biosafety of sodium hyaluronate (Healon) as an aid to electrode array insertion. Ann. Otol. Rhinol. Laryngol. **166** (Suppl.), 45–48
9. Jansson, B., Friberg, U. and Rask-Andersen, H. (1993) Endolymphatic sac morphology after instillation of hyperosmolar hyaluronan in the round window niche. Acta Otolaryngol. (Stockholm) **113**, 741–745
10. Arnbjörnsson, E.O. (1986) Sodium hyaluronate as an aid in microvascular surgery. Microsurgery **7**, 166–167
11. Braunstein, R.E., Kazim, M., and Schubert, H.D. (1995) Endoscospy and biopsy of the orbit. Ophthal. Plast. Reconstr. Surg. **11**, 269–272
12. Adams, M.E., Atkinson, M.H., Lussier, A.J., Schulz, J.I., Siminovitch, K.A., Wade, J.P. and Zummer, M. (1995) The role of viscosupplementation with hylan G-F 20 (Synvisc) in the treatment of osteoarthritis of the knee: A Canadian multicenter trial comparing hylan G-F 20 alone, hylan G-F 20 with non-steroidal anti-inflammatory drugs (NSAIDs) and NSAIDs alone. Osteoarthritis Cartilage **3**, 213–225
13. Gaustad, G. and Larsen, S. (1995) Comparison of polysulphated glycosaminoglycan and sodium hyaluronate with placebo in treatment of traumatic arthritis in horses. Equine Vet. J. **27**, 356–362
14. Lohmander, L.S., Dalén, N., Englund, G., Hamalainen, M., Jensen, E.M., Karlsson, K., Odensten, M., Ryd, L. Sernbo, I., Suomalainen, O. and Tegnander, A. (1996) Intra-articular hyaluronan injections in the treatment of osteoarthritis of the knee: a randomized, double blind, placebo controlled multicentre trial group. Ann. Rheum. Dis. **55**, 424–431
15. Balazs, E.A., Freeman, M.I., Klöti, R., Meyer-Schwickerath, G., Regnault, F. and Sweeney, D.B. (1972) Hyaluronic acid and replacement of vitreus and aqueous humor. Mod. Probl. Ophthal. **10**, 3–21
16. Larsen, N.E., Pollak, C.T., Reiner, K., Leshchiner, E. and Balazs, E.A. (1993) Hylan gel biomaterial: dermal and immunologic compatibility. J. Biomed. Mater. Res. **27**, 1129–1134
17. Lin, K., Bartlett, S.P., Matsuo, K., LiVolsi, V.A., Parry, C., Hass, B. and Whitaker, L.A. (1994) Hyaluronic acid-filled mammary implants: an experimental study. Plast. Reconstr. Surg. **94**, 306–315
18. Stenberg, A. and Läckgren, G. (1995) A new bioimplant for the endoscopic treatment of vesicoureteral reflux: experimental and short-term clinical results. J. Urol. **154**, 800–803
19. Sasaki, T. and Kawamata-Kido, H. (1995) Providing an environment for reparative dentine induction in amputated rat molar pulp by high molecular-weight hyaluronic acid. Arch. Oral Biol. **40**, 209–219
20. Sasaki, T. and Watanabe, C. (1995) Stimulation of osteoinduction in bone wound healing by high-molecular hyaluronic acid. Bone **16**, 9–15
21. Benedetti, L., Cortivo, R., Berti, T., Berti, A., Pea, F., Mazzo, M., Moras, M. and Abatangelo, G. (1993) Biocompatibility and biodegradation of different hyaluronan derivatives (Hyaff) implanted in rats. Biomaterials **14**, 1154–1160
22. Kito, H. and Matsuda, T. (1996) Biocompatible coatings for luminal and outer surfaces of small-caliber artificial grafts. J. Biomed. Mater. Res. **30**, 321–330
23. Aung. T., Inoue, K., Kogire, M., Doi, R., Kaji, H., Tun, T., Hayashi, H., Echigo, Y., Wada, M., Imamura, M., Fujisato, T., Maetani, S., Iwata, H. and Ikada, Y. (1995) Comparison of various gels for immobilization of islets in bioartificial pancreas using a mesh-reinforced polyvinyl alcohol hydrogel tube. Transplant. Proc. **27**, 619–621

24. Soon-Shiong, P., Heintz, R., Yao, Z., Yao, Q., Sanford, P., Lanza, R.P. and Meredith, N. (1992) Glucose-insulin kinetics of the extravascular bioartificial pancreas. A study using microencapsulated rat islets. Am. Soc. Artif. Int. Org. J. **38**, 851–854

25. Butnariu-Ephrat, M., Robinson, D., Mendes, D.G., Halperin, N. and Nevo, Z. (1996) Resurfacing of goat articular cartilage by chondrocytes derived from bone marrow. Clin. Orthop. **330**, 234–243

26. St. Onge, R., Weiss, C., Denlinger, J.L.and Balazs, E.A. (1980) A preliminary assessment of Na-hyaluronate injection into 'No man's land' for primary flexor tendon repair. Clin. Orthop. **146**, 269–275

27. Thomas, S.C., Jones, L.C. and Hungerford, D.S. (1986) Hyaluronic acid and its effect on postoperative adhesions in the rabbit flexor tendon. Clin. Orthop. **206**, 281-289

28. Amiel, D., Ishizue, K., Billings, E., Wiig, M., Vande Berg, J.and Akeson, W.H. (1989) Hyaluronan in flexor tendon repair. J. Hand Surg. **14A**, 837–843

29. Wiig, M.E., Amiel, D., Vande Berg, J., Kitabayashi, L., Harwood, F.L. and Arfors, K.E. (1990) The early effect of high molecular weight hyaluronan (hyaluronic acid) on anterior cruciate ligament healing: an experimental study in rabbits. J. Orthop. Res. **8**, 425-434

30. Gaughan, E.M., Nixon, A.J., Krook, L.P., Yeager, A.E., Mann, K.A., Hohammaed, H. and Bartel, D.L. (1991) Effects of sodium hyaluronate on tendon healing and adhesion formation in horses. Am. J. Vet. Res. **52**, 764–773

31. Chen, Z. and Gu, J. (1995) Experimental research on using macromolecule sodium hyaluronate to prevent flexor tendon adhesion. Chung Hua Wai Ko Tsa Chih **33**, 526-528

32. Xu, J., Gu, Y. and Wang, H. (1995) Experimental study of sodium hyaluronate products on prevention of tendon adhesion. Chung Hua Wai Ko Tsa Chih **33**, 529–531

33. Weiss, C. and Band, P. (1995) Musculoskeletal applications of hyaluronan and hylan. Potential uses in the foot and ankle. Clin. Pediatr. Med. Surg. **12**, 497–517

34. Weiss, C., Dennis, J., Suros, J.M., Denlinger, J.L., Badia, A. and Montane, I. (1989) Sodium hyaluronate for the prevention of postlaminectomy scar formation. Trans. Orthop. Res. Soc. **13**, 44

35. Songer, M.N., Ghosh, L. and Spencer, D.L. (1990) Effects of sodium hyaluronate on peridural fibrosis after lumbar laminectomy and discectomy. Spine **15**, 550–554

36. Abitbol, J-J., Lincoln, T.L., Lind, B.I., Amiel, D., Akeson, W.H. and Garfin, S.R. (1994) Preventing postlaminectomy adhesion. A new experimental model. Spine **19**, 1809–1814

37. Pfeiffer, M., Griss, P., Franke, P., Bornscheuer, C., Orth, J., Wilke, A. and Clausen, J.D. (1994) Degeneration model of the porcine lumbar motion segment: effects of various intradiscal procedures. Eur. Spine J. **3**, 8–16

38. Songer, M.N., Rauschning, W., Carson, E.W. and Pandit, S.M. (1995) Analysis of peridural scar formation and its prevention after lumbar laminectomy and discectomy in dogs. Spine **20**, 571–580

39. Wadström, J. and Tengblad A. (1993) Fibrin glue reduces the dissolution rate of sodium hyaluronate. Upsala J. Med. Sci. **98**, 159–167

40. Goldberg, E.P., Burns, J.W. and Yaacobi, Y. (1993) Prevention of postoperative adhesions by precoating tissues with dilute sodium hyaluronate solutions. Prog. Clin. Biol. Res. **381**, 191–204

41. Shushan, A., Mor-Yosef, S., Avgar, A. and Laufer N. (1994) Hyaluronic acid for preventing experimental postoperative intraperitoneal adhesions. J. Reprod. Med. **39**, 398–402

42. Holzman, S., Connolly, R.J. and Schwaitzberg, S.D. (1994) Effect of hyaluronic acid solution on healing of bowel anastomoses. J. Invest. Surg. **7**, 431–437

43. Mitchell, J.D., Lee, R., Hodakowski, G.T., Neya, K., Harringer, W., Valeri, C.R. and Vlahakes, G.J. (1994) Prevention of postoperative pericardial adhesions with a hyaluronic acid coating solution. J. Thorac. Cardiovasc. Surg. **107**, 1481–1488

44. Mitchell, J.D., Lee, R., Neya, K. and Vlahakes, G.J. (1994) Reduction in experimental pericardial adhesions using a hyaluronic acid bioabsorbable membrane. Eur. J. Cardiothorac. Surg. **8**, 149–152

45. Burns, J.W., Skinner, K., Colt, J., Sheidlin, A., Bronson, R., Yaacobi, Y. and Goldberg, E.P. (1995) Prevention of tissue injury and postsurgical adhesions by precoating tissues with hyaluronic acid solutions. J. Surg. Res. **59**, 644–652

46. Treutner, K.H., Bertram, P., Lerch, M.M., Klimaszewski, M., Petrovic-Kallholm, S., Sobesky, J., Winkeltau, G. and Schumpelick, V. (1995) Prevention of post-operative adhesions by single intraperitoneal medication. J. Surg. Res. **59**, 764–771

47. Campoccia, D., Hunt, J.A., Doherty, P.J., Zhong, S.P., O'Regan, M., Benedetti, L. and Williams, D.F. (1996) Quantitative assessment of the tissue response to films of hyaluronan derivatives. Biomaterials **17**, 963–975

48. West, J.L., Chowdhury, S.M., Sawhney, A.S., Pathak, C.P., Dunn, R.C. and Hubbell, J.A. (1996) Efficacy of adhesion barriers. Resorbable hydrogel, oxidized regenerated cellulose and hyaluronic acid. J. Reprod. Med. **41**, 149–154

49. Klein, E.S., Asculai, S.S. and Ben-Ari, G.Y. (1996) Effects of hyaluronic acid on fibroblast behavior in peritoneal injury. J. Surg. Res. **61**, 473–476

50. Becker, J.M., Dayton, M.T., Fazio, V.W., Beck, D.E., Stryker, S.J., Wexner, S.D., Wolff, B.G., Roberts, P.L., Smith, L.E., Sweeney, S.A. and Moore, M. (1996) Prevention of postoperative abdominal adhesions by a sodium hyaluronate-based bioresorbable membrane: a prospective randomized, double-blind multicenter study. J. Am. Coll. Surg. **183**, 297–306

51. Ferns, G.A.A., Konneh, M., Rutherford, C., Woolaghan, E. and Änggård, E.E. (1995) Hyaluronan (HYAL-BV 5200) inhibits neo-intimal macrophage influx after balloon-catheter induced injury in the cholesterol-fed rabbit. Atherosclerosis **114**, 157–164

52. Hellström, S.and Laurent, C. (1987) Hyaluronan and healing of tympanic membranes perforations: an experimental study. Acta Otolaryngol. (Stockholm) (Suppl.) **442**, 54–61

53. Rivas Lacarte, M.P., Casasin, T. and Alonso, A. (1992) Effects of sodium hyaluronate on tympanic membrane perforations. J. Intern. Med. Res. **20**, 353–359

54. Sugiyama, T., Miayauchi, S., Machida, A., Miyazaki, K., Tokuyasu, K. and Nakazawa, K. (1991) The effect of sodium hyaluronate on the migration of rabbit corneal epithelium. II. The effect of topical administration. J. Ocul. Pharmacol. **7**, 53–64

55. Gandolfi, S.A., Massari, A. and Orsoni, J.G. (1992) Low-molecular-weight sodium hyaluronate in the treatment of bacterial corneal ulcers. Graefe's Arch. Clin. Exp. Ophthalmol. **230**, 20–23

56. Nakamura, M., Hikida, M. and Nakano, T. (1992) Concentration and molecular weight dependency of rabbit corneal epithelial wound healing on hyaluronan. Curr. Eye Res. **11**, 981–986

57. Huang-Lee, L.L.H. and Nimni, M.E. (1994) Crosslinked CNBr-activated hyaluronan-collagen matrices: effects on fibroblast contraction. Matrix Biol. **14**, 147–157

58. Cabrera, R.C., Siebert, J.W., Eidelman, Y., Gold, L.I., Longaker, M.T. and Garg, H. (1995) The in vivo effect of hyaluronan associated protein-collagen complex on wound repair. Biochem. Mol. Biol. Int. **37**, 151–158

59. Brown, D.M., Chung, S.H., Pasia, E.N. and Khouri, R.K. (1996) Treatment of avascular ulcers with cytokine-induced tissue generation and skin grafting. Am. J. Surg. **171**, 247–250

60. Cooper, M.L., Hansbrough, J.F. and Polarek, J.W. (1996) The effect of arginine-glycine-aspartic acid peptide and hyaluronate synthetic matrix on epithelialization of meshed skin graft interstices. J. Burn Care Rehabil. **17**, 108–116

61. Hollander, D., Stein, M., Bernd, A., Windolf, J., Wagner, R. and Pannike, A. (1996) Autologe Keratinozytenkulturen auf Hyaluronsäureestermembranen: Eine Alternative in der komplizierten Wundbehandlung? Unfallchirurgie **22**, 268–272

62. Ruiz-Cardona, L., Sanzgiri, Y.D., Benedetti, L.M., Stella, V.J. and Topp, E.M. (1996) Application of benzyl hyaluronate membranes as potential wound dressings: evaluation of water vapour and gas permeabilities. Biomaterials **17**, 1639–1643

63. Glass, J.R., Dickerson, K.T., Stecker, K. and Polarek, J.W. (1996) Characterization of a hyaluronic acid-Arg-Gly-Asp peptide cell attachment matrix. Biomaterials **17**, 1101–1108

64. Seckel, B.R., Jones, D., Hekimian, K.J., Wang, K.K., Chakalis, D.P. and Costas, P.D. (1995) Hyaluronic acid through a new injectable nerve guide delivery system enhances peripheral nerve regeneration in the rat. J. Neurosci. Res. **40**, 318–324

65. Wieczorowska, K., Breborowicz, A., Martis, L. and Orepoulos, D.G. (1995) Protective effect of hyaluronic acid against peritoneal injury. Perit. Dial. Int. **15**, 81–83

66. Wang, T., Chen, C.,Heimbürger, O., Waniewski, J., Bergström, J. and Lindholm, B. (1997) Hyaluronan decreases peritoneal fluid absorbtion in peritoneal dialysis. J. Am. Soc. Nephrol., in the press

67. Balazs, E.A. and Band, P. (1984) Hyaluronic acid: its structure and use. Cosmetics Toiletries **99**, 65–72

68. Wysenbeek, Y.S., Loyo, N., Ben Sira, I., Ophir, I. and Ben Shaul, Y. (1988) The effect of sodium hyaluronate on the corneal epithelium. An ultrastructural study. Invest. Ophthalmol. Vis. Sci. **29**, 194–199

69. Itoi, M., Kim, O., Kimura, T., Kanai, A., Momose, T., Kanki, K., Yamaguchi, T., Ueno, Y., Kurokawa, M. and Komemushi, S. (1995) Effect of sodium hyaluronate ophthalmic solution on peripheral staining of rigid contact lens wearers. CLAO J. **21**, 261–264

70. Shimmura, S., Ono, M., Shinozaki, K., Toda, I., Takamura, E., Mashima, Y, and Tsubota, K. (1995) Sodium hyaluronate eyedrops in the treatment of dry eyes. Br. J. Ophthalmol. **79**, 1007–1011

71. Hamano, T., Horimoto, K., Lee, M. and Komemushi, S. (1996) Sodium hyaluronate eyedrops enhance tear film stability. Jpn. J. Ophthmol. **40**, 62–65

72. Cantor, J.O., Cerreta, J.M., Keller, S. and Turino, G.M. (1995) Modulation of airspace enlargement in elastase-induced emphysema by intratracheal instillment of hyaluronidase and hyaluronic acid. Exp. Lung Res. **21**, 423–436

73. Falanga, V., Carson, P., Greenberg, A., Hasan, A., Nichols, E. and McPherson, J. (1996) Topically applied recombinant tissue plasminogen activator for the treatment of venous ulcers. Preliminary report. Dermatol. Surg. **22**, 643–644

74. Roth, S.H. (1995) A controlled clinical investigation of 3% diclofenac/2.5% sodium hyaluronate topical gel in the treatment of uncontrolled pain in chronic oral NSAID users with osteoarthritis. Int. J. Tissue React. **17**, 129–132

75. Moore, A.R. and Willoughby, D.A. (1995) Hyaluronan as a drug delivery system for diclofenac: a hypothesis for mode of action. Int. J. Tissue React. **17**, 153–156

76. Freemantle, C., Alam, C.A., Brown, J.R., Seed, M.P. and Willoughby, D.A. (1995) The modulation of granulomatous tissue and tumour angiogenesis by diclofenac in combination with hyaluronan (HYAL EX-0001) Int. J. Tissue React. **17**, 157–166

77. Bucolo, C., Mangiafico, S. and Spadaro, A. (1996) Methylprednisolone delivery by Hyalobend corneal shields and its effects on rabbit ocular inflammation. J. Ocul. Pharmacol. Ther. **12**, 141–149

78. Nomura, M., Kubota, M.A., Kawamori, R., Yamasaki, Y., Kamada, T. and Abe, H. (1994) Effect of addition of hyaluronic acid to highly concentrated insulin on absorbtion from the conjunctiva in conscious diabetic dogs. J. Pharm. Pharmacol. **46**, 768–770

79. Cera, C., Palumbo, M., Stefanelli, S., Rassu, M. and Palù, G. (1992) Water-soluble polysaccharide-anthracycline conjugates: biological activity. Anti-Cancer Drug Design **7**, 143–151

80. Bonucci, E., Ballanti, P., Ramires, P.A., Richardson, J.L. and Benedetti, L.M. (1995) Prevention of ovariectomy osteopenia in rats after vaginal administration of Hyaff 11 microspheres containing salmon calcitonin. Calcif. Tissue Int. **56**, 274–279

81. Cascone, M.G., Sim, B. and Downes, S. (1995) Blends of synthetic and natural polymers as drug delivery systems for growth hormone. Biomaterials **16**, 569–574

82. Meyer J., Whitcomb, L., Treuheit, M. and Collins, D. (1995) Sustained *in vivo* activity of recombinant human granulocyte colony-stimulating factor (RHG-CSF) incorporated into hyaluronan. J. Control. Release **35**, 67–72

83. Payan, E., Jouzeau, J.Y., Lapicque, F., Bordji, K., Simon, G., Gillet, P., O'Regan, M. and Netter, P. (1995) *In vitro* drug release from HYC-141, a corticosteroid ester of high molecular weight hyaluronan. J. Control. Release **34**, 145–153

84. Prisell, P.T., Camber, O., Hiselius, J. and Norstedt, G. (1992) Evaluation of hyaluronan as a vehicle for peptide growth factors. Int. J. Pharm. **85**, 51–62

85. Matsumoto, Y. Yamamoto, I., Watanabe, Y. and Matsumoto, M. (1995) Enhancing effect of viscous sodium hyaluronate solution on the rectal absorbtion of morphine. Biol. Pharm. Bull. **18**, 1744–1749

86. Miyazaki, T., Yomota, C. and Okada, S. (1995) Interaction between sodium hyaluronate and tetracyclines. Yakugaku Zasshi **115**, 72–80

87. Akima, K., Ito, H., Iwata, Y., Matsuo, K., Waturo, N., Yanagi, M., Hagi, H., Oshima, K., Yagita, A., Atomi, Y. and Tatekawa, I. (1996) Evaluation of antitumor activities of hyaluronate binding antitumor drugs: synthesis, characterization and antitumor activity. J. Drug Target. **4**, 1–8

88. Maruyama, A., Asayama, S., Nogawa, M., Akaike, T. and Takei, Y. (1996) Hyaluronan as a gene carrier. Abstr. 5th World Biomaterials Congr., Toronto, Canada

89. da Cruz, L., Rakoczy, P., Perricaudet, M. and Constable, I.J. (1996) Dynamics of gene transfer to retinal pigment epithelium. Invest. Ophthalmol. Vis. Sci. **37**, 2447–2454

90. Hassan, H.G., Åkerman, B., Renck, H., Lindberg, B. and Lindquist, B. (1985) Effects of adjuvants on local anaesthetics on their duration. III. Experimental studies of hyaluronic acid. Acta Anaesthesiol. Scand. **29**, 384–388

91. Doherty, M.M., Hughes, P.J., Korszniak, N.V. and Charman, W.N. (1995) Prolongation of lidocaine-induced epidural anesthesia by medium molecular weight hyaluronic acid formulations: pharmacodynamic and pharmacokinetic studies in the rabbit. Anaesth. Analg. **80**, 740–760

92. Yerushalmi, N., Arad A. and Margalit, R. (1994) Molecular and cellular studies of hyaluronic acid-modified liposomes as bioadhesive carriers for topical drug delivery in wound-healing. Arch. Biochem. Biophys. **313**, 267–273.

93. Brown, T.J. and Fraser, J.R.E. (1995) Absorption of hyaluronan applied to the surface of the skin. In Third International Workshop on Hyaluronan in Drug Delivery. (Willoughby, D.A., ed.), Round Table Series, vol. 40, pp. 31–37, Royal Society Medicine Press, London

94. Rydell, N. (1970) Decreased granulation tissue reaction after installment of hyaluronic acid. Acta Orthop. Scand. **41**, 307–311

95. Ialenti, A. and Di Rosa, M. (1994) Hyaluronic acid modulates acute and chronic inflammation. Agents Actions **43**, 44–47

96. Larsen, N.E., Lombard, K.M., Parent, E.G. and Balazs, E.A. (1992) Effect of hylan on cartilage and chondrocyte cultures. J. Orthop. Res. **10**, 23–32

97. Kvam, B.J., Fragonas, E., Degrassi, A., Kvam C., Matulova, M., Pollesello, P., Zanetti, F. and Vittur, F. (1995) Oxygen-derived free radical (ODFR) action on hyaluronan (HA), on two HA ester derivatives, and on the metabolism of articular chondrocytes. Exp. Cell Res. **218**, 79–86

98. Sattar, A., Rooney, P., Kumar, S., Pye, D., West, D.C., Scott, I. and Ledger, P. (1994) Application of angiogenic oligosaccharides of hyaluronan increase blood vessel numbers in rat skin. J. Invest. Dermatol. **103**, 576–579

99. Lees, V.C., Fan, T.P-D. and West, D.C. (1995) Angiogenesis in a delayed revascularisation model is accelerated by low molecular weight hyaluronan fragments. Lab. Invest. **73**, 259–266

100. Alam, C.A., Seed, M.P. and Willoughby, D.A. (1995) Angiostasis and vascular regression in chronic granulomatous inflammation induced by diclofenac in combination with hyaluronan. J. Pharm. Pharmacol. **47**, 407–411

101. Montesano, R., Kumar, S., Orci, L. and Pepper, M.S. (1996) Synergistic effect of hyaluronan oligosaccharides and vascular endothelial growth factor on angiogenesis *in vitro*. Lab. Invest. **75**, 249–262

102. Han, Z.C., Bellucci, S., Shen, Z.X., Maffrand, J.P., Pascal, M., Petitou, M., Lormeau, J. and Caen, J.P. (1996) Glycosaminoglycans enhance megakaryocytopoiesis by modifying the activities of hematopoetic growth regulators. J. Cell Physiol. **168**, 97–104

103. Savani, R.C. and Turley, E.A. (1995) The role of hyaluronan and its receptors in restenosis after balloon angioplasty: development of potential therapy. Int. J. Tissue React. **17**, 141–151

104. Gustafson, S., Sangster, K., Björkman, T. and Turley, E.A. (1996) Accumulation of intravenous hyaluronan in balloon catheter-damaged rat carotid arteries. In Fourth International Workshop on Hyaluronan in Drug Delivery. (Willoughby, D.A., ed.), Round Table Series, vol. 45, pp. 27–37, Royal Society Medicine Press, London

105. Håkansson, L., Hällgren, R., Venge, P., Artursson, G. and Vedung, S. (1980) Hyaluronic acid stimulates neutrophil function *in vitro* and *in vivo*. A review of experimental results and a presentation of a preliminary clinical trial. Scand. J. Infect. Dis. (Suppl. 24), 54–57

106. Venge, P., Pedersen, B., Håkansson, L., Hällgren, R., Lindblad, G. and Dahl, R. (1996) Subcutaneous administration of hyaluronan reduces the number of infectious exacerbations in patients with chronic bronchitis. Am. J. Respir. Crit. Care Med. **153**, 312–316

107. Morales, A., Emerson, L., Nickel, J.C. and Lundie, M. (1996) Intravesical hyaluronic acid in the treatment of refractory interstitial cystitis. J. Urol. **156**, 45–48

108. Jones, L.M., Gardner, M.J., Catterall, J.B. and Turner, G.A. (1995) Hyaluronic acid secreted by mesothelial cells: a natural barrier to ovarian cancer adhesion. Clin. Exp. Metastasis **13**, 373–380

109. Nakayama, M., Arai, K., Hasegawa, K., Sato, K., Ohtsuka, K., Watanabe, H., Sakai, K., Rikiishi, H. and Abo, T. (1995) Adhesion molecules on intermediate TCR cells. II. Hepatoprotective effects of hyaluronic acid on acute liver injury. Cell. Immunol. **166**, 275–285

110. Miyazaki, K., Goto, S., Okawara, H. and Yamaguchi, T. (1984) Studies on the analgesic and anti-inflammatory effects of sodium hyaluronate. Ohyoh Yakuri **28**, 1123–1135

111. Yamashita, I., Atsuta, Y., Shimazaki, S. and Miyatsu, M. (1995) Effects of prostaglandin E2 and sodium hyaluronate on bradykinin induced knee joint pain. Nippon Seikeigeka Gakkai Zasshi **69**, 735–743

112. Wikland, M., Wik, O., Steen, S., Qvist, K., Söderlund, B. and Janson P.O. (1987) A selfmigration method for preparation of sperm for *in-vitro* fertilization. Hum. Reprod. **2**, 191–195

113. Mortimer, D., Mortimer, S.T., Shu, M.A. and Swart, R. (1990) A simplified approach to sperm-cervical mucus interaction testing using a hyaluronate migration test. Hum. Reprod. **5**, 835–841

114. Huszar, G., Willetts, M. and Corrales, M. (1990) Hyaluronic acid (Sperm Select) improves retention of sperm motility and velocity in normospermic and oligospermic specimens. Fertil. Steril. **54**, 1127–1134

115. Neuwinger, J., Cooper, T.G., Knuth, U.A. and Nieschlag, E. (1991) Hyaluronic acid as medium for human sperm migration test. Hum. Reprod. **6**, 396–400

116. Karlström, P-O., Hjelm, E. and Lundkvist, Ö. (1991) Comparison of the ability of two sperm preparation techniques to remove microbes. Hum. Reprod. **6**, 386–389

117. Aitken, R.J., Bowie, H., Buckingham, D., Harkiss, D., Richardson, D.W. and West, K.M. (1992) Sperm penetration into a hyaluronic acid polymer as a means of monitoring functional competence. J. Androl. **13**, 44–54

118. Slotte, H., Åkerlöf, E. and Pousett, Å. (1993) Separation of human spermatozoa with hyaluronic acid induces, and Percoll® inhibits, the acrosome reaction. Intern. J. Androl. **16**, 349–354

119. Zimmerman, E.R., Drobnis, E.Z., Robertson, K.R., Nakajima, S.T. and Kim, H. (1994) Semen preparation with the Sperm Select system versus a washing technique. Fertil. Steril. **61**, 269–275

120. Perry, R.L., Barratt, C.L., Warren, M.A. and Cooke, I.D. (1996) Comparative study of the effect of human cervical mucus and a cervical mucus substitute, Healonid, on capacitation and the acrosome reaction of human spermatozoa *in vitro*. Hum. Reprod. **11**, 1055–1062

121. Guo, M.M. and Hildreth, J.E. (1996) Assessment of cell binding to hyaluronic acid in a solid-phase assay. Anal. Biochem. **233**, 216–220

122. Ogino, H., Yukari, K., Terada, H. and Sawa, M. (1995) Effect of newly developed corneal storage medium on corneal endothelium – morphological study by scanning electron microscopy. Nippon Ganka Gakkai Zasshi **99**, 387–391

123. Teder, P., Versnel, M.A. and Heldin, P. (1996) Stimulatory effects of pleural fluids from mesothelioma patients on CD44 expression, hyaluronan production and cell proliferation in primary cultures of normal mesothelial and transformed cells. Int. J. Cancer **67**, 393–398

124. Repa, I., Garnic, J.D. and Hollenberg, N.K. (1990) Myocardial infarction treated with two lymphagogues, calcium dobesilate (CLS 2210) and hyaluronidase: a coded, placebo-controlled animal study. J. Cardiovasc. Pharmacol. **16**, 286–291

125. Jeanloz, R.W. (1982) Methyl derivatives of 2-acetamido-2-deoxy-3-O-(β-D-glucopyranosyl-uronic acid)-D-glucose (hyalubiouronic acid) from methylated hyaluronic acid. Carbohydr. Res. **99**, 51–58

126. Longas, M.O. and Meyer, K. (1981) Sequential hydrolysis of hyaluronate by β-glucuronidase and β-*N*-acetylhexosaminidase. Biochem. J. **197**, 275–282

127. Longas, M.O., Russell, C.S. and He, X-Y. (1986) Chemical alterations of hyaluronic acid and dermatan sulfate detected in aging human skin by infrared spectroscopy. Biochim. Biophys. Acta **884**, 265–269

128. Falk, R. (1994) Effect of hyaluronic acid on the penetration and targeting of drugs. In First International Workshop on Hyaluronan in Drug Delivery. (Willoughby, D.A., ed.), Round Table Series, vol. 33, pp. 2–10, Royal Society Medicine Press, London

129. Wigren, A., Wik, O. and Falk, J. (1975) Repeated intraarticular implantation of hyaluronic acid. An experimental study in normal and immobilized adult rabbit knee joints. Upsala J. Med. Sci. (Suppl. 17), 3–20